APPLIED DIGITAL CONTROL
Theory, Design and Implementation
Second Edition

J. R. LEIGH

Professor Associate, Brunel University, West London
Director, Intelligent Control Solutions Limited

DOVER PUBLICATIONS, INC.
Mineola, New York

Bibliographical Note

This Dover edition, first published in 2006, is an unabridged republication of the 1992 second edition of the work originally published by Prentice Hall International (UK) Ltd., London, in 1985.

Library of Congress Cataloging-in-Publication Data

Leigh, J. R. (James R.)
    Applied digital control : theory, design, and implementation / J.R. Leigh.— 2nd ed.
        p. cm.
    Originally published: New York : Prentice Hall, 1992.
    Includes bibliographical references and index.
    ISBN 0-486-45051-1 (pbk.)
    1. Digital control systems. I. Title.

TJ223.M53L45 2006
629.8'95—dc22

                                                                2006041009

Manufactured in the United States of America
Dover Publications, Inc., 31 East 2nd Street, Mineola, N.Y. 11501

To
J.L.L., A.L., W.O.B., C.M.R.B.

Contents

Preface to the Second Edition

In the seven years since the first edition, the topic of Applied Digital Control has maintained its focal position within a changing web of interacting disciplines. It has increased in importance and has witnessed some changes of emphasis, perhaps most important being the increased emphasis on robustness and realistic adaptive control. This edition has responded by including a new chapter devoted to these topics. That chapter ends with a discussion of the practicalities of achieving adaptive or robust control in an industrial situation, together with examples of commercially available systems for adaptive system implementation.

The computer systems available for control implementation have changed drastically. Large commercially available Distributed Control Systems (DCS) increasingly link technical control with business control. This new edition gives comprehensive coverage and reviews five available systems.

Personal Computers (PC) and VME-based systems have increased massively in power and industrial capability since 1985, to such an extent that, aided by ever more user-friendly software and ready made plug-in interface boards, they are being applied in 'serious' industrial situations that were previously the exclusive province of purpose-made DCS. The book has been completely revised to reflect the changed implementation scenario.

The case histories in the first edition were popular with readers and no fewer than seven new ones have therefore been added in this revision. These have been chosen to span a range of approaches (large DCS, customized microprocessor, personal computer) applied to a range of industrial environments (photographic film making, chemical processes, paper making, woodworking, metallurgical process).

The reference list has been updated and there are so many references specifically on adaptive and robust control that they merit a separate list.

Students often have difficulty establishing complete fluency of manipulation with the state variable material. To give additional help to students, I have introduced three worked examples into Chapter 8. These examples reinforce each other, deal with mainstream material and its manipulation, have links to material already familiar to students and have their solutions given in great detail.

Although control is interacting increasingly with Information Technology, the temptation to divert resources into expert systems, pattern recognition, artificial intelligence, neural nets, etc. has been resisted. In fact, all revisions to this edition have been carried out so that the book will occupy the same position in the spectrum as it did in 1985: that of a Practical Mainstream Book on Applied Digital Control: Theory, Design and Implementation.

Preface to the First Edition

Control engineering becomes ever more inextricably linked with developments in microprocessor technology. The educational implications are that control and computational aspects should be closely integrated to ensure that a student can move fluently from control design to computer implementation.

This book aims to establish a strong theoretical background to support applications material relevant to the design and implementation of digital control systems.

The material presented here has been developed from lectures given to the third year of a B.Sc. Degree Course in Control and Computer Engineering and an M.Sc. Degree in Instrumentation and Digital Systems at the Polytechnic of Central London. The strong applications theme running throughout the book rests on the author's close involvement in design of practical control systems for industry.

It is hoped that the book will be found useful and stimulating by industrial engineers and scientists as well as by students on formal courses.

Prerequisites for the book are a knowledge of mathematics, computation and classical control theory such as is possessed by most engineering undergraduates at the start of their final year. The material can be covered in a lecture time of 30 hours with, preferably, an approximately equal time allocation for supporting laboratory sessions, computer workshops and seminars.

I am pleased to acknowledge the assistance of computer manufacturers in supplying and agreeing to the publication of the material on which I have based Chapters 10 and 11. I am particularly indebted to Alain Zucho and Christian Gillet, both of Honeywell S.A. (Brussels).

I thank my colleagues, Messrs Martinho and Winter for allowing me to use some of their problems as exercises for the book. I also thank staff of the Polytechnic's Library and Computer Centre for assistance in compiling the references and bibliography. I am grateful to Alan Whittle for his help in preparing the manuscript for press.

Finally, it is my pleasure to thank Miss Sharon O'Keeffe for typing part of the first version of the manuscript, and the staff of Prentice-Hall for their help in conversion of the manuscript into the finished product.

The Structure of the Book

Chapter 1 briefly provides motivation and outlines some of the problems with which the book is concerned.

Chapters 2–7 constitute a complete course in the theory and design of individual digital control loops. Chapter 2 surveys concisely the theory of discrete time signals as required by later chapters. Chapter 3 establishes fluency in the use of the $\mathscr{Z}$ transform. Chapter 4 establishes a toolkit of design-oriented techniques. Chapter 5 is the nucleus of the first part of the book. It puts forward alternative design methods and follows these with practical design considerations such as the choice of sampling interval and word length. Chapter 6 considers the (analog and digital) hardware of the control loop with emphasis on dynamic response and cost-effective choice of elements. Chapter 7 illustrates by case histories some of the methods advocated in earlier chapters.

Chapters 8–11 constitute a concise and self-contained course in the theory, design and practical implementation of multi-loop digital systems. Chapter 8 is the nucleus of the second half of the book. It establishes theory and design techniques for multivariable systems with coverage of continuous as well as discrete systems. Chapter 9 introduces the theory of the control of large complex systems composed of many individual elements and with long- and short-term control objectives. Chapter 10 examines and rationalizes the control of large systems from a practical industrial viewpoint. The chapter contains original suggestions that may be used for characterizing distributed computer networks and matching them to a particular given industrial control requirement. Chapter 11 follows on the material of Chapter 10 by reviewing typical commercially available distributed control systems to demonstrate how manufacturers are meeting the objectives described in Chapters 9 and 10. Chapter 11 contains eight case histories, spanning a wide range of industrial situations where distributed control systems have been applied. Chapter 12 describes the principles and practice of robust, adaptive and self-tuning control.

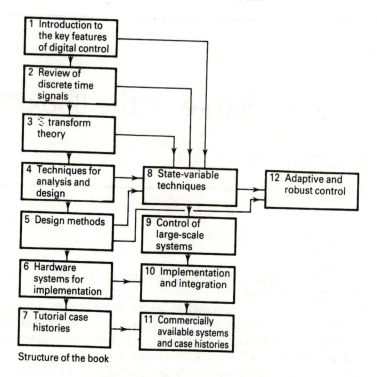

Structure of the book

Figure P1 Structure of this book.

This chapter contains descriptions of commercially available self-tuning controllers.

Appendix A gives a detailed description of a peripheral chip as used for implementation in a control loop.

Appendix B gives examples of VME bus and Personal Computer systems for control implementation.

Appendix C lists transform pairs relating time functions with $\mathscr{Z}$ transforms and Laplace transforms with $\mathscr{Z}$ transforms.

Robust and adaptive control has such a large number of interesting references that a separate reference list has been devoted to that topic.

The block diagram shows the main features of the book and their inter-relation

Acknowledgements

The awareness on which I have drawn to write this book has been built up through an energetic interchange with a wide range of professional colleagues. My indebtedness to these colleagues is gratefully acknowledged here.

Chapter 10 has benefited from a useful input from Professor Bob Malcolm of Malcolm Associates. This is gratefully acknowledged.

The ready assistance of Fisher (Provox system), Bristol Babcock, Hartmann and Brown, ASEA, Leeds and Northrup, Calex Instrumentation Ltd, Analog Devices Ltd, Honeywell, Foxboro and Jumo Ltd in supplying information on their control equipment is gratefully acknowledged.

The two case histories, C and D, describing the VME based control of a 1000 tonne press and of a fully automatic working machine, have been kindly supplied by Quarndon Electronics Ltd, UK.

Case history E, describing the Foxboro installation at Southeast Paper, has been extracted with permission from the paper by David Tobin, 'Southeast paper installs largest Foxboro distributed Control System', *Pulp & Paper*, February 1990.

Case history F, describing CIM at a 3M photographic plant has been extracted with permission from the paper, 'CIM aids 3M plant', *Control and Instrumentation*, May 1990.

Case History G, describing computer integrated manufacture applied to a Monsanto chemical plant has been extracted with permission from the paper, 'Integrated manufacture, turning vision into reality' by Stu Dutton, Monsanto Newport, *Process Industry Journal*, February 1991.

A number of errors and omissions that occurred in the first edition were pointed out by helpful readers and I take this opportunity to thank them.

Professor J R Leigh

Introduction to the Key Features of Digital Control

1.1 Important generalities

A typical digital control system involves a continuous time process, connected in closed-loop with a discrete-time computing and controlling element.

The successful design of digital control loops, therefore, requires an understanding and a bringing together of both continuous-time and discrete-time concepts as illustrated below.

1.2 A digital computer in the control loop

Figure 1.1 shows a fairly standard analog (continuous time) speed-control system in which an electric motor is controlled by a feedback loop to follow the reference signal $v(t)$.

Figure 1.2 shows the same speed-control system with a digital computer included in the loop. Notice how the computer is necessarily preceded by an analog-to-digital converter (A/D) and followed by a digital-to-analog converter (D/A).

The digital computer allows compensation and control improvement, but at this moment we are concerned with the discontinuities, errors and delays introduced by the inclusion of the computer.

Consider first the effect of the A/D converter. It will have a particular number of bits in its digital word, some finite conversion time, and it samples the signal $e(t)$ every T seconds.

1

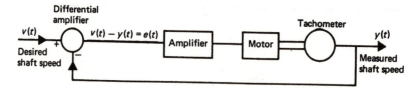

Figure 1.1   An analog speed-control loop.

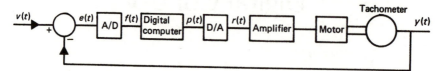

Figure 1.2   A speed-control loop containing a digital computer.

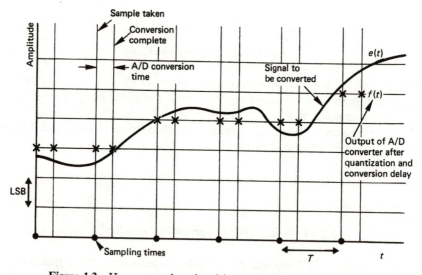

Figure 1.3   How an analog signal is converted to a digital signal.

Figure 1.3 illustrates how the A/D converter takes its sample, rounds it to the nearest least significant bit (LSB) and produces the sequence of words $f(t)$ after an appropriate conversion time.

The computer reads in the sequence of digital words $f(t)$ and, using algorithms whose accuracy depends on word length, produces the digital signal $p(t)$ after some delay to allow computation to be carried out. Notice in particular that multiplication, except of integers, always has to be followed by rounding.

Figure 1.4 shows how the most usual form of D/A converter produces the signal $r(t)$ from the word sequence $p(t)$. It can be seen that this type of converter attempts to produce a signal that is constant between sampling instants.

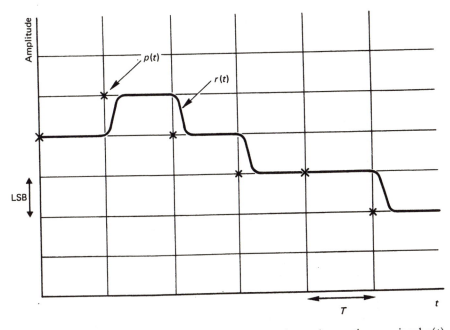

Figure 1.4 A digital-to-analog (D/A) converter produces the continuous signal $r(t)$ from the digital signal $p(t)$.

Finally, Figure 1.5 attempts to show in one diagram a summary of the main features discussed above.

Some of the questions that arise are the following:

1. The control loop contains both discrete-time and continuous-time devices. What tools can be used to determine the time and frequency-response behavior of such loops?
2. Digital control algorithms could be produced either by discretization of known analog algorithms or they could be developed, *ab initio*, as digital techniques. In the first approach, how is an analog algorithm to be discretized? This turns out to be less straightforward than might have been expected. In the second, all digital, approach, what techniques are available?
3. A high-performance control system will surely require the sampling interval T to be short. More rapid sampling will increase the computational load. How is T to be specified in relation to the other system parameters?
4. The digital devices introduce delays and quantization noise. Both can be reduced in return for additional hardware costs. What effects do these two factors have on system performance?
5. From a practical viewpoint and considering costs, what strategies, such as multiplexing of signals, would be used with advantage for particular types of application?

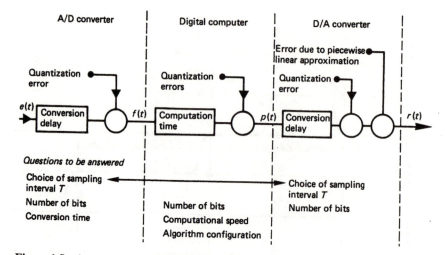

Figure 1.5 A summary of some of the questions that need to be answered and of the problems arising when a digital computer is used as a controller.

The classical tool with which to approach the problems outlined above is the $\mathscr{Z}$ transform.

In summary, milestones towards understanding and achieving competence in the design of simple digital control loops are:

1. An understanding of the nature of the sampling process and the characteristics of sampled signals.
2. Understanding and achieving fluency with the $\mathscr{Z}$ transform.
3. A complete understanding of the behavior of control loops containing a mixture of continuous-time and discrete-time devices.
4. Knowledge of control design methods for discontinuous-time controllers: Designs may be obtained by discretization of continuous-time controllers designed by continuous-time methods or they may be synthesized directly by essentially discrete-time approaches.
5. 'Knowledge' of implementation methods.

Chapters 2 to 7 provide the necessary material for the tasks outlined above.

1.3 Control of multi-loop systems

Multi-loop systems are analyzed and designed using state variable methods. These methods allow a complete systematic structural understanding of sets of interacting control loops in a way that is scarcely possible using the simpler tools of classical control.

The state-variable approach rests on the vector–matrix methods of linear algebra. State variable models give a transparent view of system structure and are the natural starting point for the design of multivariable controllers and estimators of unmeasurable process variables. This material is covered in Chapter 8.

1.4 Control of non-constant processes—robust and adaptive control

For the control of processes that vary significantly with time, a controller that also varies to match process changes is required—such a controller is called an *adaptive controller*. There are two parts to the problem: (i) Determine the current characteristics of the process, and (ii) Modify the controller automatically to match the current process.

Adaptive control is a combination of theory and pragmatism. There is tremendous interest in adaptive control but, until recently, there were relatively few convincing successful applications. This situation is changing rapidly and interesting applications are beginning to appear.

An alternative approach is to design a constant but change-tolerant controller that can cope with the envisaged variations in the process. Such a change and disturbance tolerant controller is called a *robust controller*.

Robust and adaptive control is discussed in Chapter 12.

1.5 Commercially available computer control systems

A digital controller that breaks down or gives out incorrect signals can cause enormous damage or even loss of life. For this reason, a controller, once designed, will normally be implemented through tried and tested and maintainable hardware and software.

Equipment manufacturers can supply a wide spectrum of ready made controllers that can be configured and parametrized to realize a control designer's requirements.

Every control designer who wishes to be taken seriously must have some awareness of how modern industrial control systems can be most cost-effectively implemented. Here, we satisfy the need by reviewing a number of commercially available control configurations and a range of case histories of their industrial application. Chapter 11 contains the material discussed in this section.

2

Discrete-time Signals;
Idealized Approximation;
Frequency Spectrum;
Reconstruction

2.1 Introduction

A digital controller operates on discrete time rather than continuous data, so that it is essential to obtain an understanding of the fundamental characteristics of discrete-time signals.

When a continuous signal is sampled some information may be lost, while spurious information, not present in the original signal, is created. The most important aspects of these phenomena are examined within a control-systems context.

2.2 The nature of sampled signals

When a continuous signal $f(t)$ (Figure 2.1) is sampled at regular intervals T, the resulting discrete-time signal is as shown in Figure 2.2. In the figure, q represents the time during which the sampling switch is closed.

Thus, the discrete-time signal that results when $f(t)$ is sampled is a stream of pulses of width q. The tops of the pulses are formed by the relevant section of the curve $f(t)$.

Since, in practice, the closure time q is much less than the sample interval

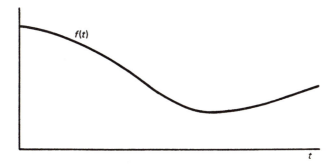

Figure 2.1   A continuous signal $f(t)$.

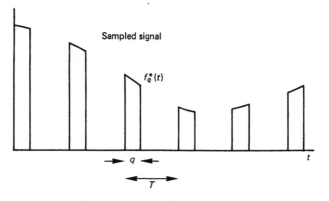

Figure 2.2   The signal f after the sampling operation—it is thendenoted f_q^*.

T, it is usual to approximate the pulses by flat-topped rectangles as shown in Figure 2.3.

    A further usually justifiable approximation is to assume that q is negligible compared with T. This leads to the idealization shown in Figure 2.4. The continuous signal $f(t)$ is represented by a set of points spaced at intervals of T. We shall refer to the signals by the notations $f_q^*(t)$ and $f^*(t)$—they are indicated on Figures 2.2 and 2.4.

2.3 Loss of information in a sampled signal

If a continuous signal $f(t)$ is considered to be carrying a certain quantity of information, then it appears highly likely that, during the sampling operation, some of the information may be lost.

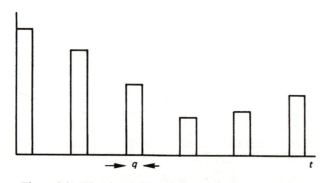

Figure 2.3 The signal f_q^* with flat-topped approximation.

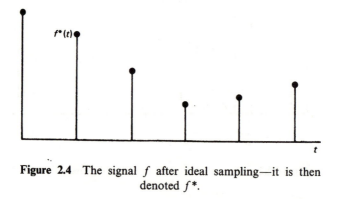

Figure 2.4 The signal f after ideal sampling—it is then denoted f^*.

For instance, the discrete-time signal in Figure 2.2 contains only a fraction of the original curve of $f(t)$. Even more extreme, the representation $f^*(t)$ has (over a finite time interval) only a finite set of points compared with the curve $f(t)$ that can be considered to be made up from an infinite set of points.

Thus, it is important to consider the question: is information lost during sampling?

We shall show that there is a minimum sampling frequency, ω_N, related to the bandwidth of the continuous signal to be sampled. If a signal is sampled at a frequency lower than ω_N, information is irretrievably lost and the signal cannot be reconstructed by any method. If a signal is sampled at a frequency greater than or equal to ω_N then no information is lost and, theoretically, the continuous signal can be reconstructed from a knowledge of the values at the sampling instants. (As we shall show in Section 2.8, such perfect reconstruction requires a knowledge of future sample values.)

2.4 Creation of spurious information in a sampled signal

It is well known that a Fourier representation of a square wave has an infinite number of terms, each successive term representing a harmonic of higher frequency. Any waveform with a zero rise time has the same characteristic. It is therefore reasonable to expect that spurious high-frequency signals will be produced when a continuous signal is replaced by an approximating sequence of pulses.

   We shall show that the process of sampling introduces harmonics. In this context they are usually called aliases. An alias is therefore a high-frequency component, not present in the original signal but apparently present in the signal that is reconstructed from samples.

2.5 The sampling process

Sampling of a continuous signal $f(t)$ at intervals T can be considered to be equivalent to the amplitude modulation of a train $p(t)$ of unit impulses of width q (see Figure 2.5). From this viewpoint,

$$f_q^*(t) = f(t)p(t). \tag{2.1}$$

Define the function $h(t - t_1)$ by the relation

$$\left.\begin{array}{ll} h(t - t_1) = 0, & t < t_1 \\ h(t - t_1) = 1, & t \geq t_1. \end{array}\right\} \tag{2.2}$$

In other words, $h(t - t_1)$ is a unit step starting at time t_1.

   Using this notation $p(t)$ can be expressed

$$p(t) = \sum_{k=-\infty}^{\infty} h(t - kT) - h(t - kT - q). \tag{2.3}$$

$p(t)$ is periodic and can be represented by a Fourier series

$$p(t) = \sum_{n=-\infty}^{\infty} C_n e^{jn\omega st}, \tag{2.4}$$

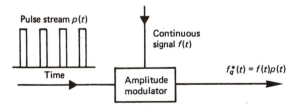

Figure 2.5   Sampling considered as amplitude modulation.

where $\omega_s = 2\pi/T$ represents the sampling frequency and the C_n are Fourier coefficients that remain to be determined.

Thus

$$f_q^*(t) = f(t)p(t) = f(t) \sum_{n=-\infty}^{\infty} C_n e^{jn\omega_s t}. \qquad (2.5)$$

Let $\mathscr{F}\{\ \}$ represent Fourier transformation, then

$$\mathscr{F}\{f_q^*(t)\} = \int_{-\infty}^{\infty} f_q^*(t) e^{-j\omega t}\, dt. \qquad (2.6)$$

We define

$$\mathscr{F}\{f_q^*(t)\} = F_q(j\omega) \qquad (2.7)$$

and recall the shift theorem,

$$\mathscr{F}\{e^{at}f(t)\} = F(j\omega - a). \qquad (2.8)$$

Hence

$$\mathscr{F}\{e^{jn\omega_s t}f(t)\} = F(j\omega - jn\omega_s). \qquad (2.9)$$

Using this result in equation (2.5) results (by linearity) in

$$F_q^*(j\omega) = \sum_{n=-\infty}^{\infty} C_n F(j\omega - jn\omega_s). \qquad (2.10)$$

This equation repays study. It shows that the frequency spectrum of the original signal, represented by $F(j\omega)$, is present, multiplied by a constant C_0. It also shows that an infinity of spurious spectra, centred at $n\omega_s, n = \pm 1, \pm 2, \ldots$ along the frequency axis and multiplied by constants $C_1, C_2, \ldots$, have been created by the sampling operation.

The constants C_n can be found from the standard Fourier formula

$$C_n = \frac{1}{T} \int_0^T p(t) e^{-jn\omega_s t}\, dt. \qquad (2.11)$$

The details are omitted, but after a little manipulation it emerges that

$$C_n = \frac{q}{T} \frac{\sin(n\omega_s q/2)}{(n\omega_s q/2)} e^{-jn\omega_s q/2}. \qquad (2.12)$$

In particular

$$C_0 = \frac{q}{T}. \qquad (2.13)$$

Figure 2.6 shows a sketch of the frequency spectrum of $f_q^*(t)$ for an assumed spectrum of $f(t)$. The central part of the spectrum is the same, within a multiplier dependent on the sampling rate, as that of the continuous signal. The remaining

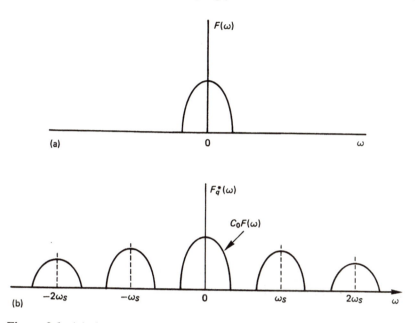

Figure 2.6 (a) An assumed spectrum for a continuous signal f; (b) The corresponding spectrum of the sampled signal.

parts of the spectrum represent spurious elements introduced by sampling. Appropriately, the spurious parts of the spectra are called aliases.

Figure 2.7 shows a sketch of the same situation as that for Figure 2.6 except that the sampling frequency ω_S is now smaller. It can be seen that the aliases are now interfering with the central part of the spectrum. To avoid this problem, it is clear that the sampling frequency must be at least twice as high as the highest frequency present in the signal $f(t)$. This can be considered to be an informal proof of Shannon's theorem, which is stated below.

SHANNON'S SAMPLING THEOREM (SOMETIMES CALLED SIMPLY THE
SAMPLING THEOREM)
A function $f(t)$ having a power spectrum restricted to a finite frequency band $(-b, b)$ is uniquely determined by a discrete set of sample values provided that the sampling frequency is greater than $2b$. The *Nyquist frequency* ω_N is defined by $\omega_N = 2b$. Further readable background on the theorem can be obtained from Reference S2.

Note that it is rare in practice to work near the limit as given by Shannon's theorem. A useful rule of thumb is to sample a signal at five to ten times the highest frequency thought to be present, if aliasing problems are to be avoided. The safety margin is required to allow for the likely presence of low-level signals at frequencies outside the defined band $(-b, b)$.

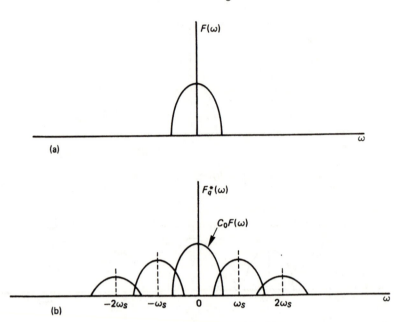

Figure 2.7 (a) An assumed spectrum for a continuous signal f; (b) The corresponding spectrum of the sampled signal when ω_s does not satisfy the sampling theorem.

Often it will be best to filter a signal before sampling it to avoid irretrievably 'folding back' the aliasing spectra. Notice in particular that high-frequency noise will be very troublesome in this connection unless it is filtered out before sampling.

ANTI-ALIASING FILTER—EXAMPLE
Let

$$f(t) = \sin \omega t + 0.2 \sin 4\omega t$$

where the terms on the right-hand side correspond to useful signal and noise respectively.

Suppose that $f(t)$ is sampled at a frequency 2ω: then the presence of the high-frequency noise leads to the generation of aliases and consequent error in the interpretation of the samples.

If $f(t)$ is processed through a simple resistance–capacitance (RC) filter of transfer function $1/(1 + sRC)$ then, approximately, the filtered signal $f'(t)$ has the form

$$f'(t) = c(\sin \omega t + 0.05 \sin (4\omega t + \phi)),$$

where c is the filter attenuation at frequency ω and ϕ is a phase angle, not relevant to the discussion. The *RC* filter can be seen to attenuate the noise in proportion to frequency—the aliasing problem is reduced but not eliminated.

RC filters are adequate for anti-aliasing application only when there is a wide frequency band separating signal from noise, to allow sufficient noise attenuation to be obtained. In other cases, sharp cut-off filters need to be used to eliminate aliases.

2.6 Ideal sampling

Usually the time q during which the sample is taken is very small compared with the sampling interval T and the following approximating assumptions can be made: (a) that the impulses are flat-topped (Figure 2.3); (b) that the impulses have zero width (Figure 2.4). The latter case is referred to as ideal sampling.

If, as in Figure 2.4, the signal $f(t)$ is subjected to ideal sampling, then the resultant discrete time signal, denoted $f^*(t)$, can be expressed as

$$f^*(t) = \sum_{k=-\infty}^{\infty} f(kT)\delta(t - kT), \tag{2.14}$$

where $\delta(t - a)$ is a Dirac impulse occurring at $t = a$ (see note at end of Section 2.6).

Fourier transforming equation (2.14) leads to

$$F^*(j\omega) = \frac{1}{T} \sum_{n=-\infty}^{\infty} F(j\omega + jn\omega_s) \tag{2.15}$$

in which $F(j\omega)$ represents the Fourier transform of the original signal $f(t)$ and $\omega_s = 2\pi/T$ is the sampling frequency.

We see that, when ideal sampling is assumed, the aliases of the central spectrum all have the same magnitude. Compare this with the situation of non-ideal sampling (equation (2.12) and Figure 2.6).

Figure 2.8 shows: (a) the spectrum of a signal $f(t)$ accompanied by a high-frequency noise component; (b) the spectrum of $f(t)$ under ideal sampling with $T = 1$ assuming that the noise has been removed before sampling by an anti-aliasing filter; (c) the spectrum of $f(t)$ and the high-frequency noise under ideal sampling with $T = 1$ with no filtering before sampling. The noise has been folded irretrievably into the low-frequency signal.

Dirac impulse function   A Dirac impulse is defined as follows.

Let $f(t)$ be the function shown in Figure 2.9(a) then the first derivative $\dot{f}(t)$ of $f(t)$ has the form shown in Figure 2.9(b).

In the limit, as $T \to 0$, $f(t)$ becomes the unit step function $h(t - a)$ while the derivative $\dot{f}(t)$ becomes the Dirac impulse function $\delta(t - a)$.

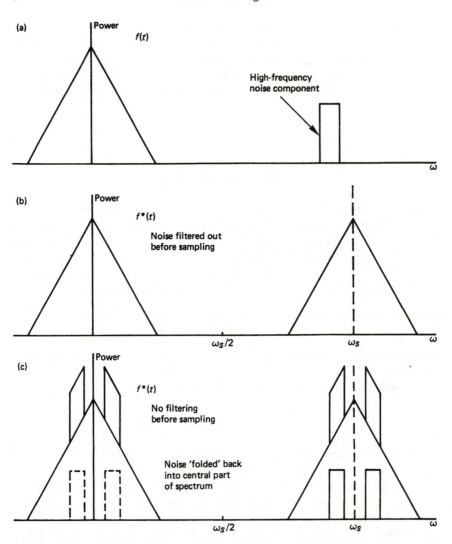

Figure 2.8 (a) An assumed spectrum for a continuous signal $f(t)$ with a high-frequency noise component; (b) The spectrum of the ideally sampled signal on the assumption that the noise is filtered out before sampling; (c) The spectrum of the ideally sampled signal when no filtering is performed before sampling.

2.7 Digital control of a single-input–single-output process

Figure 2.10 illustrates the common and important situation where a digital computer controls a single-input–single-output continuous process by means of a feedback loop.

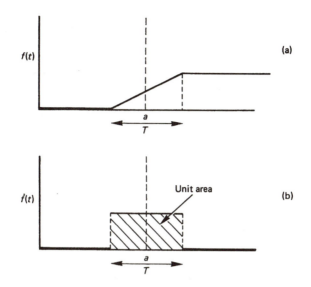

Figure 2.9   Diagrams to assist visualization of the Dirac impulse function.

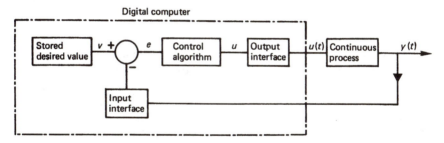

Figure 2.10   Schematic diagram showing the essential features of a digital control loop.

The dashed lines inside the digital computer indicate discrete-time signals whereas the continuous lines indicate continuous signals.

The output of the process is converted into a discrete-time signal and is subtracted from a desired value signal to form an error signal e. The control algorithm operates on the error signal to produce, through the output interface, the process input $u(t)$.

Thus, the control loop contains both continuous-time and discrete-time signals and the analytical methods that are used to investigate system behavior must deal with both types of signals. It is important to understand clearly how this is achieved.

If we refer again to Figure 2.10, it can be seen that signals v and e exist only at certain times, whereas signal y is continuously present.

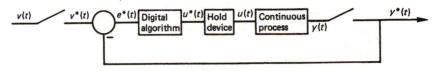

Figure 2.11   The block diagram corresponding with Figure 2.10.

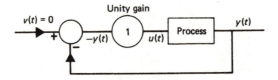

Figure 2.12   A simple continuous control loop.

The method of analysis that we are working towards does not discriminate between the cases and treats y as though it exists only at isolated instants. It thus avoids dealing with a hybrid of continuous- and discrete-time signals.

The decision to be interested in the continuous signals only at those time instants where the discrete-time signals exist is equivalent to the introduction of fictitious samplers as shown in Figure 2.11.

Here $v^*(t)$ is now assumed to be obtained by sampling some continuous signal $v(t)$. The signal $u(t)$ is obtained approximately by holding constant over each sampling instant the signal $u^*(t)$. In general, the analysis is concerned with the inter-relations between $y^*(t)$, $u^*(t)$, $v^*(t)$, the process and the algorithm.

Provided that a continuous signal $f(t)$ has been sampled at sufficiently short intervals, in accordance with Shannon's sampling theorem, then the resultant discrete time signal $f_q^*(t)$ (or its idealization $f^*(t)$) contains all the information carried by $f(t)$, together with spurious harmonics created by the sampling process.

What does this mean in practice? Consider for the sake of clarity the simple case where a computer is required to implement the control loop shown in Figure 2.12. The continuous process has input $u(t)$ and output $y(t)$, and $v(t)$ is the desired value for $y(t)$. Here $v(t) = 0$ for all values of t. The forward gain of the loop is to be unity. The computer realization is as shown in Figure 2.13. It can be seen we require that $u(t) = -y(t)$, but the computer produces $-y^*(t)$. We could set about reconstructing the signal $y(t)$ from the (known to be sufficient) information in the signal $y^*(t)$.

However, we could argue that the proper criterion for interconnecting $y^*(t)$ with the input of the process must be the effect on the process: provided that the process responds as required it is not important that the signal $u(t)$ should be a perfect reconstruction of $-y(t)$, and in practice a crude reconstruction of $-y(t)$ is usually adequate.

We examine the problem of reconstruction below.

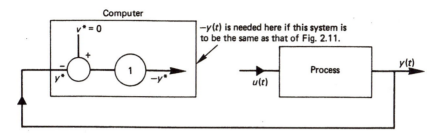

Figure 2.13 The computer realization of the loop shown in Figure 2.12 (the loop is broken to emphasize that we require $u(t)$ to be equal to $-y(t)$ if the two loops are to be equivalent).

2.8 Reconstruction of the signal f(t) from the discrete-time signal f*(t)

Assume that the signal $f(t)$ has been ideally sampled at a frequency ω_s that satisfies the sampling theorem. Then if the resultant discrete-time signal $f^*(t)$ could be passed through a filter P having the characteristics

$$
\begin{array}{cc}
\textit{Frequency} & \textit{Filter gain} \\
|\omega| < \omega_s/2 & T \\
|\omega| \geqslant \omega_s/2 & 0
\end{array} \Bigg\} \tag{2.16}
$$

the Fourier transform of the filtered discrete time signal would necessarily equal $F(\omega)$.

Assuming that such a filter, P, can be constructed, we can then write

$$F(\omega) = P(\omega)F^*(\omega). \tag{2.17}$$

By the definition of P, it follows that the filter characteristic in the time domain must be given by inverse Fourier transformation as

$$
p(t) = \frac{1}{2\pi} \int_{-\pi/T}^{\pi/T} T\,e^{j\omega t}\,d\omega
$$

$$
= \frac{T}{2\pi jt}\,(e^{j\pi t/T} - e^{-j\pi t/T}) = \frac{T}{\pi t}\sin\frac{\pi t}{T}. \tag{2.18}
$$

We also note that

$$f^*(t) = \sum_{k=-\infty}^{\infty} f(kT). \tag{2.19}$$

Now, the time-domain equivalent of equation (2.17) is

$$f(t) = p(t) * f^*(t), \tag{2.20}$$

where $*$ indicates the operation of convolution.

Hence

$$f(t) = \sum_{k=-\infty}^{\infty} f(kT) \frac{T}{\pi(t - kT)} \sin \frac{\pi(t - kT)}{T}. \qquad (2.21)$$

This equation can be regarded as the ideal against which other approaches can be judged. It allows the perfect reconstruction of $f(t)$ from its samples.

Notice however that the reconstruction requires a knowledge of future values of the sampled signal. Hence the approach cannot be used in a real-time situation where only current and past values of $f(kT)$ are available.

2.9 Approximate reconstruction of f(t) from f*(t) using hold devices

A method for approximate reconstruction of $f(t)$ that does not require future values of $f*(t)$ to be available is to fit an nth-order polynomial through the last $n + 1$ points of $f*(t)$ and extrapolate forward in time along this polynomial for each period T. Figure 2.14 illustrates the approach for $n = 2$, 1 and 0. A device that carries out this polynomial extrapolation is called an nth-order hold, where n is the order of the approximating polynomial.

The zero-order hold, although apparently crude, is the device that is normally used in practice. It has the transfer function

$$G_0(s) = \frac{1}{s} - \frac{e^{-sT}}{s} \qquad \text{or} \qquad G_0(j\omega) = \frac{1 - e^{-j\omega T}}{j\omega}. \qquad (2.22)$$

2.10 An approach to the analysis of a control loop containing a digital controller

An analog-to-digital converter approximates fairly closely the action of the ideal sampler postulated in Section 2.6. Therefore in the analysis of control loops we usually make the assumption that A/D conversion produces the signal $f*(t)$ from an input $f(t)$.

Digital-to-analog conversion, followed by a zero-order (or indeed by a higher-order) hold device cannot be assumed to yield $f(t)$ from an input of $f*(t)$. Such an assumption would be grossly inaccurate since, for instance, the output of the zero-order hold device is piecewise constant over sampling intervals.

As we have stated earlier, in a control context, it is the effect of the signal on the process that is important, rather than any consideration of accuracy of signal reconstruction. These effects are taken into account explicitly by including

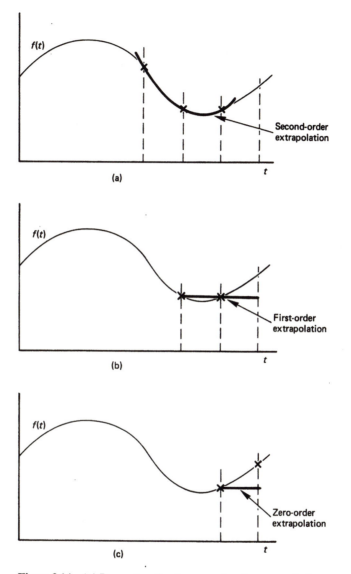

Figure 2.14   (a) Reconstruction by second-order extrapolation;
(b) Reconstruction by first-order extrapolation; (c) Reconstruction
by zero-order extrapolation.

a model of the hold device, before the process model, in the control loop. Usually, therefore, no assumption is made or needs to be made about the accuracy of reconstruction of a digital signal from the computer to a process.

Consider Figure 2.15. A continuous process with transfer function $G(s)$ is under digital control. The process output $y(s)$, after A/D conversion, becomes

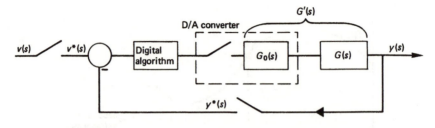

Figure 2.15 · A continuous process under digital control.

$y^*(s)$. The output from the digital algorithm passes to the dashed box marked D/A converter.

As shown in the figure, the D/A converter is modelled by an ideal sampler in series with a zero-order hold device. The zero-order hold device has the transfer function $G_0(s)$.

It is very convenient to analyze the loop by considering the two transfer functions $G_0(s)$, $G(s)$ to represent a modified process transfer function $G'(s)$. In this way the effects of non-ideal signal reconstruction by the hold device are automatically taken into account.

2.11 Other methods of sampling

It should be pointed out that several other methods of sampling are possible, apart from the method we have treated.

For instance, sampling at non-uniform intervals is sometimes encountered. Such sampling necessarily requires an extension of the analytical methods we have given.

Pulse-width modulation is used in certain applications. Here, the pulses are of fixed height and information is carried by the variable width of the pulses.

This type of modulation is particularly useful for achieving variable-speed control of a d.c. motor supplied from a constant voltage source. Despite the discontinuous nature of the applied voltage, the inertia of motor and load is sufficient in most applications to ensure that a virtually constant speed corresponds with each choice of pulse width.

These variations in sampling technique do not really enter the mainstream of digital control theory and they are not pursued further in this text. See Reference C2 for further details.

Exercises

2.1 A disk marked by a single spot rotates at 256 revolutions per second. It is illuminated by a stroboscope that flashes 2^k times per second. What is the least integer value for k that will reveal that the disk is rotating?

2.2 A circular potentiometer is used to measure shaft position. In a computer control system, the potentiometer output is sampled every T seconds. Calculate the longest value of T that can be used without introducing ambiguity of position measurement if the maximum shaft velocity is $100\,\text{rad/s}$. (Note that as the shaft rotates, the potentiometer produces a sawtooth output voltage.) Observe that the reasoning required parallels that of Shannon's theorem.

2.3 A function generator is known to produce a voltage $v(t) = \sin \omega t$, yet samples taken every T seconds indicate a constant voltage. What is the smallest value of T for which this is possible?

2.4 Explain the principal reasons for loss of information and generation of

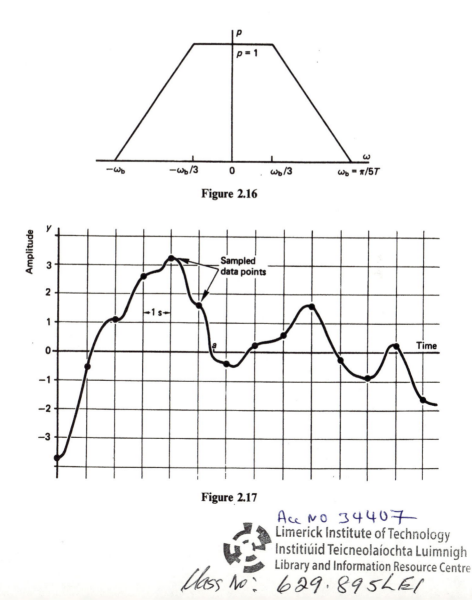

Figure 2.16

Figure 2.17

errors when a continuous signal is subject to sampling followed by reconstruction.

2.5 A continuous signal has the spectrum shown in Figure 2.16. Plot the spectrum of the sampled signal for the two cases; (a) $q/T = 0.1$, $T = 1$ (q is the duration of the sample); (b) ideal sampling with $T = 1$.

(*Note*: You may wish to make use of equations (2.10), (2.12) and (2.15).)

2.6 To obtain a feeling for equation (2.21) and for its operation as a reconstructor, write a computer program that will solve the equation for summations of k over the range (a) $(-2, 2)$, (b) $(-5, 5)$. Test the algorithm on the sampled data points of the graph in Figure 2.17 to obtain an estimate for the value of y at point a. Comment critically on the accuracy of reconstruction and the computational practicability of the algorithm. (Notice how reconstruction in real time suffers from the disadvantage that future data points are necessarily not available—an algorithm based on equation (2.21) can only be applied after some modification to make it asymmetrically dependent only on past sample values.) If the reconstruction is poor, explain why this is so and repeat the exercise under modified conditions to obtain better results.

3

$\mathscr{Z}$ Transform Techniques

3.1 Introduction

The $\mathscr{Z}$ transform is the principal analytical tool for single-loop discrete-time systems.

This chapter is concerned with establishing a good understanding of the properties of the transform and giving sufficient practice in its manipulation. On this latter point, it is felt that although readers may be fully familiar with the Laplace transform (with which the $\mathscr{Z}$ transform has many similarities), they should nevertheless carefully work the examples at the end of the chapter since there are important differences in the methods of manipulation.

The fundamental theorems of this chapter are given without proof. They rest on a foundation of complex variable theory that can be found in texts on complex analysis. Readers might with advantage review the foundations by reading the chapter on complex integrals in Reference K8.

3.2 The $\mathscr{Z}$ transform

The $\mathscr{Z}$ transform is to discrete-time systems what the Laplace transform is to continuous-time systems, the symbol z being the analog of the symbol s. It allows concise representation, convenient manipulation and time solution and leads into methods for system analysis and design. There are many similarities between the two approaches and a familiarity with the Laplace transform will assist in understanding the $\mathscr{Z}$ transform. Conceptually, the symbol z can be associated with time shifting in a difference equation in the same way that s can be associated with differentiation in a differential equation.

3.2.1 The $\mathscr{Z}$ transform of a time function

A discrete-time function $f*(t)$ can be Laplace transformed

$$\mathscr{L}\{f*(t)\} = \int_0^\infty f*(t)\,e^{-st}\,dt. \tag{3.1}$$

We shall sometimes use the notation

$$\mathscr{L}\{f*(t)\} = F*(s). \tag{3.2}$$

The function $f*(t)$ is regarded as taking the value zero except at instants where $t = kT$, where k takes only integral values.

Hence

$$\int_0^\infty f*(t)\,e^{-st}\,dt = \sum_{k=0}^\infty f(kT)\,e^{-skT} \tag{3.3}$$

$$= \sum_{k=0}^\infty f(kT)(e^{st})^{-k} = F*(s). \tag{3.4}$$

The $\mathscr{Z}$ transform of $f(t)$ is defined as being equal to the Laplace transform of $f*(t)$: that is,

$$\mathscr{Z}\{f(t)\} \triangleq \mathscr{L}\{f*(t)\}. \tag{3.5}$$

If $\mathscr{Z}\{f(t)\}$ is designated $F(z)$ then, equivalently to equation (3.4), we have

$$F(z) = F*(s) = \sum_{k=0}^\infty f(kT)(e^{st})^{-k}. \tag{3.6}$$

Finally, if we define $z = e^{sT}$, then

$$F(z) = \sum_{k=0}^\infty f(kT)z^{-k}. \tag{3.7}$$

Notice carefully that in the calculation of $F(z)$, the signal $f*(t)$ (rather than $f(t)$) is used. We can imagine that equation (3.7), by choosing to ignore the signal except at sampling instants, has introduced a fictitious (i.e. not physically present) sampler into the system. Clearly both $f(t)$ and $f*(t)$ have the same $\mathscr{Z}$ transform (see Figure 3.1).

Equation (3.7) is the fundamental, defining relation for the $\mathscr{Z}$ transform of a time function and it must be memorized. It is the preferred starting point for proving 'from first principles' properties of the $\mathscr{Z}$ transform. It also furnishes a numerical method for the calculation of $\mathscr{Z}$ transforms of particular time functions.

Example of the use of equation (3.7) The sequence $\{1, 2, 4, 8, 16, \ldots\}$ may be written as 2^k, $k = 0, \ldots$. Find the $\mathscr{Z}$ transform of this sequence using the defining equation (equation (3.7)).

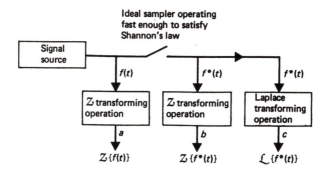

Figure 3.1 A diagram to assist the visualization of Section 3.2.1—the three signals a, b, c are identical.

Notice that a time interval T is not stated in the problem and here we can think of k as indexing the sequence or, if we insist on a time series visualization, we can suppose that $T = 1$. In either case we obtain

$$F(z) = \sum_{k=0}^{\infty} f(k)z^{-k} = \sum_{k=0}^{\infty} 2^k z^{-k}$$
$$= 1 + 2z^{-1} + 4z^{-2} + 8z^{-3}.$$

Seeking a closed-form solution (that is, one not involving an infinite series), we note that

$$\frac{1}{z-2} = z^{-1} + 2z^{-2} + 4z^{-3} + 8z^{-4}.$$

Therefore $F(z) = z/(z-2)$ is the required closed-form solution.

3.2.2 The $\mathscr{Z}$ transform corresponding with a particular Laplace transform

It is often required to determine the $\mathscr{Z}$ transform corresponding with a particular Laplace transform. Be very careful to note that, although we may denote the $\mathscr{Z}$ transform equivalent of $G(s)$ by $G(z)$, $G(z)$ is *not obtained by simply substituting z for s in $G(s)$*. The basic approach is to follow the sequence $G(s) \rightarrow g(t) \rightarrow G(z)$, i.e. to inverse Laplace transform $G(s)$ into a time function $g(t)$, which is then $\mathscr{Z}$ transformed. If partial fractions are used before inverse Laplace transforming, then the elements of $g(t)$ are usually simple enough to be found in a table such as that in Appendix C and the $\mathscr{Z}$ transformation is easily achieved. The requirement to move from s to z representation occurs so frequently that a table of equivalent S and $\mathscr{Z}$ transforms will be found convenient, and such a table is included in Appendix C.

Another method of converting a transform $G(s)$ into the equivalent transform $G(z)$ is the following.

Assume that the Laplace transform $G(s)$ is available and that $G(s)$ can be expressed in the form $G(s) = N(s)/D(s)$, where $D(s)$ has a finite number of distinct roots. Then

$$G(z) = \sum_{n=1}^{p} \frac{N(x_n)}{D'(x_n)} \frac{1}{1 - e^{x_n T} z^{-1}}, \qquad (3.8)$$

where $D' = \partial D / \partial s$ and x_n, $n = 1, \ldots, p$ are the roots of the equation $D(s) = 0$.

(The derivation of equation (3.8) is omitted. The equation is obtained by application of Cauchy's residue theorem using contour integration around the poles and zeros of $G(s)$ in the complex plane.)

Example Let

$$G(s) = \frac{1}{s^2 + (a + b)s + ab}.$$

Determine $G(z)$ by the two alternative methods given earlier.

$$G(s) = \frac{1}{(s + a)(s + b)}$$

$$g(t) = \mathscr{L}^{-1}\{G(s)\} = \frac{1}{b - a}(e^{-at} - e^{-bt}).$$

From the basic definition,

$$G(z) = \sum_{k=0}^{\infty} \frac{1}{b - a}(e^{-akT} - e^{-bkT})z^{-k}$$

$$= \frac{1}{b - a}[(1 + e^{-aT}z^{-1} + e^{-2aT}z^{-1} + \cdots)$$

$$- (1 + e^{-bT}z^{-1} + e^{-2bT}z^{-2} + \cdots)]$$

$$= \frac{1}{b - a}\left[\frac{z(e^{-aT} - e^{-bT})}{(z - e^{-aT})(z - e^{-bT})}\right].$$

Using the second method

$$G(s) = \frac{N(s)}{D(s)},$$

where

$$N(s) = 1, \qquad D(s) = (s + a)(s + b), \qquad D'(s) = 2s + a + b;$$

$$G(z) = \sum_{n=1}^{p} \frac{N(x_n)}{D'(x_n)} \frac{1}{1 - e^{x_n T} z^{-1}}.$$

Putting $x_1 = -a$, $x_2 = -b$ in (3.8),

$$G(z) = \frac{1}{-a+b}\frac{1}{1-e^{-aT}z^{-1}} + \frac{1}{a-b}\frac{1}{1-e^{-bT}z^{-1}}$$

$$= \frac{1}{b-a}\left[\frac{z(e^{-aT}-e^{-bT})}{(z-e^{-aT})(z-e^{-bT})}\right].$$

3.2.3 Considering z as an operator

It may be helpful to consider z as a shift operator that moves a time function one discrete time step forward. Thus

$$zF(z) = \mathscr{L}\{f[(k+1)T]\}, \tag{3.9}$$

$$z^nF(z) = \mathscr{L}\{f[(k+n)T]\}, \tag{3.10}$$

$$z^{-n}F(z) = \mathscr{L}\{f[(k-n)T]\}. \tag{3.11}$$

This shift property is listed again together with other properties of z, in Section 3.2.4. It will be seen in that section that for completeness, equations (3.9), (3.10) need to include initial condition effects. Essentially the $\mathscr{L}$ transform operates on an infinite sequence to yield an infinite series, as the following examples make clear.

Example Let $u(k)$ represent the sequence $u(k) = \{0, 0, 1, 0, 0, \ldots\}$. Then

$$\mathscr{L}\{u(k)\} = \sum_{k=0}^{\infty} u(k)z^{-k} = z^{-2}.$$

Example Let $u(k)$ represent the sequence $u(k) = \{a^0, a^1, a^2, \ldots\}$. Then

$$\mathscr{L}\{u(k)\} = \sum_{k=0}^{\infty} u(k)z^{-k} = \sum_{k=0}^{\infty} a^k z^{-k} = \sum_{k=0}^{\infty} \left(\frac{a}{z}\right)^k.$$

Probably the most useful concrete visualization of z is to consider z^{-1} as representing a unit delay.

Simple illustrative example of the use of equation (3.8): $\mathscr{L}$ transform of the unit step function $h(t)$ The unit step function $h(t)$ is defined by

$$\left.\begin{array}{ll} h(t) = 0, & t < 0 \\ h(t) = 1, & t \geq 0. \end{array}\right\} \tag{3.12}$$

Let $H(s) = \mathscr{L}\{h(t)\}$ then $H(s) = 1/s$ and we have for substitution into equation. (3.8),

$$H(s) = \frac{1}{s}, \qquad N(s) = 1, \qquad D(s) = s, \qquad D'(s) = 1, \qquad x_1 = 0,$$

$$H(z) = \frac{1}{1 - e^0 z^{-1}} = \frac{z}{z - 1}. \tag{3.13}$$

Check from first principles, using equation (3.7)   Substitution of equation (3.12) into equation (3.7) yields

$$H(z) = \sum_{k=0}^{\infty} z^{-k}.$$

$\mathscr{L}$ transforms are wherever possible put into closed forms. Here we can express $H(z)$ in the form

$$H(z) = \frac{1}{1 - z^{-1}} = \frac{z}{z - 1}, \tag{3.14}$$

agreeing with equation (3.13).

3.2.4 Properties of the $\mathscr{L}$ transform

1. *Linearity*

$$\mathscr{L}\{f_1(t) + f_2(t)\} = \mathscr{L}\{f_1(t)\} + \mathscr{L}\{f_2(t)\} = F_1(z) + F_2(z) \tag{3.15}$$

(for any transformable functions f_1, f_2);

$$\mathscr{L}\{\alpha f(t)\} = \alpha \mathscr{L}\{f(t)\} = \alpha F(z) \tag{3.16}$$

(for any scalar α).

2. *Shift property*

$$\mathscr{L}\{f(t + nT)\} = z^n \left[F(z) - \sum_{k=0}^{n-1} f(kT) z^{-k} \right]. \tag{3.17}$$

3.

$$\mathscr{L}\{e^{-at} f(t)\} = F(e^{aT} z). \tag{3.18}$$

4. *Final-value theorem*

$$\lim_{k \to \infty} f(kT) = \lim_{z \to 1} (1 - z^{-1}) F(z) \tag{3.19}$$

$$= \lim_{z \to 1} (z - 1) F(z)$$

provided that all poles of $(1 - z^{-1}) F(z)$ are inside the unit circle.

5. *Convolution*

$$F_1(z)F_2(z) = \mathcal{L}\left\{\sum_{n=0}^{k} f_1(nT)f_2((k-n)T)\right\}$$

$$= \mathcal{L}\{f_1(t) * f_2(t)\} \qquad (3.20)$$

where $*$ indicates convolution.

PROOF OF THE PROPERTIES

The statements above can be proved by use of the fundamental definition of the $\mathcal{L}$ transform.

Example   Proof of property 2

$$\mathcal{L}\{f(t + T)\} = \mathcal{L}\{f((k + 1)T)\}$$

where $kT = t$. Then

$$\mathcal{L}\{f((k + 1)T)\} = \sum_{k=0}^{\infty} f((k+1)T)z^{-k}$$

$$= z \sum_{k=0}^{\infty} f((k+1)T)z^{-(k+1)}$$

$$= z\left(f(0) + \sum_{k=0}^{\infty} f((k+1)T)z^{-(k+1)} - f(0)\right)$$

$$= z \sum_{k'=0}^{\infty} f(k'T)z^{-k'} - zf(0)$$

where $k' = k + 1$. Thus

$$\mathcal{L}\{f((k + 1)T)\} = z\mathcal{L}\{f(kT)\} - zf(0).$$

(Note that, here and elsewhere in the book, $f(0)$ denotes the value of f at 0^+.)

$$\mathcal{L}\{f(t + nT)\} = z^n\left[F(z) - \sum_{k=0}^{n-1} f(kT)z^{-k}\right].$$

Proof

$$\mathcal{L}\{f(t + nT)\} = \sum_{k=0}^{\infty} f(kT + nT)z^{-k}$$

$$= z^n \sum_{k=0}^{\infty} f(kT + nT)z^{-(k+n)}$$

$$= z^n\left[F(z) - \sum_{k=0}^{n-1} f(kT)z^{-k}\right].$$

Note that since it is usual to define $f(t) = 0$ for $t < 0$,

$$\mathscr{L}\{f(t - nT)\} = z^{-n}F(z).$$

Example   Outline of the proof of property 4   We argue plausibly that:

(a) If all poles of $F(z)$ are inside the unit circle, then $f(kT) \to 0$ as $k \to \infty$.
(b) If $F(z)$ has any pole outside the unit circle then $f(kT)$ is unbounded and this case is ruled out by the conditions stated with Property 4.
(c) If $F(z)$ has a pole on the unit circle except at $z = 1$, the solution is oscillatory and never settles.
(d) The only case remaining where $\lim_{k \to \infty} f(kT)$ may be finite and constant and which satisfies the conditions of the test is when $F(z)$ has a pole at $z = 1$. Thus the value of $f(kT)$ is equal to the constant term generated by the pole at $z = 1$.

This can be found by partial fraction expansion of $F(z)$, when the required quantity will be given as the numerator of the term in $1/(z-1)$; that is,

$$\lim_{k \to \infty} f(kT) = \lim_{z \to 1} (z - 1)F(z).$$

Numerical example of the final-value theorem   Let

$$F(z) = \frac{z}{(z - 0.2)(z - 1)}.$$

Applying rule 4 (the final-value theorem),

$$\lim_{k \to \infty} f(kT) = \lim_{z \to 1} (1 - z^{-1})F(z)$$

$$= \frac{(1 - z^{-1})z}{(z - 0.2)(z - 1)}\bigg|_{z \to 1}$$

$$= \frac{(z - 1)}{(z - 0.2)(z - 1)}\bigg|_{z \to 1} = \frac{1}{0.8} = 1.25.$$

(The complete solution is found by noting that

$$F(z) = \frac{-0.25}{(z - 0.2)} + \frac{1.25}{(z - 1)}.$$

Then from tables (as will be explained in Section 3.5)

$$f(kT) = -0.25\,e^{-akT} + 1.25,$$

where a satisfies $e^{-aT} = 0.2$.)

3.3 The pulse transfer function

Modified notation   In the interests of clear exposition we shall now start to use lower case symbols $f(z)$, $y(z)$, $u(z)$ etc. to represent signals instead of $F(z)$, $Y(z)$, $U(z)$ etc. We shall reserve capital symbols such as $G(z)$, $H(z)$ to represent system elements.

The pulse transfer function of a dynamic element is denoted $G(z)$ and is defined by the relation

$$G(z) = \frac{y(z)}{u(z)}, \tag{3.21}$$

where u, y are the input and output variables respectively of the dynamic element.

Note that $G(z)$ relates the input and output signals of a dynamic element only at discrete times separated by intervals T. The function $G(z)$ establishes a relation between $u(z)$ and $y(z)$ or equivalently between $u^*(s)$ and $y^*(s)$ as Figure 3.2 makes clear.

We shall usually refer to $G(z)$ simply as a transfer function, the symbol z making it clear that we are discussing the pulse transfer function.

3.3.1 Manipulation of block diagrams

The presence or absence of a sampler in a signal line affects the $\mathscr{Z}$ transform of the system very significantly. The rules given below will help the manipulation, which in general, needs to be carried out in terms of s rather than z (why is this?).

Some helpful rules and comments

1. Taking the $\mathscr{Z}$ transform of a time function involves the assumption that a sampler is present. If no such sampler is actually present we refer to this as fictitious sampling.
2. Let $y(s) = G(s)u(s)$; then

$$y(z) = y^*(s) = (G(s)u(s))^* \neq G^*(s)u^*(s) = G(z)u(z). \tag{3.22}$$

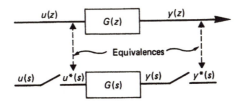

Figure 3.2   Illustrating the relation between $G(z)$ and $G(s)$.

3. Let $y(s) = G(s)u^*(s)$; then

$$y^*(s) = (G(s)u^*(s))^* = G^*(s)u^*(s) = G(z)u(z) \qquad (3.23)$$

(compare with 2).

Sketch of Proof of equation (3.23)   Let $y(s) = G(s)u^*(s)$; then

$$y^*(s) = \frac{1}{T}\sum_{n=-\infty}^{\infty} y(s + jn\omega_s) = \frac{1}{T}\sum_{n=-\infty}^{\infty} g(s + jn\omega_s)u^*(s + jn\omega_s).$$

Now $u^*(s + jn\omega_s) = u^*(s)$; therefore

$$y^*(s) = u^*(s)\frac{1}{T}\sum_{n=-\infty}^{\infty} G(s + jn\omega_s) = G^*(s)u^*(s).$$

Intuitively, equation (3.22) tells us that once two continuous functions have been combined by convolution, they must thenceforth be treated as an entity for purposes of $\mathscr{L}$ transformation.

Equation (3.23) tells us that if at least one of the continuous functions has been sampled before continuous convolution, then $\mathscr{L}$ transformation of the product is equal to the product of the $\mathscr{L}$ transformations. (Note that of course $(u^*(s))^* = u^*(s)$, since sampling an already-sampled signal has no further effect.)

It is very important to become proficient in manipulation of block diagrams with different dispositions of samplers. In this connection it is suggested that the following examples be worked in detail and that all the block diagram manipulation exercises at the end of the chapter be worked through.

$\mathscr{L}$ **transfer functions in series**   With reference to Figure 3.3,

$$y^*(s) = G_2^*(s)x^*(s) = G_2^*(s)G_1^*(s)u^*(s),$$

$$\frac{y(s)}{u(z)} = G_2(z)G_1(z). \qquad (3.24)$$

Note carefully that this configuration implies the existence of a sampling operation between the two dynamic elements. Compare with the situation below.

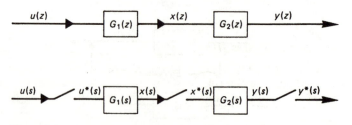

Figure 3.3 $\mathscr{L}$ transfer functions in series—the two diagrams are equivalent.

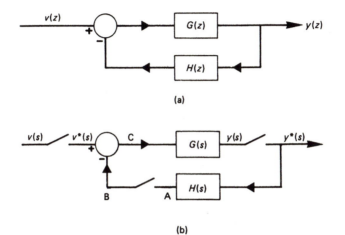

Figure 3.4 S transfer functions in series preceded by a sampler.

(a)

(b)

Figure 3.5 $\mathscr{Z}$ transfer functions in a feedback loop—the two
diagrams are equivalent.

S transfer functions in series, preceded by a sampler   Since $\mathscr{Z}$ transformation
gives results only at sampling instants, the technique can be considered to
impose a fictitious sampler on the system output (see Figure 3.4):

$$\frac{y^*(s)}{u^*(s)} = \frac{y(z)}{u(z)} = \mathscr{Z}\{G_1(s)G_2(s)\} \triangleq G_1G_2(z). \qquad (3.25)$$

$\mathscr{Z}$ transfer functions in a feedback loop   For this type of loop we have

$$\frac{y(z)}{v(z)} = \frac{G(z)}{1 + G(z)H(z)}. \qquad (3.26)$$

The derivation is obtained as follows (refer to Figure 3.5):

At point A the signal is $H(s)y^*(s)$

At point B the signal is $(H(s)y^*(s))^* = H^*(s)y^*(s)$. (Be sure to understand
this step, which uses the rule described by equation (3.23).)

At point C the signal is $v^*(s) - H^*(s)y^*(s)$.

Therefore

$$y(s) = G(s)(v^*(s) - H^*(s)y^*(s))$$

and

$$y^*(s) = (G(s)(v^*(s) - H^*(s)y^*(s)))^*$$
$$= G^*(s)v^*(s) - G^*(s)H^*(s)y^*(s).$$

Then

$$y(z) = G(z)v(z) - G(z)H(z)y(z)$$
$$y(z)(1 + G(z)H(z)) = G(z)v(z).$$

Finally

$$\frac{y(z)}{v(z)} = \frac{G(z)}{1 + G(z)H(z)}$$

as required.

Example In Figure 3.5(b) let $G(s) = 1/(s + 1)$ and $H(s) = 1/s$. Determine the overall transfer function of the configuration.

From the table in Appendix B we have

$$G(z) = z/(z - e^{-T}), \qquad H(z) = z/(z - 1).$$

The overall transfer function is given by

$$\frac{G(z)}{1 + G(z)H(z)} = \frac{z/(z - e^{-T})}{1 + [z/(z - e^{-T})][z/(z - 1)]}$$
$$= \frac{z(z - 1)}{(z - e^{-T})(z - 1) + z^2} = \frac{z(z - 1)}{2z^2 - z(1 + e^{-T}) + e^{-T}}.$$

3.4 A brief description of three methods for obtaining information on system behavior between sampling instants

A continuous process whose output is sampled periodically may appear to be performing satisfactorily according to the sampled information even though large deviations are occurring in the intervals between sampling instants. Unless the sampling theorem can be invoked with a guarantee that such behavior is impossible, it may be necessary to investigate the behavior between sampling instants.

Suppose that it is required to investigate the behavior of the output of the system shown in Figure 3.6 between sampling instants.

Method 1 (Refer to Figure 3.7) Here the input sampler is preceded by a fictitious sampler operating n times faster than the actual input sampler. We

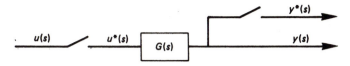

Figure 3.6　The system whose behavior between sampling instants is to be investigated.

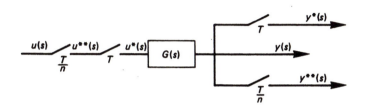

Figure 3.7　Determination of behavior between sampling instants, Method 1.

denote the more rapidly sampled signal by $u^{**}(s)$. Note that since n is an integer, $u^*(s)$ is not modified by the presence of the additional sampler.

The output $y^{**}(s)$ from the additional rapid sampler at the output is able to furnish information on the output response between sampling instants.

The procedure for applying the method can be derived from first principles or more quickly as follows. We argue that we can obtain the information we want by leaving the input signal alone and introducing fictitious rapid samplers into the process transfer function $G(z)$. Suppose we require to obtain output information every T/n seconds. We denote the transfer function $G(z)$ with rapid fictitious samplers by the symbol $G(z)_n$ and argue that the equation

$$y(z)_n = G(z)_n u(z)$$

will furnish the information we need. Now

$$G(z)_n = \sum_{k=0}^{\infty} g\left(\frac{kT}{n}\right) z^{-k/n}$$

from the definition of the $\mathscr{Z}$ transform, hence

$$G(z)_n = G(z)\Big|_{\substack{z=z^{1/n} \\ T=T/n}}$$

The example at the end of this section makes the method clear. (Note that when $G(s)$ contains the quantity T, as it will do when a hold device is incorporated into the process, then no substitution of $T = T/n$ must be made for any such T. To make such a substitution would erroneously redefine the period of the hold.)

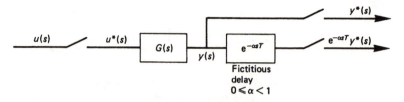

Figure 3.8   Determination of behavior between sampling instants, Method 2.

Method 2   (Refer to Figure 3.8)   A fictitious variable delay is introduced at the output. As α runs through values between 0 and 1, values of the output between sampling instants are generated. The approach is sometimes called the *modified $\mathscr{L}$ transform method*, since the method as usually applied involves modifying the system transfer functions so that they explicitly contain the variable delay term. We do not explore the method further here, since methods 1 and 3 offer adequate mechanisms for inter-sample information to be determined.

Method 3   Method 3 is not so much a method as the application of straightforward reasoning to each particular situation that is to be investigated.
   For instance:

(a) The output of a continuous system preceded by a sampler can be calculated (continuously) numerically by superimposing impulse responses.
(b) The output of a continuous system preceded by a sample and zero-order hold can be calculated (continuously) numerically by superimposing step responses.
(c) The behavior of a continuous system in a discrete-time closed loop can be calculated (continuously) numerically one sample interval at a time, the system input being updated at every sampling instant.

Example   Determination of behavior between sampling instants   A process of transfer function $G(s) = 1/(s + 1)$ is preceded by a sampler operating with a period $T = 3.4657$ seconds. It is subjected to a step input.

(a) Calculate the system output using the normal $\mathscr{L}$ transform;
(b) calculate the system output at intervals of $T/5$ using the method of rapid fictitious sampling;
(c) confirm the result in (b) by classical transform methods applied over each interval T.

   (a) *Normal $\mathscr{L}$ transform analysis*

$$y(z) = G(z)u(z) = \frac{z}{z - e^{-T}} u(z) = \frac{z}{z - 0.03125} u(z).$$

In the time domain, the step response is generated by the equation

$$y(k) = 0.03125y(k-1) + 1,$$

which yields the solution

k	0	1	2	3	4
y(k)	1	1.03125	1.03223	1.03226	1.03226

(Note how, in this solution, we have preferred a difference equation route instead of using long division.)

(b) *Calculation every T/5 second*

$$y(z) = G(z)_5 u(z) = \frac{z^{1/5}}{z^{1/5} - e^{-T/5}} \frac{z}{z-1}.$$

Setting $\gamma = z^{1/5}$ and evaluating $e^{-T/5}$ yields

$$y(z)_5 = \frac{\gamma^6}{(\gamma - 0.5)(\gamma^5 - 1)}.$$

$$y(z)_5 = \frac{\gamma^6}{\gamma^6 - 0.5\gamma^5 - \gamma + 0.5}.$$

Then by long division (why did we not use the difference equation method here?),

$$y(z)_5 = 1 + 0.5\gamma^{-1} + 0.25\gamma^{-2} + 0.125\gamma^{-3} + 0.0625\gamma^{-4} + 1.03125\gamma^{-5} + \cdots.$$

The result over the first period T is therefore given as

k	0	0.2	0.4	0.6	0.8	1
y(k)	1	0.5	0.25	0.125	0.0625	1.03125

Table 3.1

k	y(k)	k	y(k)	k	y(k)
0	1	1	1.03125	2	1.03223
0.1	0.707	1.1	0.729	2.1	0.730
0.2	0.5	1.2	0.515	2.2	0.516
0.3	0.354	1.3	0.365	2.3	0.365
0.4	0.25	1.4	0.258	2.4	0.258
0.5	0.176	1.5	0.182˙	2.5	0.182
0.6	0.125	1.6	0.129	2.6	0.129
0.7	0.088	1.7	0.091	2.7	0.091
0.8	0.0625	1.8	0.065	2.8	0.065
0.9	0.044	1.9	0.046	2.9	0.046
				3	1.03226

We note immediately how the $\mathscr{L}$ transform analysis of (a) failed to indicate the true nature of the response that is now revealed.

(c) *Classical continuous analysis over each period T*
The process has the response within each sampling period

$$y(kT + t) = y(kT)e^{-(t-kT)}.$$

We use the equation to calculate the response every 0.1 second over the first three sampling intervals. The results are given in Table 3.1.

3.5 Inverse $\mathscr{L}$ transformation to yield time solutions

3.5.1 Inversion by means of the method of residues

The fundamental relation connecting a $\mathscr{L}$ transform representation with its time response is

$$f_k = \frac{1}{2\pi j} \oint_C f(z)z^{k-1}\, dz, \tag{3.27}$$

where C is a closed path that encloses all the singularities of the integrand.

Note that a discussion of the mathematical background to equation (3.27) would include consideration of the analyticity of the integrand, the question of k taking on negative values and the convergence of the integral.

For most physically meaningful applications the function $f(z)$ is rational and causal and, under these restricted conditions, the expression is easily evaluated using the theory of residues. The *residue* of a function $f(z)$ at a particular pole p_1, is defined in terms of a Laurent expansion of the function in the region of p_1. Let the Laurent series be written

$$f(z) = A_{-n}(z - p_1)^{-n} + \cdots + A_{-1}(z - p_1)^{-1} + A_0 + A_1(z - p_1) + \cdots \tag{3.28}$$

then the residue of $f(z)$ at the pole p_1 is defined to be equal to A_{-1}.

Complex variable theory then tells us (see for instance K4) that provided that each of the poles of $f(z)$ is simple (i.e. not repeated), then

$$\frac{1}{2\pi j} \oint f(z)z^{k-1}\, dz = \sum_{i=1}^{n} r_i, \tag{3.29}$$

where r_i is the residue of the integrand at the ith pole. Provided that the integrand has simple poles, then each of the residues r_i is determined from the relation

$$r_i = \lim_{z \to p_i} (z - p_i)f(z)z^{k-1}. \tag{3.30}$$

The above expressions allow inverse transformation of particular functions $f(z)$ to be carried out.

Example Let

$$f(z) = \frac{z}{(z-1)(z-2)(z-3)}. \tag{3.31}$$

The poles are at $p_1 = 1$, $p_2 = 2$, $p_3 = 3$

$$f(k) = \sum_{\lambda=1}^{3} \lim_{z \to p_i} (z - p_i) \frac{z}{(z-1)(z-2)(z-3)} z^{k-1}$$

$$= \frac{1}{2} - 2^k + \frac{3^k}{2}.$$

Example

$$f(z) = \frac{(1-c)z}{(z-1)(z-c)}$$

where $c = e^{aT}$. Inverse transformation by the residue method proceeds as follows.

$$f(kT) = \text{sum of residues of } f(z)z^{k-1}$$

$$= \frac{(1-c)z^k}{z-c} \bigg|_{z \to 1} + \frac{(1-c)z^k}{z-1} \bigg|_{z \to c}$$

$$= 1 - c^k = 1 - e^{akY}$$

(agreeing with an entry in the table in Appendix B).

The residue method of inversion for a denominator with repeated roots
Where $f(z)$ has a repeated pole of order m at z_0, the residue at that pole is found by using the following formula:

$$\text{Residue at } z_0 = \frac{1}{(m-1)!} \lim_{z \to z_0} \frac{d^{m-1}[(z-z_0)^m F(z)z^{k-1}]}{dz^{m-1}}. \tag{3.32}$$

Do not be put off by the somewhat forbidding nature of the formula—it is in fact simple and convenient to use, as will be demonstrated by an example.

Derivation of the formula is omitted since it would take us too far into complex variable theory, involving Cauchy's integral formula and the Laurent expansion.

Example

$$f(z) = \frac{cTz}{(z-c)^2}$$

where $c = e^{-aT}$. Invert the transform by the method of residues.

Here we note the repeated pole and apply the rule

$$\text{Residue at } z = c = \frac{1}{(m-1)!} \frac{d^{m-1}[(z-c)^m F(z)z^{k-1}]}{dz^{m-1}}\Bigg|_{z \to c},$$

where m is the order of the pole. Since $m = 2$, we obtain

$$f(kT) = \frac{d}{dz}((z-c)^2 F(z)z^{k-1})\Bigg|_{z \to c}$$

$$= \frac{d}{dz}(cTz^k)\Bigg|_{z \to c} = kTc^k$$

$$= kT\,e^{-akT}.$$

(This transform pair will be found in the tables, Appendix C.)

3.5.2 Transform inversion by partial fraction expansion followed by recognition of inverse transforms in tables

The rules for expansion of a function in partial fractions are assumed known. (If not, almost any text treating. Laplace transform applications, for example K8, can be consulted.) It is not difficult to see how the method could be derived from the inversion integral (equation (3.27)). The method is adequately explained by an example.

Example   Let

$$f(z) = \frac{z}{(z-1)(z-2)(z-3)} = \frac{1}{2(z-1)} - \frac{2}{(z-2)} + \frac{3}{2(z-3)}.$$

Notice that most of the $\mathscr{L}$ transforms in the table, Appendix C, have a z term in the numerator. So that we can use the table, we write

$$zf(z) = \frac{z}{2(z-1)} - \frac{2z}{(z-1)} + \frac{3z}{2(z-3)};$$

then from the table

$$\mathscr{L}^{-1}\{zf(z)\} = \frac{1}{2} - 2 \times 2^k + \frac{3 \times 3^k}{2}.$$

Finally, using the shift property of the $\mathscr{L}$ operator,

$$\mathscr{L}^{-1}\{f(z)\} = \frac{1}{2} - 2 \times 2^{k-1} + \frac{3 \times 3^{k-1}}{2} = \frac{1}{2} - 2^k + \frac{3^k}{2}.$$

Notice also that a transform of the type $z/(z-a)$ can sometimes with advantage be written in the form $z/(z - e^{CT})$, for some constant C. For instance, some sets of tables use (without explanation) the first form when $a = \pm 1$ and the second form otherwise.

3.5.3 Transform inversion by long division

Direct division of the numerator by the denominator of a transform results in an infinite sequence representing the time response. (Where a closed-form solution is required, this method is not applicable.)

Again let

$$f(z) = \frac{z}{(z - 1)(z - 2)(z - 3)}$$

then long division yields

$$f(z) = z^{-2} + 6z^{-3} + 25z^{-4} + 90z^{-5} + \cdots$$

and by direct interpretation the solution has the form

k	1	2	3	4	5
$f(k)$	0	1	6	25	90

The reader is asked to check that the solution

$$\frac{1}{2} - 2^k + \frac{3^k}{2},$$

generated earlier, produces the same numerical solution.

3.6 Difference equations

3.6.1 Introduction

An nth-order difference equation is a relation of the general form

$$f(k, y(k), \Delta y(k), \dots, \Delta^n y(k)) = 0 \tag{3.33}$$

where $\Delta y(k)$ is defined as $y(k + 1) - y(k)$; $\Delta^2 y(k)$ is defined as $\Delta y(k + 1) - \Delta y(k)$, etc.

Equation (3.33) can be transformed into the alternative form

$$g(k, y(k), y(k + 1), \dots, y(k + n)) = 0. \tag{3.34}$$

This latter representation is called the *standard form* of a difference equation. In this book we use only the standard form and make no use of the first representation.

Difference equations arise in their own right. For instance, the Fibonacci equation, representing growth of a species, has the form

$$y(k) = y(k-1) + y(k-2).$$

The operation of a computer algorithm can often with advantage be represented by a difference equation. The variables in the difference equation are constant until updated by operation of another cycle of calculation, exactly as in the operation of a cyclic computer program.

Often, in this book, a difference equation will be a *discrete approximation* of some original differential equation representing a physical process. The variables in such an approximation are allowed to take on any magnitude but changes in magnitude are only allowed to take place at discrete time instants.

3.6.2 Solution by means of the *ℒ* transform

ℒ transform solution of a difference equation is the analog of Laplace transform solution of a differential equation.

The procedure involves:

(a) *ℒ* transformation of the difference equation;
(b) manipulation of the transformed expression;
(c) inverse *ℒ* transformation to yield the time solution.

The approach is illustrated by reference to a first-order equation.

A first-order difference equation is of the form

$$y(k+1) = ay(k) + bu(k), \qquad y(0) = y_0. \tag{3.35}$$

ℒ transformation yields

$$zy(z) - zy(0) = ay(z) + bu(z),$$
$$y(z) = y(0) + z^{-1}(ay(z) + bu(z)), \tag{3.36}$$
$$y(k) = \mathscr{L}^{-1}\{y(0) + z^{-1}(ay(z) + bu(z))\},$$

and a time solution of the difference equation can be obtained provided that the inverse transformation can be carried out. To carry the development further we take a particular second-order difference equation

$$y(k+2) - 5y(k+1) + 6y(k) = u(k) \tag{3.37}$$

in which $u(k) = 1$, for all $k \geq 0$, and where the initial conditions $y(0)$, $y(1)$ are both zero. *ℒ* transforming leads to

$$(z^2 - 5z + 6)y(z) = u(z) = \frac{z}{z-1},$$

and since the initial conditions are zero, the difference equation can be represented equivalently by the transfer function

$$G(z) = \frac{y(z)}{u(z)} = \frac{1}{z^2 - 5z + 6}. \tag{3.38}$$

Inversion to obtain the time solution

$$y(z) = G(z)u(z) = \frac{1}{(z-2)(z-3)}\frac{z}{z-1};$$

$y(k) =$ the sum of the residue of $G(z)u(z)$

$$= \frac{zz^{k-1}}{(z-1)(z-3)}\bigg|_{z\to2} + \frac{zz^{k-1}}{(z-1)(z-2)}\bigg|_{z\to3} + \frac{zz^{k-1}}{(z-2)(z-3)}\bigg|_{z\to1}$$

$$= \frac{2^k}{-1} + \frac{3^k}{2} + \frac{1}{2} = \frac{1}{2} - 2^k + \frac{3^k}{2}. \tag{3.39}$$

Example   The Fibonacci equation is

$$y(k+2) = y(k+1) + y(k), \tag{3.40}$$

with initial conditions $y(1) = y(0) = 1$, $\mathscr{L}$ transforming yields (see Section 3.2.4 for justification of the next step)

$$z^2 y(z) - z^2 y(0) - zy(1) - [zy(z) - zy(0)] - y(z) = 0,$$

$$y(z) = \frac{z^2 - z}{z^2 - z - 1} y(0) + \frac{z}{z^2 - z - 1} y(1).$$

Let α, β be the roots of the denominator; then the solution is given in terms of residues as

$$y(k) = \frac{(z^2 - z)z^{k-1}(z - \alpha)}{(z^2 - z - 1)}\bigg|_{z\to\alpha} + \frac{(z^2 - z)z^{k-1}(z - \beta)}{(z^2 - z - 1)}\bigg|_{z\to\beta}$$

$$= \left(\frac{z^{k+1} - z^k}{z - \beta} + \frac{z^k}{z - \beta}\right)\bigg|_{z\to\alpha} + \left(\frac{z^{k+1} - z^k}{z - \alpha} + \frac{z^k}{z - \alpha}\right)\bigg|_{z\to\beta}$$

$$= \frac{\alpha^{k+1}}{\alpha - \beta} + \frac{\beta^{k+1}}{\beta - \alpha} = \frac{\alpha^{k+1} - \beta^{k+1}}{\alpha - \beta}.$$

Now

$$\alpha = \frac{1}{2} + \frac{\sqrt{5}}{2}, \qquad \beta = \frac{1}{2} - \frac{\sqrt{5}}{2}.$$

Substituting the values for α, β leads to the solution

$$y(k) = \frac{(1 + \sqrt{5})^{k+1} - (1 - \sqrt{5})^{k+1}}{2^{k+1}\sqrt{5}}. \tag{3.41}$$

Table 3.2   Numerical solution of equation (3.42)

k	$u(k)$	$y(k)$	$y(k+1)$	$y(k+2)$
0	1	0	0	1
1	1	0	1	6
2	1	1	6	25
3	1	6	25	90

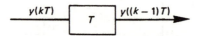

Figure 3.9   The block diagram representing a unit delay.

3.6.3 Numerical solution of difference equations

The difference equation (3.37) can be solved numerically as follows. Write the equation in the form

$$y(k+2) = u(k) + 5y(k+1) - 6y(k) \qquad (3.42)$$

and recall that $u(k) = 1$ for all k and that the equation has zero initial conditions. We set out the calculation in tabular form. Such calculations are ideally suited to computer calculation.

Table 3.2 will be seen to yield results identical to those of Section 3.5.3, where the transffer function $G(z)$ corresponding with equation (3.42) was inverted by long division.

3.6.4 Computer realization of difference equations by means of delay blocks

Define delay blocks as shown in Figure 3.9. Then a difference equation can be mechanized by means of a combination of such blocks together with interlinking algebraic operations. For instance, let the difference equation be

$$y((k+3)T) = u(kT) - y((k+2)T) - 3y((k+1)T) - 5y(kT). \qquad (3.43)$$

The equation can be mechanized as shown in Figure 3.10. Initial conditions have to be given to

$$y(kT), \qquad y((k+1)T), \qquad y((k+2)T).$$

The difference equation can then be solved for time moving forward at intervals of T.

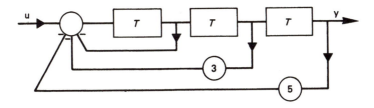

Figure 3.10 Representation of a difference equation by delay blocks.

The realization shown in Figure 3.10 is called the direct method of realization. Other methods of realization may be preferable for computer implementation. For instance, a particular method of realization may be chosen because of noise considerations. Other methods of realization are discussed in Chapter 5.

3.7 The z plane and its relation with the s plane

The utility of the s plane in continuous-time control analysis and design is well established. This section describes the parallel development of the z plane in relation to discrete-time systems and establishes useful correspondences between s plane and z plane representations.

3.7.1 Representation of poles and zeros of $G(z)$ in the z plane

The behavior of discrete-time systems can be investigated in the z *plane*. The z plane is the standard complex number plane taking values $z = a + jb$.

Let $G(z) = N(z)/D(z)$; then the poles and zeros of $G(z)$ are defined as the complex-valued roots of the equations $D(z) = 0$ and $N(z) = 0$, respectively.

Example

$$G(z) = \frac{z}{(z-1)(z^2 + z + 5/16)}$$

has one zero at $z = 0$ and poles at $z = 1$ and $z = -\frac{1}{2} \pm j\frac{1}{4}$. These poles and zeros are plotted in Figure 3.11.

The transfer function $G(z)$ can also be written

$$G(z) = G^*(s)$$

and the poles and zeros of $G^*(s)$ can be displayed in the s plane.

It is convenient to replace T by $2\pi/\omega_s$, where ω_s is the sampling frequency and, as usual in s plane investigations, s is replaced by $s = \sigma + j\omega$.

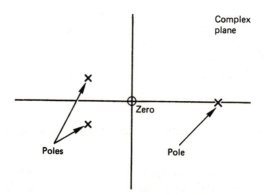

Figure 3.11   The poles and zeros of the function

$$G(z) = \frac{z}{(z-1)(z^2 + z + 5/16)}.$$

As the poles and zeros of a transfer function $G(z)$ move in the z plane, so the corresponding poles and zeros of $G^*(s)$ move in the s plane. An understanding of the relations between loci in the two planes gives a valuable insight into system behavior, and such relations are investigated below.

3.7.2 Relation between the z plane and the s plane

The relation $z = e^{sT}$ provides a mapping from s to z, which we denote by M. The inverse mapping M^{-1} then maps from z to s. Given any s, $M(s) = e^{sT}$ represents the corresponding value of z.

As usual, let $s = \sigma \pm j\omega$ and $T = 2\pi/\omega_s$, where ω_s is the sampling frequency; then

$$M(s) = M(\sigma \pm j\omega) = e^{(\sigma \pm j\omega)(2\pi/\omega_s)} = \exp\left(\frac{2\pi\sigma}{\omega_s}\right) \pm \exp\left(\frac{j2\pi\omega}{\omega_s}\right). \quad (3.44)$$

$M(s)$ is a vector in the z plane. Its magnitude is given by $e^{2\pi\sigma/\omega_s}$ and its direction is governed by $e^{\pm j2\pi\omega/\omega_s}$. The mapping between σ in the s plane and $e^{2\pi\sigma/\omega_s}$ in the z plane can be seen to be one-to-one. This means that there is a unique two-way relationship between values of σ in the s plane and the distance from the origin of the corresponding point in the z plane.

Complex poles in the s plane occur in conjugate pairs. Let $\pm\omega_1$ be the imaginary components of any pair of poles in the s plane; then the corresponding angles generated in the z plane are given by $e^{\pm j2\pi\omega_1/\omega_s}$. However,

$$e^{\pm j2\pi(\omega_1 \pm n\omega_s)/\omega_s} = e^{\pm j2\pi\omega_1/\omega_s}. \quad (3.45)$$

Thus an infinite sequence of points

$$\sigma \pm j\omega_1 \pm jn\omega_s, \qquad n = 0, \pm 1, \pm 2, \ldots$$

in the s plane generates the same pair of angles, $\pm 2\pi\omega_1/\omega_s$, in the z plane.

Given any particular pair of complex conjugate points in the z plane, the mapping M^{-1} from the z plane to the s plane produces in general an infinite sequence of points.

To understand the complete behavior of the mapping M we must represent it by

$$M(s) = e^{2\pi\sigma/\omega_s} e^{\pm j2\pi(\omega \pm n\omega_s)/\omega_s}, \qquad n = 0, \pm 1, \pm 2, \ldots. \qquad (3.46)$$

When $n = 0$, M^{-1} maps the whole of the z plane into a strip in the s plane of infinite width and height ω_s, symmetrical about the $\omega = 0$ axis. This strip is called the *primary strip* in the s plane.

For other values of n, M^{-1} maps the whole of the z plane into other infinite strips of height ω_s centered at the line $\omega = n\omega_s$. These infinite strips, roughly speaking, tell us that poles that have the same imaginary value, modulo ω_s, in the s plane, are identical when transformed into the z plane.

Clearly the further strips that are generated by M^{-1} when $n \neq 0$ can be closely identified with the aliasing spectra of Section 2.4.

3.7.3 Establishment of some specific relations between s and z planes

Most readers will already have some knowledge of the s plane and will appreciate the value of being able to correlate particular freatures in the s plane with particular types of system behavior. It will be recalled that some of the more useful s plane features are the following:

1. A representative point moving along the imaginary axis in the s plane can be considered to generate the system frequency response.
2. Let $s = \sigma + j\omega$; then the real part, σ, governs the rate of growth or decay of transient solutions in the time domain.
3. } Loci of constant damping factor and of constant undamped natural
4. } frequency are very useful in system design work.
5. (Related to (2)) Poles in the right-half s plane indicate instability.
6. Poles nearest to the origin are called dominant poles since they largely govern system behavior.

Using the basic relation $z = e^{sT}$, where T is some chosen sampling interval, relations between the primary strip in the s plane and the z plane can be established. Below we concentrate our attention on the useful properties 1 to 5 of the s plane that we have just discussed.

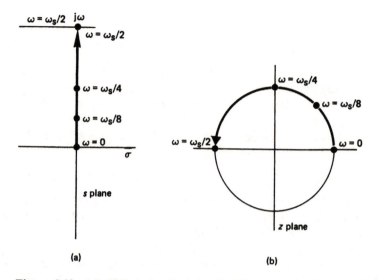

Figure 3.12 (a) The path of a point in the s plane; (b) The corresponding path in the z plane.

Define the sampling frequency ω_S by

$$\omega_S = 2\pi/T,$$

and as usual let $s + \sigma + j\omega$. Then corresponding to each point in the primary strip of the s plane is the point

$$z = e^{(\sigma+j\omega)2\pi/\omega_S} \tag{3.47}$$

in the z plane.

1. *Let $\sigma = 0$ and let ω move from $\omega = 0$ to $\omega = \omega_S/2$* (Figure 3.12(a)). The points in the z plane are given by $z = e^{j2\pi\omega/\omega_S}$.

Suppose $\omega = \omega_S/4$; then $z = j$. This and other correspondences are sketched in Figure 3.12(b).

It also follows immediately that horizontal lines in the s plane map into radial lines in the z plane (Figure 3.13).

We can also see that z plane poles located on the half-line from 0 to $+\infty$ (Figure 3.14) correspond to a non-oscillatory response. Poles located at any other location in the z plane correspond to oscillatory responses. Let a pair of poles be situated on the lines making angles θ and $-\theta$ with the real axis (see Figure 3.15); then the oscillatory response will have $2\pi/\theta$ discrete data points in each oscillation period.

2. *Let σ be fixed at $\sigma = \alpha > 0$ and let ω move from $\omega = 0$ to $\omega = \pm\omega_S/2$.* The values of z are $z = e^{j(\omega/\omega_S)2\pi} \cdot e^{2\pi\sigma/\omega_S}$.

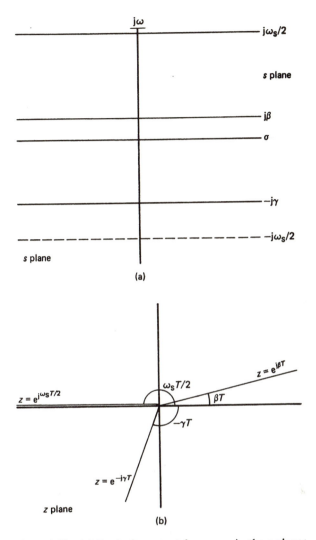

Figure 3.13 (a) Loci of constant frequency in the s plane; (b) Corresponding loci in the z plane.

The locus in the z plane is a circle of radius $r = e^{\alpha T}$. Figure 3.16 shows this relationship and the further locus corresponding to fixing σ at the value $\sigma = -\alpha$.

3. *Let L be the lines of constant damping factor in the s plane* (Figure 3.17(a)). The damping factor ζ is equal to $\cos \alpha$. (This can be confirmed in any classical control text.)

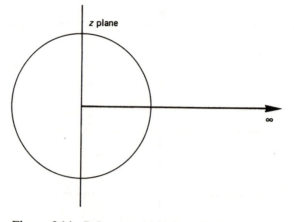

Figure 3.14 Poles on the infinite half line $(0, \infty)$ correspond to non-oscillatory responses.

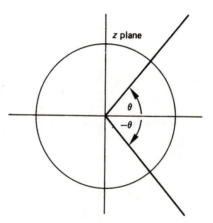

Figure 3.15 Poles located on the marked locus have $2\pi/\theta$ discrete data points in each oscillation period.

The locus in the z plane is generated by the equation

$$z = \exp\left[-2\pi \frac{\omega}{\omega_s} \tan\left(\frac{\pi}{2} - \cos^{-1}\zeta \right) \right] \exp\left(\pm j2\pi \frac{\omega}{\omega_s} \right)$$

as ω varies from $\omega = 0$ to $\omega = \pm\omega_s/2$.

The z plane locus is formed by the first parts of two logarithmic spirals (see Figure 3.17(b)). The lines of constant damping factor for a number of different values of ζ are plotted in Figure 3.18.

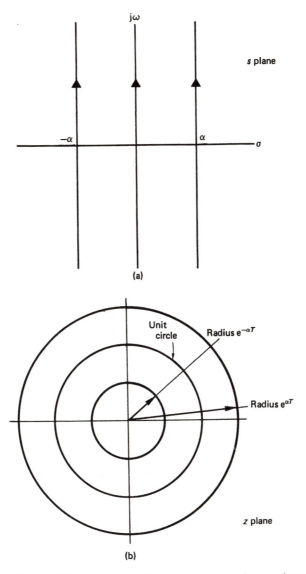

Figure 3.16 (a) Loci of constant σ in the s plane; (b) Corresponding loci in the z plane.

4. *The locus of constant undamped natural frequency*. The decaying response of a second order, undamped, continuous system is of the form

$$y(t) = A\,e^{-\zeta\omega_n t}\sin\omega_n(1-\zeta^2)^{1/2}t$$

where ζ is the damping factor, ω_n is the undamped natural frequency and A is a constant.

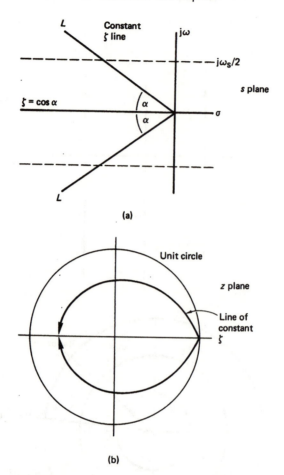

Figure 3.17 (a) Locus of constant damping factor
in the s plane; (b) Corresponding locus in the z plane.

Let $\sigma_1 = \zeta\omega_n$. The damped natural frequency of the decaying sinusoid will be denoted ω_1, where

$$\omega_1 = \omega_n(1 - \zeta^2)^{1/2}.$$

σ_1 can be eliminated from the equations to yield

$$\omega_1^2 + \sigma_1^2 = \omega_n^2.$$

In other words, the locus of constant ω_n in the s plane is a circle of radius ω_n centered at the origin (Figure 3.19).

The loci of constant undamped natural frequency ω_n in the z plane are

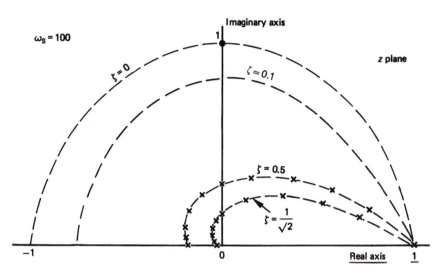

Figure 3.18 Loci in the z plane for a number of selected damping factors (ζ values).

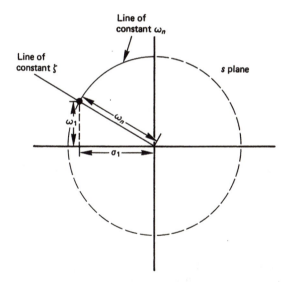

Figure 3.19 Locus of constant ω_n in the s plane.

given by the equation

$$z = e^{[-\sqrt{(\omega_n^2 - \omega^2)} + j\omega]T}$$

as ω varies. The derivation of the expression is given below.

Specimen loci for four different values of ω_n are given in Figure 3.20.

Derivation of the locus of constant ω_n in the z plane. In the z plane we need

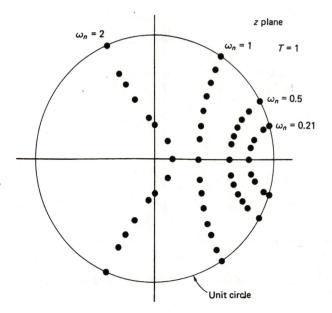

Figure 3.20   Loci of constant ω_n in the z plane.

to plot points

$$e^{sT} = e^{(\sigma + j\omega)T}$$

for points $(\sigma, j\omega)$ on the ω_n locus in the s plane.

From earlier,

$$\sigma = \pm \sqrt{(\omega_n^2 - \omega^2)}$$

and we take the minus sign for σ, since this will correspond to a left-half plane position in the s plane.

Thus, the locus of constant ω_n in the z plane is given by points

$$e^{(-\sqrt{(\omega_n^2 - \omega^2)} + j\omega)T}$$

as ω varies.

To illustrate the shape of the loci, we have fixed T at $T = 1$ and plotted loci in Figure 3.20, for some different values of ω_n. The values used are shown in Table 3.3 for $\omega_n = 0.5$.

It can be shown that the loci of constant ω_n are everywhere orthogonal to the loci of constant damping factor ζ. In the section on root locus in Chapter 4, it is suggested how a knowledge of the ω_n and damping factor loci can be used for controller design.

5. *Stability determination in the z plane.* The whole of the left half of the primary strip in the s plane maps into the interior of the unit circle in the z

Table 3.3   Specimen calculation for the information plotted in Figure 3.20

$$\omega_n = 0.5, \; T = 1$$

ω rad/s	Real part	Imaginary part $\pm$
0.1	0.61	0.061
0.2	0.62	0.126
0.3	0.64	0.198
0.4	0.68	0.288
0.45	0.724	0.34
0.48	0.77	0.40
0.49	0.79	0.43
0.499	0.851	0.46
0.4999	0.869	0.475
0.5	0.877	0.479

plane. Hence systems whose poles are all inside the unit circle in the z plane are asymptotically stable. Systems with any poles outside the unit circle are unstable—their output increases without limit. Poles on the unit circle are analogous to s plane poles on the imaginary axis (see Figure 3.12).

(A system is asymptotically stable if, with no input applied, its output $y(t)$ satisfies the condition $y(t) \to 0$ as $t \to \infty$ for any initial condition $y(0)$.)

3.8 Correspondence between pole locations in the z plane and system time responses

It is now possible to make sketches of a system's (unforced) time response corresponding to particular pole locations. The necessary theoretical background has already been established and without further derivation we give a number of examples of equivalences in Figure 3.21.

3.9 Analysis of a simple loop containing a discrete-time controller

3.9.1 Example

The destabilizing effect of including a discrete-time element in a continuous control loop is well illustrated by the following example. The example is so

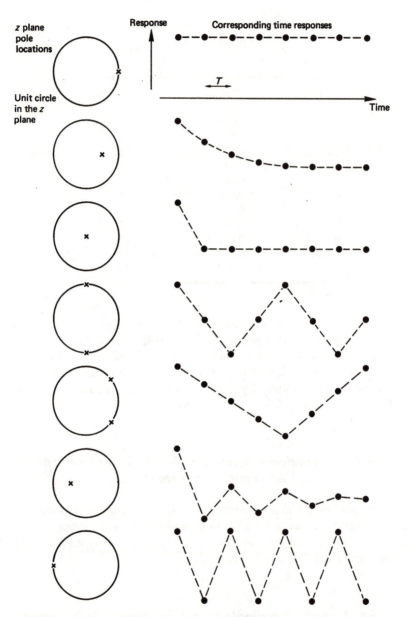

Figure 3.21   Pole locations in the z plane and the corresponding time responses.

simple that the behavior can be found by intuitive reasoning as well as by $\mathscr{L}$ transform analysis.

The approach through intuitive reasoning is valuable in that it allows the nature of the response of the discrete-time system to be understood geometrically as well as analytically.

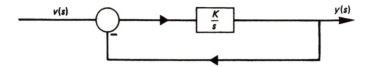

Figure 3.22   A simple continuous-time closed-loop system.

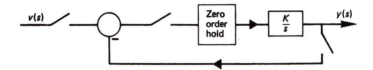

Figure 3.23   The system of Figure 3.22 with the addition of discrete-time components.

Example   First we establish a continuous-time system as a basis for comparison. Then we go on to investigate the equivalent discrete-time system.

Figure 3.22 shows the continuous-time closed-loop system, and Figure 3.23 shows the same system under the simplest possible discrete-time control. The closed-loop transfer function of Figure 3.22 is

$$G(s) = \frac{K}{s + K}$$

and the response to the initial condition $y(0) = -1$ is given by

$$y(t) = -e^{-Kt}.$$

The response is plotted in Figure 3.24 against normalized time. The system is clearly stable for all positive values of K. It has a non-oscillatory exponential response.

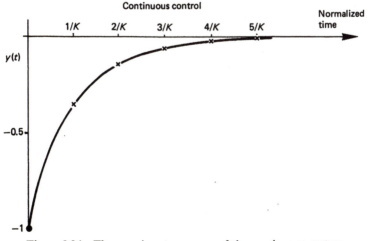

Figure 3.24   The zero-input response of the continuous system.

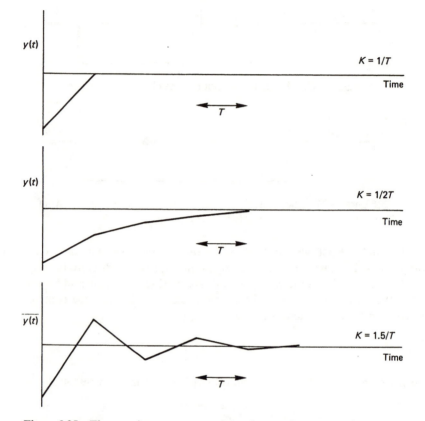

Figure 3.25 The zero-input response of the discrete-time system for different values of K.

3.9.2 Intuitive analysis of the equivalent discrete-time system

Let us turn now to the discrete-time system shown in Figure 3.23. Over each period T, the input to the integrator is held constant at some value, say e, where e indicates the error between input and output of the closed loop system.

The output of the integrator is therefore a series of ramps, extending over time periods T and of slope Ke, where e is the system error at the start of the relevant time period T.

For any particular choice of K, the system response can be constructed graphically. Responses have been constructed in Figure 3.25 for

$$K = -\frac{1}{2T}, \quad \frac{1}{T}, \quad \frac{1.5}{T}, \quad \frac{2}{T}, \quad \frac{3.}{T}.$$

The responses range through analogs of the continuous cases: overdamped, critically damped, oscillatory, in continuous oscillation and unstable.

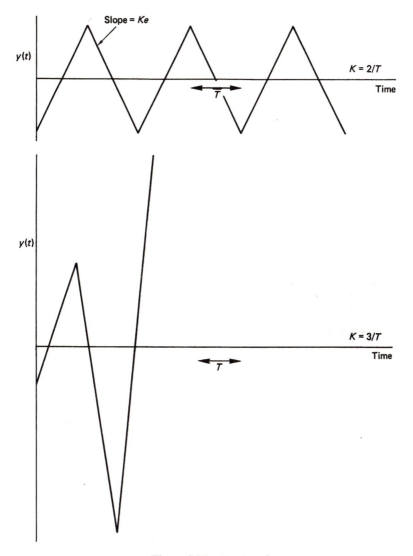

Figure 3.25   *Continued.*

3.9.3 $\mathscr{L}$ transform analysis of the discrete-time system

The continuous transfer function $G_0(s)$ of the zero-order hold element is

$$G_0(s) = \frac{1 - e^{-sT}}{s},$$

where T is the length of the hold period. The open-loop discrete transfer function of the system is

$$\mathscr{L}\{G_0(s)\} = \mathscr{L}\left\{\frac{K}{s}\frac{1-\mathrm{e}^{-sT}}{s}\right\}$$

$$= (1-z^{-1})\mathscr{L}\left\{\frac{K}{s^2}\right\}$$

$$= (1-z^{-1})\frac{KTz}{(z-1)^2} = \frac{KT}{(z-1)}.$$

The closed-loop transfer function is

$$\frac{KT/(z-1)}{1+KT/(z-1)} = \frac{KT}{z-1+KT}.$$

K is constrained to take positive values and we see that as K is increased from zero, the closed-loop pole, taking the value $1-KT$, passes outside the unit circle when $KT = 2$. This represents the stability limit and at $KT = 2$, the pole position predicts continuous oscillation at a frequency $\omega = \omega_s/2$, corresponding to a period of $2T$ in agreement with Figure 3.25.

The z plane pole lies in the right half-plane for $1 - KT > 0$ corresponding to non-oscillatory response, whereas for $2 < KT < 1$ the pole position corresponds to oscillatory response—it would be easy to specify the value of KT to obtain a particular damping factor using a knowledge of the loci of constant ζ.

For instance (Section 3.7.3) a damping factor of ζ will be obtained if the pole is situated left of the origin on the real axis of the z plane at a position

$$-\mathrm{e}^{-2\pi\omega\tan(\pi/2-\cos^{-1}\zeta)/\omega_s}.$$

On the left-hand real axis in the z plane, $\omega = \omega_s/2$; hence the pole must be at

$$-\mathrm{e}^{-\pi\tan(\pi/2-\cos^{-1}\zeta)}.$$

Suppose that we require that $\zeta = 1/\sqrt{2}$; then the pole must be situated at

$$-\mathrm{e}^{-\pi\tan(\pi/2-\pi/4)} = -\mathrm{e}^{-\pi} = -0.043;$$

that is, we require that

$$-0.043 - 1 + KT = 0,$$

$$K = \frac{1.043}{T}.$$

Finally, when $KT = 1$, so-called dead-beat response is obtained (see below). Notice that if the sampling period T is very short, then K will need to be correspondingly high—in practice, constraints may be violated and the system will then become nonlinear. Physically, such violation of constraints corresponds with driving motors or actuators into saturation or in exceeding the permissible limits of signal levels.

A *dead-beat response* is one where, after the application of step change in desired value, the output value moves without overshoot in one sampling interval to its new value and remains there.

Let a system have input u and output y, then if

$$\frac{y(z)}{u(z)} = z^{-1}$$

a dead-beat response will be obtained. To see this let $u(z) = z/(z-1)$; that is, $u(t)$ is a unit step, then

$$y(z) = \frac{1}{z-1} = z^{-1} + z^{-2} + z^{-3} + \cdots$$

and the time response is

k	0	1	2	3
$y(k)$	0	1	1	1

Exercises

3.1 Define the $\mathscr{Z}$ transform and show that the following relations hold

(a) $\mathscr{Z}\{u(k)\} = \dfrac{z}{z-1}$

where $u(k) = 1, k \geqslant 0; u(k) = 0, k < 0$

(b) $\mathscr{Z}\{c^k\} = \dfrac{z}{z-c}$

(c) $\mathscr{Z}\{1 - c^k\} = \dfrac{(1-c)z}{(z-1)(z-c)}$

(d) $\mathscr{Z}\{\sin ak\} = \dfrac{\sin a}{z^2 - 2z \cos a + 1}$

(e) $\mathscr{Z}\{c^k \cos ak\} = \dfrac{(z - c\cos a)z}{z^2 - 2cz \cos a + c^2}$

(f) $\mathscr{Z}\{f(k+1)\} = zF(z) - zf(0)$
     where $F(z)$ denotes $\mathscr{Z}\{f(k)\}$

(g) $\mathscr{L}\{e^{-at}f(t)\} = F(e^{aT}z)$

(h) $\mathscr{L}\{kT\} = \dfrac{Tz}{(z-1)^2}$

3.2 Invert the following $\mathscr{L}$ transforms to obtain time functions. Check your solutions either in the $\mathscr{L}$ transform tables in the Appendix or by long division and numerical comparison of the first few terms.

(a) $\dfrac{z(z-2)}{z^2 - 4z + 16}$

(b) $\dfrac{z^2}{z^2 - 4z + 3}$

(c) z^{-3}

(d) $\dfrac{Tz}{(z-1)^2}$

(e) $\dfrac{T^2 z(z+1)}{2(z-1)^3}$

(f) $\dfrac{z(a-b)}{(z-a)(z-b)}$

(g) $\dfrac{aTz}{(z-a)^2}$

3.3 (An aside on the convergence of the $\mathscr{L}$ transform)

As we have noted earlier, the $\mathscr{L}$ transform of the unit step function is given by

$$\sum_{k=0}^{\infty} z^{-k} = \frac{z}{z-1}$$

Observe that the sum is finite only if $|z| > 1$.

Show that a $\mathscr{L}$ transform that converges for $|z| < 1$ will necessarily correspond with a time function that is finite for $k < 0$ and zero for $k \geqslant 0$.

(Experience of convergence questions, connected with the integral

$$\int_0^{\infty} e^{-st} f(t)\, dt$$

will be found relevant in this exercise.)

3.4 Derive $\mathscr{L}$ transforms corresponding to the following Laplace transforms.

(a) $\dfrac{1}{s^2}$

(b) $\dfrac{1}{s(s+a)}$

(c) $\dfrac{1}{(s+2)(s+3)}$

(d) $\dfrac{1}{s(s+10)^2}$

(e) $\dfrac{ab}{s(s+a)(s+b)}$

3.5 State the transfer function $G_0(s)$ and the pulse transfer function $G_0(z)$ of the zero-order hold. Determine the step response in the time domain of $G_0(s)$ by Laplace and of $G_0(z)$ by $\mathscr{Z}$ transform methods. Comment on the results obtained.

Repeat for first- and second-order hold devices after first deriving their transfer functions.

3.6 Show the validity of the two important relations.

$$(g(s)u^*(s))^* = g^*(s)u^*(s)$$

$$(g(s)u(s))^* \neq g^*(s)u^*(s)$$

Explain the physical significance of the relations.

3.7 A process has the transfer function

$$G(s) = \frac{1}{s+a}.$$

(a) Obtain the $\mathscr{Z}$ transform $G(z)$ of $G(s)$.
(b) The process is preceded by a zero-order hold. Write down $G'(s)$, the transfer function of the process with the zero-order hold.
(c) Obtain the $\mathscr{Z}$ transform $G'(z)$ of $G'(s)$.
(d) Write down the $\mathscr{Z}$ transform of

$$\frac{G'(z)}{1+G'(z)}.$$

3.8 A process of transfer function

$$G(s) = \frac{10}{s(s+2)}$$

is preceded by a zero-order hold with sampling interval 1. Determine the transfer function $G'(z)$ of the combination of $G(s)$ and zero-order hold. Investigate by computation how the step response of $G'(z)$ behaves as T is made smaller. Do you find evidence that, in the limit as $T \to 0$, the step

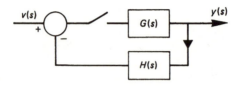

Figure 3.26

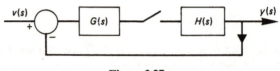

Figure 3.27

response of $G'(z)$ will approach that of $G(s)$? Summarize what you have learned from this exercise.

3.9 Calculate $y(z)/v(z)$ for the closed-loop shown in Figure 3.26.

3.10 Derive a relation between $y(z)$ and $v(z)$ for the system shown in Figure 3.27. (Notice that it will not be possible to produce an expression for $y(z)/v(z)$ in this case.)

3.11 For each of the four configurations in Figure 3.28, determine the overall $\mathscr{L}$ transfer function or the relation between input and output.

3.12 Determine the $\mathscr{L}$ transfer functions of the two systems shown in Figure 3.29. Let $a = 1$ and the sampling interval $T = 0.693$. Insert these numerical values and simplify the transfer functions.

3.13 Figure 3.30 shows three loci in the s plane. Derive and sketch the corresponding z plane loci.

3.14 (a) A second-order discrete-time system has a sampling frequency ω_s. Derive an expression for the locus in the z plane for the poles of the system with fixed damping factor. (b) Show that the loci of constant undamped natural frequency in the s plane constitute a family of concentric circles. Derive and sketch their z plane equivalents.

3.15 (a) Give an example of a transfer function $G(z)$ of first order, whose step response is oscillatory. (b) Explain how such behavior is possible for $G(z)$, although it is not possible for any continuous first-order transfer function $G(s)$. (c) Complete your explanation by sketching the s plane equivalent of your example from (a).

3.16 A system has the transfer function

$$G(z) = \frac{1 + 3z^{-1}}{b + cz^{-1} + dz^{-2}}.$$

State the equivalent difference equation and show that its solution depends on the roots of the auxiliary equation $bm^2 + cm + d = 0$.

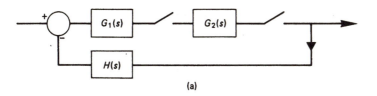

(a)

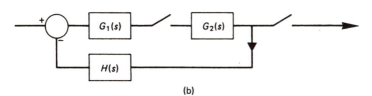

(b)

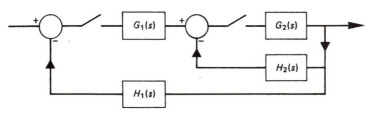

(c)

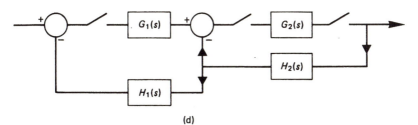

(d)

Figure 3.28

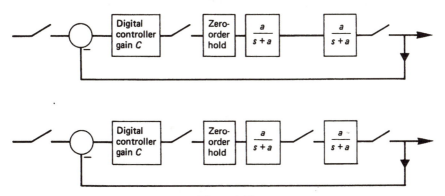

Figure 3.29

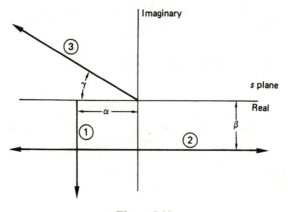

Figure 3.30

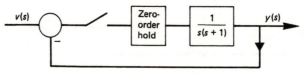

Figure 3.31

3.17 $v(t)$ is a unit step input. The sampler operates with an interval of $T = 0.5$ seconds. Calculate and plot the time response $y(kT)$ for Figure 3.31.

3.18 In the difference equation

$$y(k + 2) - 3y(k + 1) + 2y(k) = 0,$$

$$y(0) = 0, \qquad y(1) = 1.$$

(a) Solve the equation numerically to calculate $y(9)$.
(b) Obtain an analytic solution of the equation and from it confirm the value for $y(9)$ found in (a).

3.19 Sketch the unit step responses of
(a) a first-order system of transfer function

$$G_1(s) = \frac{1}{s + 1};$$

(b) a second-order system of transfer function

$$G_2(s) = \frac{100}{s^2 + 12s + 100}.$$

Investigate the practicability of identifying the equivalent pulse transfer functions $G_1(z)$, $G_2(z)$, by using the relations $G(z) = y(z)/u(z)$ in conjunction

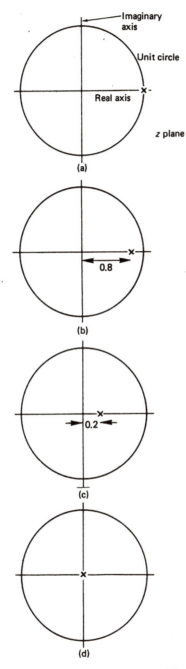

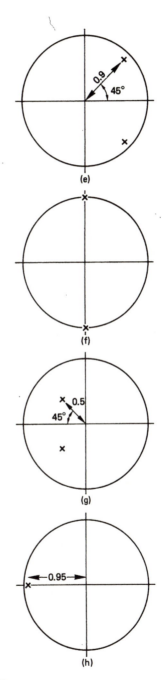

Figure 3.32

with the step responses plotted earlier. We already know that $u(z) = z/(z-1)$, representing the unit step. It remains to read a sequence of samples representing $y(z)$ from the appropriate step response curve and to attempt calculation of $G_1(z)$, $G_2(z)$, followed by comparison with the values for $G_1(z)$, $G_2(z)$ predicted by theory.

Comment on the practicability of the approach as a means of modelling a (continuous-time) dynamic element whose step response is available.

3.20 Each of the systems whose pole(s) are plotted in the z plane in Figures 3.32(a)–(h) has zero input and initial conditions $y(k) = 1$, $k = 0$; $y(k) = 0$ otherwise. Calculate and plot the time response for each system.

3.21 Examine the table in Appendix C of Laplace transforms $G(s)$ and corresponding $\mathscr{L}$ transforms $G(z)$. This table was obtained by inverse Laplace transforming $G(s)$ and $\mathscr{L}$ transforming the resulting time function.

Note the similarities and differences between the $G(z)$ listed here and those that would be obtained from $G(s)$ by the pole-mapping method.

3.22 A system has the transfer function

$$G(z) = \frac{y(z)}{u(z)} = \frac{1}{z^2 + 1}.$$

Show that when the input u is a unit step function the output $y(k)$ has the form

$$y(k) = \{0, 0, 1, 1, 0, 0, 1, 1, 0, 0, 1, 1, \ldots\}.$$

Show that when the input has the form

$$u(k) = \{1, 0, -1, 0, 1, 0, -1, \ldots\},$$

then the output is

$$y(k) = \{0, 0, 1, 0, -2, 0, 3, 0, -4, 0, 5, 0, -6, 0, 7, \ldots\}.$$

Noting that the poles of $G(z)$ are on the unit circle, explain why a bounded input produces a bounded output in the first case but an unbounded output in the second case. Summarize your findings in the most general form possible (i.e. your explanation should consider cases other than that quoted).

4

Methods of Analysis and Design

4.1 Introduction

This chapter establishes the main tools that will be used in the design of single-input–single-output digital control loops. The background tool is the $\mathscr{L}$ transform—it underpins all the material to be presented here.

Discretization is important both from a practical and a theoretical viewpoint, and therefore several alternative approaches are juxtaposed to allow them to be judged in context. (Recall that, since we are largely concerned with the discontinuous control of continuous-time devices, discretization will be required at some stage of the design procedure.)

The material on root-locus and frequency-response techniques will be found very familiar by readers with a background in continuous-time control systems. Both techniques will be used in Chapter 5 to support algorithm design.

Stability tests are sufficiently useful to be given a short section to themselves, even though stability information is also generated by the root-locus and frequency-response techniques.

4.2 Discretization

4.2.1 Introduction: discretization requirements for digital control ·

First we explore the similarities and differences between digital algorithms and digital simulations. (Control engineers are heavily involved with both algorithms and simulations.)

Figure 4.1 shows the schematic outline of a *digitally implemented algorithm*.

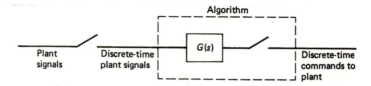

Figure 4.1   Illustrating the discretization of a continuous algorithm.

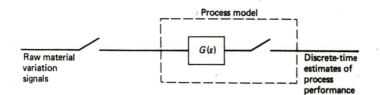

Figure 4.2   Illustrating the discretization of a continuous process.

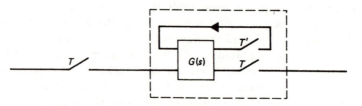

Figure 4.3   Illustrating the availability of outputs at intermediate times between sampling instants.

The algorithm receives discrete-time plant signals and produces discrete-time commands.

Figure 4.2 shows the schematic outline of a mathematical model of a continuous process as implemented in a digital computer (a *digital simulation*). The model receives logged discrete-time data describing (say) raw material variations and produces discrete-time estimates of the corresponding performance of the process.

Apart from the different labelling of the diagrams, Figures 4.1 and 4.2 are identical—the algorithm-implementation problem and the modelling problem are basically the same.

The two problems usually differ in one respect—that of time simultaneity. A real-time algorithm is required to produce commands with as little delay as possible, whereas the process model, operating on recorded data, may have access to 'future' values of the input signals.

Figure 4.3 shows an important feature common to both problems: values of the output signal can be generated, should they be required, corresponding to any time instant. We can imagine that the continuous output signal is internally available and may be sampled at will. This is shown schematically in Figure 4.3 by the loop with fictitious sampler of interval T' ($T' < T$).

Note particularly that the input signals are available only at sampling instants—if intermediate values are needed they can be estimated by some interpolation technique.

Time discretization is the mechanism by which a differential equation or continuous-time transfer function $G(s)$ is converted to a difference equation or a discrete-time transfer function $G(z)$. Such discretization is required whenever a continuous-time dynamic algorithm is to be implemented within a digital computer.

Most of the industrial processes that we are called upon to control are continuous in nature. (Of course, they may embody important discontinuous aspects associated with start-up, batch sequencing or imposed by certain types of actuators or measuring devices.) Mathematical models of continuous processes are usually based around differential equations or, equivalently, around transfer functions in the operator s.

A very extensive range of well-tried methods for control systems analysis and design, together with the associated algorithms for implementation, are in continuous-time form.

To move from continuous-time forms to discrete-time forms requires some mechanism for time discretization (we shall refer to this mechanism simply as discretization).

Discretization turns out to be more interesting than might have been expected. Suppose that the continuous system is characterized by its time response, frequency response and pole–zero locations. None of the discretization methods to be described preserves all these characteristics exactly. The topic of discretization is pursued at some depth not only because of its practical usefulness but also because it links nicely a number of different approaches.

4.2.2 Principles of discretization

Let $f(t)$ be a continuous function. It is unknown but its values, $f(1), f(2,), \ldots$, spaced at intervals T are given.

Suppose that we wish to integrate the function $f(t)$ over one time interval T. Although $f(t)$ is unknown, we can fit a polynomial $p(t)$ through a number of points $f(i)$, $i \geq 1$, and then integrate $p(t)$. We can write

$$\int_0^T f(t)\, \mathrm{d}t \simeq \int_0^T p(t)\, \mathrm{d}t. \tag{4.1}$$

The polynomial $p(t)$ is expressed in terms of values of the function $f(t)$ at sampling instants kT, $k = 0, 1, \ldots$.

Let us take the simplest case where the integral is to be approximated by

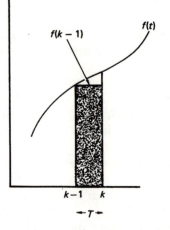

Figure 4.4 Forward rectangular approximation.

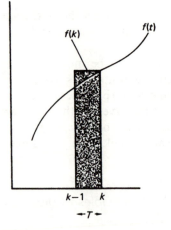

Figure 4.5 Backward rectangular approximation.

a flat-topped rectangle (Figure 4.4 or Figure 4.5). Define

$$q(k) = \int_0^{kT} f(t)\, \mathrm{d}t. \tag{4.2}$$

This rectangular approximation is equivalent to choosing $p(t)$ to be polynomial of order zero and leads to two alternative formulae corresponding to Figures 4.4 and 4.5. Table 4.1 outlines these alternatives comparatively.

Table 4.1

Approximation of Figure 4.4	Approximation of Figure 4.5
$q(k) = q(k-1) + Tf(k-1)$ (4.3)	$q(k) = q(k-1) + Tf(k)$ (4.4)
$q(z) = z^{-1}q(z) + Tz^{-1}f(z)$	$q(z) = z^{-1}q(z) + Tf(z)$
$D_1(z) = Tz^{-1}/(1 - z^{-1})$	$D_2(z) = T/(1 - z^{-1})$
$D_1(z) = T/(z - 1)$ (4.5)	$D_2(z) = Tz/(z - 1)$ (4.6)

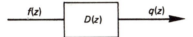

Figure 4.6 Discrete approximation to integration considered as an operation in z.

We can consider the operations of Table 4.1 to be representation by functions $D_1(z)$, $D_2(z)$ (see Figure 4.6).

Trapezoidal (mid-ordinate) integration   In trapezoidal integration, the area over each interval T is approximated by the trapezium shaded in Figure 4.7, and $p(t)$ is a first-order polynomial. This leads to the rule

$$q(k) = q(k-1) + \frac{T}{2}(f(k) + f(k-1))$$

$$q(z) = z^{-1}q(z) + \frac{T}{2}(f(z) + z^{-1}f(z))$$

(4.7)

$$D_3(z) = \frac{q(z)}{f(z)} = \frac{T}{2}\left(\frac{1 + z^{-1}}{1 - z^{-1}}\right).$$

(4.8)

This equation is strongly linked with the w transformation which will be described in Section 4.5.2.

4.2.3 Approximation of Differential Equations by Difference Equations (Differential Mapping Methods)

Suppose that an algorithm, in the form of a differential equation or a transfer function $G(s)$, is to be implemented in a digital computer. The algorithm must

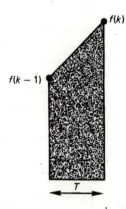

Figure 4.7 Trapezoidal
integration.

be written as a difference equation approximating the original differential
equation.

Perhaps the most obvious strategy is to use the forward difference
approximation

$$\left.\frac{dy}{dt}\right|_{t=kT} = \frac{y(k+1) - y(k)}{T}. \tag{4.9}$$

This, when used to discretize a continuous transfer function $G(s)$, leads to the
discrete approximation

$$G_1(z) = G(s)|_{s=(z-1)/T}. \tag{4.10}$$

The backward difference approximation leads to the discrete approximation

$$G_2(z) = G(s)|_{s=(1-z^{-1})/T} \tag{4.11}$$

and the trapezoidal approximation leads to the discrete approximation

$$G_3(z) = G(s)|_{s=(2/T)[(z-1)/(z+1)]}. \tag{4.12}$$

The three substitutions (4.10)–(4.12) are easily seen to be exactly equivalent to
the three integration techniques D_1, D_2, D_3 of Section 4.2.2. These techniques,
when used for discretization, are often referred to as differential mapping
techniques.

Although the three approximation strategies differ only in apparently trivial
details, the dynamic characteristics of the resultant difference equations vary
significantly and are by no means in faithful correspondence with those of the
original continuous system.

We illustrate by an example some of the problems that can arise when a continuous algorithm is converted to discrete form. (This example is continued into Exercise 4.1.)

Example The continuous algorithm

$$\frac{d^2u}{dt^2} + 0.4\frac{du}{dt} + 0.68u = e \tag{4.13}$$

is to be converted into a discrete-time algorithm by application of the three approximation methods of this section. Determine the poles of the continuous system and of its discrete approximations $G_1(z)$, $G_2(z)$, $G_3(z)$ as in equations (4.10)–(4.12). (Let the sampling interval $T = 1$ for simplicity.)

The continuous algorithm can be represented as a transfer function

$$G(s) = \frac{1}{s^2 + 0.4s + 0.68} = \frac{1}{(s + 0.2 + j0.8)(s + 0.2 - j0.8)}. \tag{4.14}$$

The forward difference approximation is equivalent to setting

$$s = \frac{z - 1}{T} = z - 1$$

in $G(s)$. Therefore

$$G_1(z) = \frac{1}{(z - 1)^2 + 0.4(z - 1) + 0.68};$$

similarly for the backward difference approximation

$$G_2(z) = \frac{1}{(1 - z^{-1})^2 + 0.4(1 - z^{-1}) + 0.68}.$$

The poles of $G(s)$ are at $s = -0.2 \pm j0.8$.
The poles of $G_1(z)$ are at $z = 0.8 \pm j0.8 = 1.13 \angle \pm\pi/4$.
The poles of $G_2(z)$ are found from the equation

$$\frac{z - 1}{z} + 0.2 \pm j0.8 = 0$$

leading to

$$z = 0.576 \pm j0.384$$
$$= 0.693 \angle \pm0.588$$
$$= 0.693 \angle \pm33.7°.$$

Notice particularly that G and G_2 are stable transfer functions whereas G_1 is an unstable transfer function.

For the trapezoidal approximation

$$G_3(z) = \frac{1}{\{(2/T)[(z-1)/(z+1)]\}^2 + 0.4\{(2/T)[(z-1)/(z+1)]\} + 0.68}.$$

(4.15)

The denominator is

$$\frac{4(z-1)^2 + 0.8(z-1)(z+1) + 0.68(z+1)^2}{(z+1)^2}$$

(4.16)

and the poles satisfy the equation

$$5.48z^2 - 6.64z + 3.88 = 0.$$

The values are found to be

$$z = 0.606 \pm j0.584$$

$$= 0.841 \angle \pm 0.767$$

$$= 0.841 \angle 43.94°.$$

The mapping $s \to e^{sT}$ would produce the transfer function $G_0(z)$ by letting

$$z = e^{-0.2 \pm j0.8} \qquad \text{(recall that } T = 1\text{)}$$

with poles at $z = 0.819 \angle \pm 0.8 = 0.819 \angle \pm 45.84°$.

The pole positions for the four approximating transfer functions are plotted in Figure 4.8.

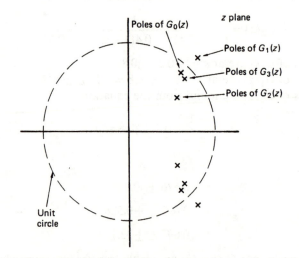

Figure 4.8 Pole positions found in the example of Section 4.2.3.

The trapezoidal method can be seen to be quite accurate in terms of pole placement—this is typical and in general the trapezoidal method offers much higher accuracy in pole placement than the two simpler (forward and backward approximation) methods.

4.2.4 Discretization by matching poles and zeros (sometimes called pole mapping or pole matching)

Section 4.2.3 has made clear that conversion of a transfer function to discrete time by simple difference approximations in general results in significant distortion of the dynamic characteristics. A stable continuous transfer function may even be transformed into an unstable discrete-time function.

An alternative approach is to argue that the transformation from continuous to discrete time should ideally be performed by the relation $z = e^{sT}$.

Direct substitution of the relation into $G(s)$ will produce a rather intractable ratio of polynomials in $\ln z$. However, the relation $z = e^{sT}$ can be used to map the poles and zeros of $G(s)$ into the z plane. It then remains to synthesize a discrete-time transfer function $G(z)$ having the stipulated poles and zeros and to provide a constant multiplier to give the correct steady-state gain.

The only difficulty in carrying the plan through is concerned with the mapping of the zeros of $G(s)$.

Let $G(s) = P(s)/Q(s)$ and suppose that the orders of the polynomials $P(s)$, $Q(s)$ are m, n respectively. For physical realizability, $m \leqslant n$, and if $m < n$ then $G(s)$ has $n - m$ zeros at infinity.

It can be argued that each zero at infinity in $G(s)$ should be represented as a zero at $z = -1$ in the z plane.

The basis for this argument is that such placing of the zeros causes the system gain to be zero at the frequency $\omega = \pi/T = \omega_N$, which, as the highest permissible frequency that does not violate the sampling theorem, is in a sense the analog of $\omega = \infty$ in the continuous case. This is not the only possible approach and at times it may be preferable to map zeros at infinity into $z = 0$ or to omit to map them altogether. A further factor that influences the decision on how to map zeros concerns the time requirement for computation of a real-time algorithm—if an algorithm of transfer function $D(z)$ has equal degree in numerator and denominator, then zero time is being allowed for computation. Where the time required for computation is not negligible compared with the sampling interval T, it may be necessary to reduce the order of the numerator by omitting a zero, so as to allow time for computation.

Example   Given

$$G(s) = \frac{(s + 1)}{(s + 2)(s + 3)},$$ (4.17)

determine an equivalent transfer function $G(z)$ by the method advocated above.

There is one zero at infinity in $G(s)$. This leads to

$$G(z) = \frac{(z+1)(z-e^{-T})}{(z-e^{-2T})(z-e^{-3T})}. \tag{4.18}$$

Let $G(z)$, $y(z)/u(z)$ and note, as discussed above, how (since the orders of numerator and denominator are equal) the difference equation linking y with u will imply that a change in u is transmitted instantaneously to y.

Example Let

$$G(s) = \frac{1}{(s+0.2+j0.8)(s+0.2-j0.8)}. \tag{4.19}$$

$G(s)$ has two zeros at infinity, therefore $G(z)$ must have two zeros at $z = -1$. Hence

$$G(z) = \frac{(z+1)^2}{(z-e^{-(0.2+j0.8)T})(z-e^{-(0.2-j0.8)T})}.$$

Set $T = 1$; then

$$G(z) = \frac{(z+1)^2}{z^2 - 1.14z + 0.67}. \tag{4.20}$$

The poles of $G(z)$ are at

$$z = 0.57 \pm j0.588$$

$$= 0.818 \angle \pm 0.8$$

as obtained in Section 4.2.3 for $G_0(z)$.

The difference equation corresponding to $G(z)$ is

$$u(k+2) - 1.14u(k+1) + 0.67u(k) = e(k+2) + 2e(k+1) + e(k)$$

and when a scaling factor of $1.47/7.55$ is included, the time response shown in Figure 4.9 is generated.

4.2.5 Discretization by substitution of the first few terms in the series for $(1/T) \ln z$

$G(s)$ could be converted to an equivalent transfer function $G(z)$ by setting $s = (1/T) \ln z$ in $G(s)$, but this results in an unwieldy expression for $G(z)$. An alternative is to approximate $\ln z$ by the first few terms of its series.

The series for $\ln z$ is

$$\ln z = 2\left(r + \frac{r^3}{3} + \frac{r^5}{5} + \cdots\right), \qquad |r| < 1 \tag{4.21}$$

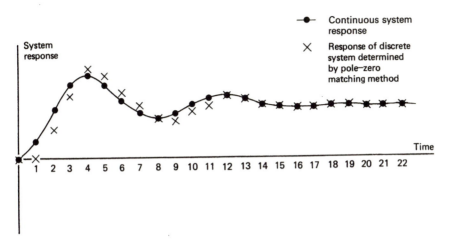

Figure 4.9 Plot of exact and approximate solutions for the example of Section 4.2.4.

where

$$r = \frac{1 - z^{-1}}{1 + z^{-1}}.$$

Then

$$s^{-1} = \frac{T}{1\left(r + \dfrac{r^3}{3} + \dfrac{r^5}{5} + \cdots\right)}$$

$$= \frac{T}{2}\left(\frac{1}{r} - \frac{r}{3} - \frac{4}{45}r^3 - \cdots\right). \tag{4.22}$$

The series can be truncated and used in a practicable discretization mechanism. Similar truncated series can be produced for s^{-2}, s^{-3}, etc. Suggested expressions for substitution in $G(s)$ are:

$$s^{-1} \simeq \frac{T}{2r} = \frac{T(1 + z^{-1})}{2(1 - z^{-1})}, \tag{4.23}$$

$$s^{-2} \simeq \frac{T^2}{4}\left(\frac{1}{r^2} - \frac{2}{3}\right) = \frac{T^2}{12}\frac{1 + 10z^{-1} + z^{-2}}{(1 - z^{-1})^2}, \tag{4.24}$$

$$s^{-3} \simeq \frac{T^3}{2}\left(\frac{z^{-1} + z^{-2}}{(1 - z^{-1})^3}\right). \tag{4.25}$$

In use, the expression for $G(s)$ is multiplied out and arranged in terms of s^{-1}, s^{-2}, s^{-3}. The substitutions (4.23), (4.24),... are made and the resulting expression in z is simplified. An example makes the method clear.

Example   Let

$$G(s) = \frac{1}{s^2 + 0.4s + 0.68} \tag{4.26}$$

be converted to an equivalent transfer function $G(z)$ by the method of truncated approximations to $\ln z$ (use $T = 1$).

$$G(s) = \frac{s^{-2}}{1 + 0.4s^{-1} + 0.68s^{-2}},$$

and the substitutions produce

$$G(z) = \frac{z^2 + 10z + 1}{12(1.257z^2 - 1.43z + 0.857)}$$

$$= \frac{(z + 0.1)(z + 9.9)}{12(z - 0.57 + j0.6)(z - 0.57 - j0.6)}. \tag{4.27}$$

4.2.6 Discretization by z transformation of G(s) (sometimes referred to as the impulse-invariant response method of discretization)

Given a transfer function $G(s)$ that is to be discretized, it would seem timely to ask: why not simply say that

$$G(z) = G^*(s) = \mathscr{Z}\{G(s)\} \tag{4.28}$$

is a discrete equivalent of $G(s)$?

This point will be returned to shortly. In the meantime, assume (incorrectly) that there exists a perfect device for reconstructing continuous signals from a finite number of samples. Let the device have transfer function $R(s)$.

A little thought then shows that

$$G(z) = \mathscr{Z}\{GR(s)\} \tag{4.29}$$

would furnish perfect discretization, since then $G(s)$ receives and gives out exactly the same signals as it did in the continuous case (see Figure 4.10).

We can now see why equation (4.28) may be non-ideal. $G(z) = G^*(s)$ contains the implicit assumption that the input signal is zero except at sampling

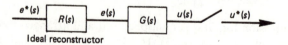

Figure 4.10   Ideal discretization.

instants. We therefore require a fictitious reconstruction device that approximates the effect of $R(s)$ in equation (4.29).

A zero-order hold of transfer function $(1 - e^{-sT})/s$ is the simplest reconstruction but, of course, its non-ideal reconstruction will introduce error and its type of action will inevitably produce a time delay.

This section therefore proposes two discretization methods:

(a) As in equation (4.28), to use the transformation

$$G(z) = \mathscr{Z}\{G(s)\},$$

where the actual mechanism of transformation is that described in Section 3.2.2. The impulse responses of $G(s)$ and $G(z)$ are then identical.

(b) As described in this section, to allow for a reconstructor, in the form of a zero-order hold, to arrive at the discretization mechanism

$$G(z) = \mathscr{Z}\left\{\frac{1 - e^{-sT}}{s}G(s)\right\}.$$

4.2.7 Discretization by matching the step response

Let $u(t)$ be the (known) unit step response of the continuous system that is to be discretized. Let $u_0, u_1, \ldots$ be the values of $u(t)$ at $t = kT, k = 0, 1, \ldots$.

Then the z transform $u(z)$ of the step response is known in the form of a sequence.

The z transform $h(z)$ of a unit step applied at time $kT, k = 0$ is given by

$$h(z) = \frac{z}{z - 1}. \tag{4.30}$$

$G(z)$, the required discretized transfer function, can then be obtained directly from

$$G(z) = \frac{u(z)}{h(z)}.$$

By this method $G(z)$ will be defined numerically in the form

$$G(z) = \left(\frac{z - 1}{z}\right)(0 + u_1 z^{-1} + u_2 z^{-2} + \cdots).$$

In general $G(z)$ is represented by a long sequence that cannot easily be converted to a closed form and the method of step response matching is not advocated for discretization.

However, this section has introduced the idea of obtaining a discrete-time model of a continuous system from input–output data only—making no use of knowledge of $G(s)$.

Very frequently we are faced with the problem of designing a control system for a plant whose characteristics are unknown but from which input–output data are available. The topic is treated in Chapter 5.

4.2.8 Discretization by solution of the continuous equation over each time step

The principle is most easily explained with the aid of a simple example. Let the equation that is to be discretized be

$$\frac{du}{dt} = au, \tag{4.31}$$

where a is a constant.

The equation has the solution

$$u(t) = e^{a(t - t_0)}u(t_0), \tag{4.32}$$

where t_0 is some initial time.

Let $t_0 = kT$ and $t = (k + 1)T$; then

$$u(k + 1) = e^{aT}u(k), \tag{4.33}$$

e^{aT} is a constant and the discretization is both simple and perfectly accurate.

If equation (4.31) is allowed to have a forcing term to become

$$\frac{du}{dt} = au + v,$$

the solution now involves convolution:

$$u(t) = e^{a(t - t_0)}u(t_0) + \int_{t_0}^{t} e^{a(t - \tau)}v(\tau)\, d\tau.$$

Again let $t_0 = kT$:

$$t = (k + 1)T,$$

$$u(k + 1) = e^{aT}u(k) + \int_{kT}^{(k + 1)T} e^{a((k + 1)T - \tau)}v(\tau)\, d\tau, \tag{4.34}$$

$$u(k + 1) = e^{aT}u(k) + e^{a(k + 1)T} \int_{kT}^{(k + 1)T} e^{-a\tau}v(\tau)\, d\tau, \tag{4.35}$$

Suppose temporarily that v is constant on each sampling instant—as it would be if, for instance, it were generated by a zero-order hold. Then

$$u(k+1) = e^{aT}u(k) + e^{a(k+1)T}\left(-\frac{1}{a}\right)e^{-a\tau}\bigg|_{kT}^{(k+1)T}v(k), \qquad (4.36)$$

$$u(k+1) = e^{aT}u(k) - \frac{1}{a}e^{a(k+1)T}[e^{-a(k+1)T} - e^{-akT}]v(k),$$

$$u(k+1) = e^{aT}u(k) + \frac{1}{a}[e^{aT} - 1]v(k),$$

$$u(k+1) = \phi u(k) + \psi v(k), \qquad (4.37)$$

where

$$\phi = e^{aT}, \qquad \psi = \frac{1}{a}[e^{aT} - 1]. \qquad (4.38)$$

This leads to the z transfer function

$$\frac{u(z)}{v(z)} = \frac{\psi}{z - \phi}. \qquad (4.39)$$

Equation (4.39) generates the same solution as equation (4.32).

We now remove the assumption that v is constant over each period of length T. Here we have to proceed by approximation of v. Two simple possibilies are to use the approximations

$$v(t) = v(k) + (v(k) - v(k-1))_2, \qquad kT \leqslant \tau < (k+1)T \qquad (4.40)$$

or

$$v(\tau) = \frac{(k) + v(k+1)}{2}, \qquad kT \leqslant \tau < (k+1)T. \qquad (4.41)$$

The second algorithm will be more accurate but will impose a delay of T on a real-time algorithm.

Equations of order 2 or more The approach can be applied to equations of order 2 or higher using the transition-matrix approach. Here we give a minimal treatment and reserve a complete discussion until Chapter 8.

Let the equation to be solved be

$$\frac{d^2y}{dt^2} + a\frac{dy}{dt} + by = 0, \qquad y(0) = \alpha, \qquad \frac{dy}{dt}(0) = \beta. \qquad (4.42)$$

The procedure is as follows:

1. Represent the original equation as n first-order equations (where n is the order of the original equation) by introducing dummy variables $x_1, \ldots, x_n$.

2. Represent the equations in matrix form.
3. Solve by Laplace and inverse Laplace transformation, involving, implicitly, the transition matrix.
4. Discretize the time in the solution.

The steps appear as follows for equation (4.42):

1. Let $x_1 = y$, $x_2 = \dfrac{dx_1}{dt}$; then $\dfrac{dx_2}{dt} = -ax_2 - bx_1$.

2. $\begin{bmatrix} \dot{x}_1 \\ \dot{x}_2 \end{bmatrix} = \begin{bmatrix} 0 & 1 \\ -b & -a \end{bmatrix} \begin{bmatrix} x_1 \\ x_2 \end{bmatrix} = (\text{say})\dot{x} = Ax.$

3. $sx(s) - x(0) = Ax(s)$,

$x(t) = \mathcal{L}^{-1}\{(sI - A)^{-1}x(0)\} \triangleq \Phi(t)x(0),$

where $\Phi(t) = \mathcal{L}^{-1}\{(sI - A)^{-1}\}$ can be defined at this point as the *transition matrix*

4. $x(k + 1) = \Phi(T)x(k).$ (4.43)

Example Discretize the equation (for a time interval $T = 1$)

$$\frac{d^2 y}{dt^2} + 0.4\frac{dy}{dt} + 0.68y = 0 \qquad (4.44)$$

using the method explained above.

We shall number the steps in the example to correspond with the steps in the explanation.

1. Let

$$x_1 = y, \qquad x_2 = \frac{dx_1}{dt} = \frac{dy}{dt}.$$

2. Define

$$x = \begin{bmatrix} x_1 \\ x_2 \end{bmatrix};$$

then the matrix representation is

$$\dot{x} = \begin{bmatrix} 0 & 1 \\ -0.68 & -0.4 \end{bmatrix} x = Ax.$$

3. $\Phi(t) = \mathcal{L}^{-1}\{(sI - A)^{-1}\}$

$= \begin{bmatrix} e^{-0.2t}(\cos 0.8t + \frac{1}{4}\sin 0.8t) & 1.25e^{-0.2t}\sin 0.8t \\ -(0.68/8)e^{-0.2t}\sin 0.8t & e^{-0.2t}(\cos 0.8t - \frac{1}{4}\sin 0.8t) \end{bmatrix}.$

4. $\Phi(T)|_{T=1} = \begin{bmatrix} 0.7172 & 0.73415 \\ -0.4992 & 0.4235 \end{bmatrix}.$

Hence

$$\begin{bmatrix} y(k+1) \\ [dy(k+1)/dt] \end{bmatrix} = \begin{bmatrix} 0.7172 & 0.73415 \\ -0.4992 & 0.4235 \end{bmatrix} \begin{bmatrix} y(k) \\ [dy(k)]/dt \end{bmatrix}. \qquad (4.45)$$

4.2.9 Discussion

We have devoted considerable space to the topic of discretization because an understanding of the various possible approaches helps the formation of a good theoretical foundation for the study of digital systems.

The main point is to be aware of the significant features of discretization and to have a rough quantitative understanding of the errors that are likely to be introduced by the various methods.

The method of discretization must be chosen to match the problem at hand and it is not possible to recommend a 'best method' for all situations. However, as a general guideline, it can be said that the trapezoidal method of Section 4.2.3 or the pole–zero matching method of Section 4.2.4 are the most useful, based on criteria of the preservation of time and frequency response and ease of implementation.

It is strongly recommended that as many as possible of the exercises related to discretization be worked through. Exercise 4.5 suggests comparative tests based on step response. Note however that preservation of (gain and phase) frequency response and pole locations may be as important in application as preservation of time response.

4.3 The root-locus diagram

Consider the system of Figure 4.11. The poles of the system move as C is varied and the root locus consists of a display of the path of the poles in the z plane.

Clearly the poles of any usable system must not be outside the unit circle, on stability grounds. The system response may also be required to satisfy certain

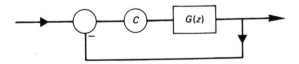

Figure 4.11 The closed-loop system used for root-locus investigation. The gain C is regarded as a design variable.

damping and frequency criteria and these can be interpreted in terms of conditions on the location of poles within the unit circle.

More subtle but just as important is the effect of choice of sampling interval on system performance. This can also be interpreted in terms of the location of poles in the z plane.

4.3.1 Some general rules for sketching the root locus

Although a computer package will be used routinely to generate root loci, it will still be useful to know the basic rules for approximate sketching of loci. Such an outline knowledge will help, amongst other things, intelligent interaction with a root-locus computer package.

One helpful point: the construction of the root-locus diagram for $G(z)$ is identical to that for the continuous transfer function $G(s)$; only the interpretation of the pole positions is different in the two cases.

Root locus branches start and terminate on singularities of $G(z)$ Every point on the root locus satisfies

$$1 + CG(z) = 0 \qquad (4.46)$$

so that, as $C \to 0$, $|G(z)| \to \infty$. Thus, the branches of the loci start from poles of $G(z)$.

Similarly, as $C \to \infty$, $|G(z)| \to 0$ and the branches of the loci terminate on zeros of $G(z)$.

The number of branches Clearly the number of branches in the root locus must equal the number of poles of $G(z)$ where, in counting the number of poles or zeros, their multiplicities are taken into account.

The number of branches seeking infinity Let $G(z)$ have m zeros and n poles with $n > m$; then $n - m$ branches of the root locus seek infinity. m branches terminate on the zeros.

The asymptotes of the infinity-seeking branches are arranged symmetrically and they intersect at an easily determined point on the real axis It is shown in Reference D2, Section 3.5, that the asymptotes for the infinity-seeking branches are as shown in Figure 4.12. (If positive feedback is used, the angles are different.) Again let m and n be the order of the numerator and denominator respectively. The intersection of the asymptotes is given by

$$-a = \frac{\Sigma \, \text{pole coordinates} - \Sigma \, \text{zero coordinates}}{n - m}.$$

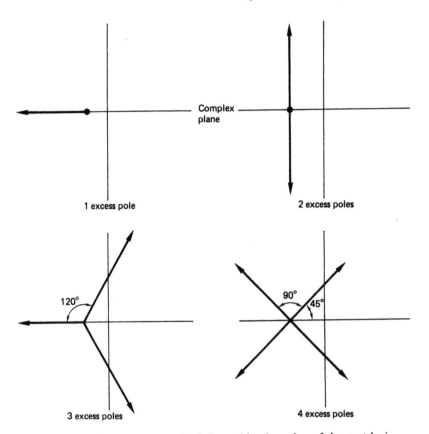

Figure 4.12 Asymptotes of infinite seeking branches of the root loci.

The segment of the real axis to the left of an odd number of singularities of $G(z)$ is part of the root locus This is proved in Reference D2. (Again the assumption is that the loop is providing negative feedback and the rule will not apply for a positive-feedback loop.)

4.3.2 Example

Root-locus solution by sketching Sketch the root locus for the system shown in Figure 4.13.

The continuous-time system has the transfer function

$$G(s) = \frac{C}{s(s + 2)},$$

$$\mathscr{L}^{-1}\{G(s)\} = \mathscr{L}^{-1}\left\{\frac{C}{2}\left(\frac{1}{s} - \frac{1}{s + 2}\right)\right\} = \frac{C}{2}(1 - e^{-2t}).$$

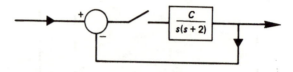

Figure 4.13 The closed-loop system for the example
of Section 4.3.2.

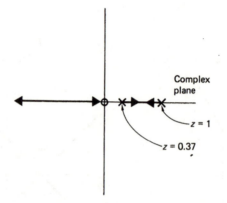

Figure 4.14 The root loci for the example
must occupy the sections of the real axis
shown.

Hence the impulse invariant equivalent of $G(s)$ is

$$G(z) = \frac{C}{2} \frac{z(1 - e^{-2T})}{(z - 1)(z - e^{-2T})}.$$

The open-loop transfer function has one zero at $z = 0$ and two poles at $z = 1$ and $z = e^{-2T}$ respectively. Letting $t = 0.5$ for definiteness, the second pole is then at $z = e^{-1} = 0.37$.

From the general rules above, the locus has two branches starting at the poles. One branch terminates at the zero, while the second branch seeks infinity along the negative real axis as an asymptote. The segment of the real axis to the left of 1 or 3 singularities (poles or zeros) is part of the locus. Putting this information together, we can sketch in part of the locus, together with directional arrows (Figure 4.14).

We now reason as follows: Two connecting pieces of locus need to be drawn between the two disconnected sections already drawn (for if not, the rules about loci starting on poles and terminating on zero cannot be satisfied). Such connections must be off the real axis (why?) and by the complex conjugate symmetry that must always be preserved, one connection is above the axis and

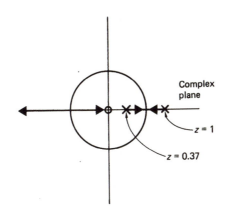

Figure 4.15 Completion of the diagram
of Figure 4.14 by 'informed guesswork'.

one below. By this informed guesswork, we complete the diagram as in
Figure 4.15.

We hasten to add that we have no information (apart from some possible
inherited wisdom) on the location and shape of the interconnection added in
between Figures 4.14 and 4.15. However, we can be reasonably certain that we
have the correct topography for the locus. For some purposes this will suffice.
For other cases, a more detailed delineation of the precise location of the locus
will be undertaken by computer package. (It used to be common to learn the
'eleven rules for root-locus sketching' and should the reader wish, he can enhance
the brief sketching rules given here by extracting and condensing from early
(pre-computer) texts such as Reference D2.)

Returning to a consideration of the root-locus sketch (Figure 4.15), we
would expect the roots of the closed loop system to be first real, then complex,
and then real again as the gain C is increased from zero. At some particular
gain, C_{max}, the closed-loop system will reach its stability limit.

The inter-relation of sampling interval T with maximum gain C_{max} We
have allowed the sampling interval T to vary and for each case have determined
C_{max}. The results are listed in Table 4.2. (We have also listed in Table 4.2 the
position of the second closed-loop pole at the gain C_{max}.) The data from the
table are plotted in Figure 4.16 to show the inter-relation of C_{max} with T.

Detailed construction of the root-locus diagram For comparison purposes,
we construct the root-locus for the example of Figure 4.13.

The closed-loop transfer function is given by

$$\frac{G(z)}{1+G(z)} = \frac{Cz(1-e^{-2T})}{2z^2 + (C(1-e^{-2T}) - 2(1+e^{-2T}))z + 2e^{-2T}}. \quad (4.47)$$

Table 4.2

T	C_{max}	Position of second pole
0.0001	40 000	$\simeq -1$
0.001	4000	$\simeq -1$
0.01	400	$\simeq -1$
0.05	80	$\simeq -0.9$
0.1	40	$\simeq -0.8$
0.2	20	$\simeq -0.7$
0.5	8.65	$\simeq -0.37$
1	5.25	$\simeq -0.135$
2	4.15	$\simeq 0$
5	4	$\simeq 0$
10	4	$\simeq 0$
100	4	$\simeq 0$

Setting $t = 0.5$ and finding the roots of the quadratic denominator by the usual quadratic formula, the root-locus diagram, Figure 4.17 is easily plotted. The correctness of the sketch, Figure 4.15, is confirmed and quantitative information is now available. In particular we can see that $C_{max} = 8.65$ (read off from the point where the locus reaches the unit circle).

Comments on the results of the example We notice that as the sampling interval $T \to 0$ so the gain $C_{max} \to \infty$. In other words, in the limit, the continuous case is approached. Note carefully in a digital (as opposed to a sampled data) situation, the amplitude is quantized to an accuracy set by the word length, and as $T \to 0$ the continuous case is not then approached.

We also notice, perhaps to our surprise, that as T becomes large, so $C_{max} \to 4$. In other words, provided that C satisfies $C < 4$, the system is not unstable for any value of T. Since the system is so simple, we can reason this through as follows. Assume that, in Figure 4.13, $e(t)$ has an initial value $e(t)|_{t=0} = 1$. At the first operation of the sampler a unit impulse is input to $G(s)$. The response of $G(s)$ to this unit impulse is given by $y(t) = \frac{1}{2}(1 - e^{-2t})$ (convince yourself of this). The time constant of $G(s)$ is 0.5 s and for $T \gg 0.5$, we shall have

$$y(t)|_{t=T} \simeq \frac{C}{2}.$$

At the second operation of the sampler, an impulse scaled by the factor $-C/2$ is input to $G(s)$. The steady-state output then becomes

$$y(t)|_{t=2T} \simeq \frac{C}{2} - \left(\frac{C}{2}\right)^2.$$

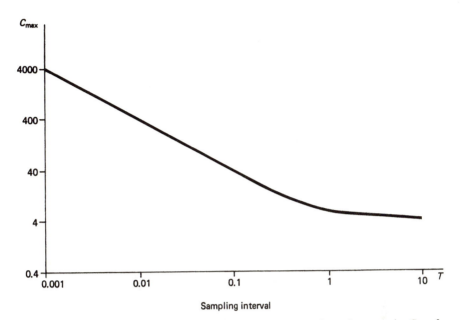

Figure 4.16 The relation between sampling interval T and maximum gain C_{max} for the example of Section 4.3.2.

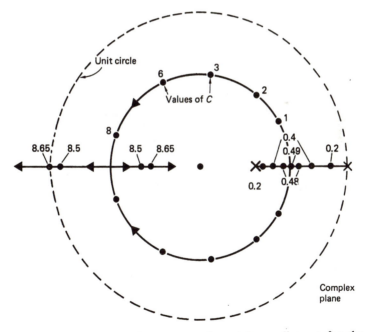

Figure 4.17 The exactly constructed root locus diagram for the example of Section 4.3.2

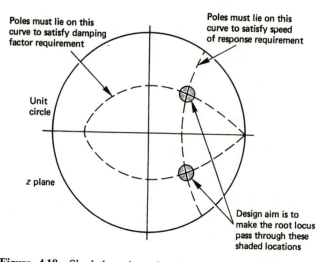

Figure 4.18 Shaded regions in the complex plane denoting (typical) preferred regions for the system poles.

We can see that if $C = 4$, the steady-state outputs oscillate between the values $y = \pm 2$ and the system is at the limit of stability.

4.3.3 Use of the root locus as a design tool

We have seen in Section 3.8 how the time response of a system depends on the locations of the system poles in the z plane.

This knowledge forms the basis for a control-design method. A desirable region for pole location can be selected in the z plane and the root locus modified by informed trial and error until it passes through the desired region. The system gain is a parameter along the locus and it can be read off where the locus passes through the desired region.

For a typical second-order system, the closed-loop poles might be located to satisfy, simultaneously, conditions on speed of response and damping factor. We illustrate qualitatively in Figure 4.18 how preferred pole locations might be chosen.

4.3.4 Two difficulties in the interpretation of the root locus in the z plane

A small change in the loation of a pole situated close to $z = 1$ often corresponds to a large change in system behavior. The diagram therefore needs to be interpreted with special care in the region of $z = 1$.

System behavior is strongly dependent on the location of zeros, particularly when the zeros are in the region of $z = 1$. This makes the exact design of systems through the root-locus technique more complex than if only root locations were involved. This point is taken up again in Section 4.4.2.

4.4 Frequency-response methods

Frequency-response methods are very valuable tools in the analysis and design of continuous systems. Here we consider the frequency response of discrete-time systems described by their transfer functions $G(z)$.

By the frequency response of a system we mean the steady-state response of the system to a sinusoidal input. Recall that a continuous linear system subjected to a sinusoidal input has an output of the same frequency. All that can change are the magnitude and phase angle: these two quantities vary as a function of frequency and characterize the frequency response.

In considering the frequency response of discrete-time systems, two special features are encountered.

1. If ω is increased above the limiting frequency $\omega_s/2$, then input sampling will not uniquely characterize the input signal $a \sin \omega t$.
2. The output of $G(z)$ is a stream of impulses that generates an infinite number of aliasing harmonic spectra.

4.4.1 Direct method of frequency-response determination

To determine the response of a continuous system of transfer function $G(s)$ to a steady sinusoidal input, we replace s by $j\omega$ in $G(s)$ and operate with the restricted transfer function $G(\omega)$.

In a similar manner, the frequency response of a transfer function $G(z)$ can be obtained from the restricted transfer function $G(z)|_{z = e^{j\omega T}}$.

We can consider that as s moves in the s plane from 0 to $\omega_s/2$, generating the frequency response of $G^*(s)$, so z moves around the unit circle in the z plane, generating the frequency response of $G(z)$.

Notice particularly that as z repeatedly traverses the unit circle, the same frequency response will be generated repeatedly. Thus, the frequency response of $G(z)$ over the interval $\omega T = [-\pi, \pi]$ is repeated over the intervals $[\pm \pi, \pm 3\pi]$, $[\pm 3\pi, \pm 5\pi]$, ...

The direct method is illustrated through the following example.

Example The example uses the frequency-response method to demonstrate the destabilizing effect of discrete-time control. Our starting point is the continuous-point control system of Figure 4.19.

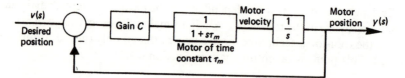

Figure 4.19 Closed-loop system to be investigated by frequency-response methods.

The open-loop transfer function of the continuous system is

$$G(s) = \frac{C}{(1 + s\tau_m)s}.$$

Let $\tau_m = 0.5$; then $G(s) = 2C/[s(s + 2)]$, and using the impulse invariant transformation, the corresponding function $G(z)$ is found to be

$$G(z) = \frac{Cz(1 - e^{-2T})}{(z - 1)(z - e^{-2T})} \qquad (4.48)$$

(this transfer function (differing only by a constant factor) was derived in Section 4.3.2).

Setting $z = e^{j\omega T} = \cos \omega T + j \sin \omega T = (\text{say})a + jb$ and setting $e^{-2T} = d$, then

$$G(z)|_{z = e^{j\omega T}} = \frac{C(1 - d)(a + jb)}{(a - 1)(a - d) - b^2 + jb(2a - 1 - d)}, \qquad (4.49)$$

which is easily evaluated for Nyquist plotting with the aid of a small computer. The Nyquist plot is shown in Figure 4.20 for a value of $C = 5$. The Nyquist plot for the equivalent continuous system $G(s)$ is shown for comparison on the same diagram. The destabilizing effect of sampling is clearly seen.

4.4.2 The effect of z plane zeros on frequency response

The location of zeros has an important effect on system dynamic behavior as mentioned in Section 4.3.4. Here we examine by means of an example the effect of zero locations on frequency response.

Example Let

$$G(z) = \frac{z + a}{z - 0.5},$$

where a is allowed to vary.

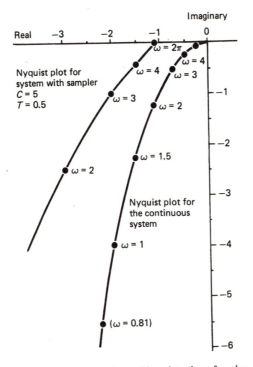

Figure 4.20 Open-loop Nyquist plots for the system of Figure 4.19 with and without the inclusion of a sampler before the motor.

The frequency response is given by the expression

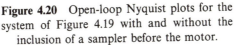

$$\frac{a + \cos \omega T + \mathrm{j} \sin \omega T}{-0.5 + \cos \omega T + \mathrm{j} \sin \omega T}.$$

Figure 4.21 shows plots of frequency response for some different values of a. It is clear that the zero frequency gain and the frequency response plots generally are drastically modified by the location of the zero.

4.4.3 Approximate frequency response through transformation of $G(z)$ to the w plane

THE W TRANSFORMATION

The relation $z = e^{sT}$ provides the connection between the z plane and the s plane. However, because of the irrationality of e, a transformation based on this relation leads to mathematically awkward expressions.

For certain problems (frequency response, stability determination) it is

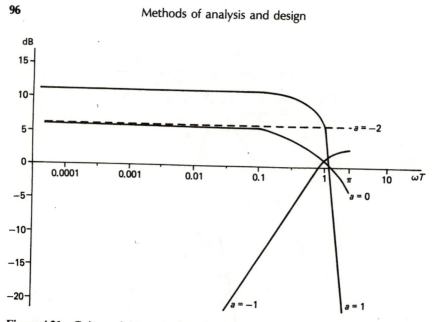

Figure 4.21   Gain vs. frequency plots for the element $(z + a)/(z + 0.5)$ for different values of a.

convenient to set

$$z = \frac{1 + w}{1 - w}$$

to transform from the z plane into the w plane.

The w plane can be regarded as, in a broad sense, an analog of the s plane. We agree to set

$$w = \sigma_w + j\omega_w \tag{4.50}$$

to emphasize the similarity with the s plane so that ω_w is the analog of ω in the s plane.

Along the imaginary axis in the w plane, $\sigma_w = 0$ and the equivalent locus in the z plane is given by

$$z = \frac{1 + j\omega_w}{1 - j\omega_w}$$

as ω_w moves from $-\infty$ to ∞, z takes on the values

$$z = 1 \angle (\tan^{-1} \omega_w - \tan^{-1}(-\omega_w))$$
$$= 1 \angle 2 \tan^{-1} \omega_w.$$

The imaginary axis in the w plane maps into the unit circle in the z plane. For $\omega \leqslant \omega_s/2$, ω_w has a one-to-one relation with frequency ω. In fact we can show that

$$\omega_w = \tan\left(\frac{\omega T}{2}\right). \tag{4.51}$$

Frequency-response plots, similar to those for continuous systems, can be drawn in the w plane, allowance being made at the interpretation stage for the relation between ω_w and true frequency ω.

Example Let

$$G(z) = \frac{z}{(z-1)(z-0.4)}. \tag{4.52}$$

Determine the frequency response by the W transformation method.
Setting

$$z = \frac{1+w}{1-w}$$

yields

$$G(w) = \frac{(1+w)}{[1+w-(1-w)](1+w-(1-w)0.4)}$$

$$= \frac{1+w}{1.2w(1 \pm w)}$$

Setting $w = j\omega_w$ yields

$$G(j\omega_w) = \frac{1+j\omega_w}{1.2j\omega_w - 1.2\omega_w^2},$$

$$|G(j\omega_w)| = \frac{(1+\omega_w^2)^{1/2}}{[(1.2\omega_w^2)^2 + (1.2\omega_w)^2]^{1/2}},$$

$$\angle\, G(j\omega_w) = \tan^{-1}\omega_w - \tan^{-1}\frac{1.2\omega_w}{1.2\omega_w^2} = \tan^{-1}\omega_w - \tan^{-1}\frac{1}{\omega_w}$$

where ω_w is related to ω by the expression

$$\omega_w = \tan\left(\frac{\omega T}{2}\right)$$

By reference to Section 4.2.2, it will be realized that the W transformation, achieved by setting $z = (1+w)/(1-w)$, is closely related to trapezoidal integration. The W transformation (often referred to as the bilinear transformation) is widely used, particularly in the design of digital filters.

Note that for algorithm design, to be described in Chapter 5, we shall prefer to use not the w plane but a slightly modified version called the w' plane, which has the merit of more nearly preserving true frequency. The details will be explained in Section 5.2.6.

4.4.4 Discussion

The direct frequency-response method, where z is replaced in $G(z)$ by $e^{j\omega T}$, is the recommended method whenever computer assistance is available to complete the calculation.

The w plane transformation has the advantage that the transform $G(w)$ tends to be much simpler than $G(z)|_{z=e^{j\omega T}}$, so that for hand sketching of Bode diagrams (for instance) this form would be preferred.

Finally, a place will still be found for the classical continuous frequency-response technique as the example below illustrates.

Example   The zero-order hold device (Section 2.9) has the transfer function

$$G(s) = \frac{1 - e^{-sT}}{s};$$

determine its frequency response.

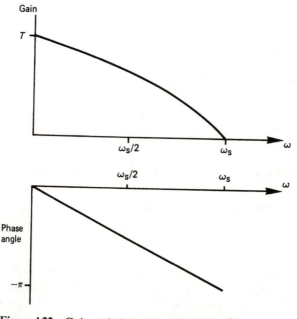

Figure 4.22   Gain and phase characteristics of a zero-order hold.

Replacing s by $j\omega$, we have

$$G(j\omega) = \frac{1 - e^{-j\omega T}}{j\omega},$$

and after manipulation, $G(j\omega)$ can be expressed in polar coordinates:

$$G(j\omega) = T\left|\frac{\sin\theta}{\theta}\right| \angle -\theta$$

where $\theta = \omega T/2$.

The gain and phase angle are sketched against ω in Figure 4.22.

4.5 Stability tests

Given a closed-loop transfer function

$$G(z) = N(z)/D(z)$$

stability depends on the location of the roots of the equation $D(z) = 0$. These are defined as the closed-loop poles. Poles outside the unit circle indicate an unstable system.

Whereas the Hurwitz or Routh tests will indicate whether poles are on the imaginary axis or in the right half complex plane, a different test is required to check whether any root of the equation $D(z) = 0$ satisfies

$$|z| \geqslant 1.$$

The following methods are available to check for roots on or outside the unit circle in the z plane:

(a) to factorize the equation $D(z) = 0$ and hence obtain specific locations for the poles;
(b) to determine the specific locations of the roots using a computer-implemented numerical method;
(c) to apply a criterion that will give a direct yes/no answer to the stability question without determining the location of the roots (considered in Section 4.5.1);
(d) to transform the problem into the s plane (or an analog of the s plane) and apply the Hurwitz or Routh criterion (considered in Section 4.5.2).

Methods (a) and (b) do not need to be discussed further. Turning to method (c), a number of possible different approaches are possible. These include Jury's test and the Shur–Cohn criterion. These tests are difficult to derive, difficult to remember and cumbersome to apply—their complexity increases rapidly with the order of the system being tested. However, for equations of order 2 or 3 Jury's test is simple. It is quoted below.

Methods of analysis and design

4.5.1 *Jury's test for equations of order 2 and 3*

Jury's test can be applied to equations of any order. However, it rapidly increases in complexity with increasing order and is usually only worth considering as a practical test for equations of very low order. Accordingly, we describe the test as applicable to equations of order 2 or 3.

Order 2
Given

$$f(z) = a_2 z^2 + a_1 z + a_0 = 0, \tag{4.53}$$

where $a_2 > 0$, no roof of f is on or outside the unit circle provided that

$$f(1) > 0, \qquad f(-1) > 0, \qquad |a_0| < a_2.$$

Order 3
Given

$$f(z) = a_3 z^3 + a_2 z^2 + a_1 z + a_0 = 0, \tag{4.54}$$

where $a_3 > 0$, no root of f is on or outside the unit circle provided that

$$f(1) > 0, \qquad f(-1) < 0, \qquad |a_0| < a_3,$$

$$\left| \det \begin{bmatrix} a_0 & a_3 \\ a_3 & a_0 \end{bmatrix} \right| > \left| \det \begin{bmatrix} a_0 & a_1 \\ a_3 & a_2 \end{bmatrix} \right|. \tag{4.55}$$

Example The simple feedback system of Figure 4.13 has a continuous-time element of transfer function

$$G(s) = \frac{C}{s(s+2)}$$

preceded by a sampler.

The closed-loop transfer function is given by

$$\frac{G(z)}{1 + G(z)}.$$

First $G(z)$ is obtained from $G(s)$ by the impulse invariant transformation as

$$G(z) = \frac{C}{2} \left(\frac{z(1 - e^{-2T})}{(z-1)(z - e^{-2T})} \right)$$

(see Section 4.3.2 where this expression was derived).

We check to ensure that all roots are inside the unit circle:

$$1 + G(z) = 1 + \frac{C}{2} \left(\frac{z(1 - e^{-2T})}{(z-1)(z - e^{-2T})} \right).$$

Then equivalent to $1 + G(z) = 0$ we have

$$2z^2 + (C(1 - e^{-2T}) - 2(1 + e^{-2T}))z + 2e^{-2T} = 0.$$

Applying the three conditions of Jury's test

(a) $f(1) > 0 \Rightarrow C(1 - e^{-2T}) > 0$

since $C > 0$ and $T > 0$ this condition is always satisfied.

(b) $f(-1) > 0 \Rightarrow C(e^{-2T} - 1) + 4(1 + e^{-2T}) > 0.$

This relation shows the interaction between gain C and sampling interval T. As $T \to 0$, the continuous case is approached and the condition is satisfied for any finite gain C.

(c) $|a_0| < a_2 \Rightarrow e^{-2T} < 1,$

which is satisfied for all positive T. (Notice that the example is artificial, to the extent that normally a hold device would have followed the sampler in a normal control application. The hold was omitted in this case to allow a direct comparison between the discrete and continuous loops.)

4.5.2 Stability investigation by transforming to the W plane

Transformation of a polynomial in z into a polynomial in s could be achieved in theory at least by replacing z by e^{sT}. In practice it is much easier to use an alternative transformation. The mapping $P: z \to w$ described by

$$P(z) = \frac{1 + w}{1 - w} \tag{4.56}$$

maps the exterior of the unit circle into the right half w plane. Thus $|z| > 1$ in the z plane implies that $w > 0$ in the w plane and a stability test can be carried out on the transformed problem in the w plane by methods usually used in the s plane.

Example To test the polynomial equation

$$z^4 - 2z^3 + 1.75z^2 - 0.75z + 0.125 = 0 \tag{4.57}$$

for roots on or outside the unit circle. Set

$$z = \frac{1 + w}{1 - w}$$

to produce the transformed polynomial

$$5.625w^4 + 6w^3 + 3.25w^2 + w + 0.125 = 0.$$

The Hurwitz test is applied as follows. Let the equation whose root locations are to be tested be in the form

$$a_0 w^n + a_1 w^{n-1} + \cdots + a_n = 0, \qquad \text{where } a_0 > 0.$$

The following matrices Δ_j are constructed

$$\Delta_j = \begin{bmatrix} a_1 & a_3 & a_5 & a_7 & \cdots & \\ a_0 & a_2 & a_4 & a_6 & \cdots & \\ 0 & a_1 & a_3 & a_5 & \cdots & \\ 0 & a_0 & a_2 & a_4 & \cdots & \\ \vdots & \vdots & \vdots & \vdots & & \\ 0 & 0 & 0 & 0 & \cdots & a_{j-4}, a_{j-2}, a_j \end{bmatrix}$$

The Hurwitz test then states that the equation has no roots on the imaginary axis or in the right half plane provided that either condition below is satisfied:

$$\det \Delta_j > 0 \begin{cases} \text{for all even } j \leqslant n \\ \text{for all odd } j \leqslant n \end{cases}.$$

For the equation under test

$$\Delta_3 = \begin{bmatrix} 6 & 1 & 0 \\ \vdots & \vdots & \vdots \\ 5.625 & 3.25 & 0.125 \\ \vdots & \vdots & \vdots \\ 0 & 6 & 1 \end{bmatrix}.$$

And the stability requirement is that

$$19.5 - 5.625 > 0,$$

which is clearly satisfied.

The Hurwitz test shows that no roots of the equation in w are in the right half plane and this confirms that the original polynomial in z has all roots inside the unit circle. Note that the roots of the original polynomial in z are at

$$z = \tfrac{1}{2}, \tfrac{1}{2}, \tfrac{1}{2} \pm j\tfrac{1}{2}.$$

Comment Further tests for system stability will be met in Chapter 8, dealing with multi-variable systems. Of course, stability information is also furnished, *inter alia*, by frequency-response and root-locus techniques.

Exercises

4.1 Convert each of the transfer functions $G_1(z)$, $G_2(z)$ and $G_3(z)$ from the example of Section 4.2.3 into a difference equation and, for each, determine numerically the response to a unit step using $T = 1$.

The original system from which the three discrete-time transfer functions above were obtained has the unit step response

$$u(t) = 1.47[-e^{-0.2t}(\cos 0.8t + \tfrac{1}{4} \sin 0.8t)].$$

Plot the three approximate step responses and the response of the original system on the same time axis for comparison.

4.2 Show that if y is the integral of a function u then, by trapezoidal approximation,

$$y(z) = \frac{T}{2} \frac{z+1}{z-1} u(z).$$

Given the differential equation

$$\frac{d^3 y}{dt^3} + \frac{d^2 y}{dt^2} + \frac{dy}{dt} + y = u$$

with

$$y(0) = \frac{dy}{dt}(0) = \frac{d^2 y}{dt^2}(0) = 0$$

and the equivalent set of first-order differential equations

$$\left.\begin{aligned}
\frac{dx_1}{dt} &= x_2 \\[4pt]
\frac{dx_2}{dt} &= x_3 \\[4pt]
\frac{dx_3}{dt} &= -x_1 - x_2 - x_3 + u
\end{aligned}\right\}$$

(where $x_1 = y$).

Decide on an appropriate time step and compute numerically, after trapezoidal discretization, the step responses of the two equivalent systems. Comment on the relative merits of the two approaches to solution.

4.3 Show that the amplitude and phase characteristics of the discrete

Table 4.3

Transfer function	Amplitude	Phase angle
D_2	$\left\|\dfrac{T}{2}\operatorname{cosec}\dfrac{\omega T}{2}\right\|$	$-\dfrac{\pi}{2}+\dfrac{\omega T}{2}$
D_3	$\left\|\dfrac{T}{2}\cot\dfrac{\omega T}{2}\right\|$	$-\dfrac{\pi}{2}$

approximations to integration

$$D_2(z) = \frac{T}{1-z^{-1}},$$

$$D_3(z) = \frac{T}{2}\left(\frac{1+z^{-1}}{1-z^{-1}}\right)$$

are given by Table 4.3.

For a continuous-time integrator, the amplitude is $1/\omega$ and the phase angle $-\pi/2$. Set $T=1$ and plot the amplitudes and phase angles for D_2, D_3 and the continuous integrator over the frequency range 0 to 6 rad/s.

4.4 Use the pole–zero matching method to obtain a discrete-time equivalent to the continuous transfer function

$$G(s) = \frac{4}{s^2 + 2s + 4}$$

when $T = \pi/5$ seconds.

4.5 Discretize the transfer function

$$G(s) = \frac{1}{(s+2)(s+3)} = \frac{y(s)}{u(s)}$$

by the methods:

    Forward rectangular approximation (Section 4.2.3)
    Backward rectangular approximation (Section 4.2.3)
    Trapezoidal approximation (Section 4.2.3)
    Pole matching (Section 4.2.4)
    Substitution of the first terms of the series for $\ln z$ (Section 4.2.5)
    $\mathscr{Z}$ transformation of $G(s)$, no zero-order hold (Section 4.2.6)
    $\mathscr{Z}$ transformation of $G(s)$, with zero-order hold (Section 4.2.6)

For each case, compute the solution for $u = 0$, $y(0) = 100$, $\dot{y}(0) = 0$, using the time step $T = 0.1$ second. Compare with the exact solution and comment critically on the results.

4.6 Not all differential equations met in practice are linear. Investigate and report on the discretization of the equation

$$\frac{d^2 y}{dt^2} + (y^2 - 1)\frac{dy}{dt} + y = 0.$$

4.7 Work through and confirm the solution in the examples of Sections 4.2.3 and 4.2.4.

4.8 It always seems to be remarkable that a non-periodic oscillatory solution can be generated by repeated multiplication by the same constant matrix.

In equation (4.44) set $y(0) = 1$, $dy/dt = 0$ and generate time solutions using the four discretization intervals $T = 0.1, 1, 2, 5$ seconds. Comment on your results. (You will need to go through the steps 1 to 4 as in Section 4.2.8.)

4.9 Investigate how well the frequency response of the lightly damped system

$$G(s) = \frac{100}{s^2 + 2s + 100}$$

is preserved under the three discretization schemes: impulse invariant transformation, pole mapping method, trapezoidal discretization.

4.10 Why is discretization of $G(s)$ not performed simply by replacing s in $G(s)$ by $(1/T)\ln z$? What are the disadvantages of simple, rectangular approximation approaches to discretization? Do these disadvantages lessen as the sampling interval T is shortened?

'Zeros at infinity' in $G(s)$ are usually represented in $G(z)$ by zeros at $z = -1$ in the pole–zero mapping approach to discretization. Can you justify such action?

All numerical methods for the solution of differential equations (and there are many) are nothing more or less than discretization methods! Discuss the validity and implications of this assertion.

4.11 A closed-loop system has the differential equation

$$\frac{d^2 y}{dt^2} + 2\zeta\omega_n\frac{dy}{dt} + \omega_n^2 y = \omega_n^2$$

where ζ denotes damping factor and ω_n denotes undamped natural frequency.

(a) Convert the equation to a transfer function $G(s)$.
(b) Transform $G(s)$ into an equivalent $G(z)$ by the impulse invariant method.
(c) Set $\zeta = 0.5$ and plot the z plane root locus as ω_n varies (use $T = 0.1$).
(d) Set $\zeta = 1.5$ and repeat (c).
(e) Comment on the results obtained.

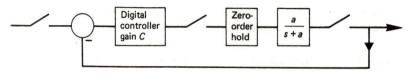

Figure 4.23

4.12 Determine the open-loop and closed-loop $\mathscr{Z}$ transfer functions of the system shown in Figure 4.23. Let $a = 1$, $T = 0.693$. Determing how the closed-loop system behavior varies as C is increased by plotting the root-locus diagram. Next fix C at $C = 1$ and with $a = 1$ allow T to vary to produce another root-locus diagram. Finally, let $a = 1$ and let T take on the values $T = 0.01, 0.1, 1, 10$. For each value of T, determine the gain C that brings the closed-loop system to the stability limit. Sketch the relation between T and C_{max}.

4.13 Use the known properties of pole locations in the z plane to synthesize a transfer function, $H(z)$, satisfying $\zeta = 0.6$, $\omega_n = 3\pi/(10T)$. In a control-design problem, $H(z)$ will represent the desired closed-loop transfer function of the completed system. Show how a controller $D(z)$ might be designed to control a process $G(s)$ such that the resultant closed-loop system has the overall transfer function $H(z)$. (Control design approaches are pursued in the next chapter.)

4.14 Explain carefully what you understand by the frequency response of a dynamic element of transfer function $G(z)$. What assumptions did you make about:

(a) The sampling of the input signal?
(b) The reconstitution of the output signal?

4.15 Consider the backward rectangular method of discretization as a mapping from the s plane to the z plane. Show that this mapping transforms the left half s plane into a circle in the z plane. Is the size of the circle dependent on the choice of sampling interval?

4.16 Investigate the effect of an added zero on the behavior of a closed-loop system by sketching the root locus for the following transfer functions:

$$G_1(z) = \frac{1}{(z-1)(z-0.5)} \qquad G_2(z) = \frac{(z-0.5)}{(z-1)(z-0.5)}$$

$$G_3(z) = \frac{(z-0.75)}{(z-1)(z-0.5)} \qquad G_4(z) = \frac{(z-0.4)}{(z-1)(z-0.5)}$$

$$G_5(z) = \frac{(z-0.1)}{(z-1)(z-0.5)} \qquad G_6(z) = \frac{(z+1)}{(z-1)(z-0.5)}.$$

4.17 Show that the transfer functions of a zero-order hold and a first-order hold device are given respectively by:

$$G_0(s) = \frac{1 - e^{-sT}}{s},$$

$$G_1(s) = (1 + sT)T\left(\frac{1 - e^{-sT}}{sT}\right)^2.$$

Plot the magnitude and phase characteristics of both devices against frequency. Confirm from the plots:

(a) that the zero-frequency gain for both devices is equal to T, just as for the perfect reconstructor of Section 2.8;
(b) that both devices have finite gains at frequencies above π/T, differing in this respect from the ideal reconstructor;
(c) that the first-order hold strategy is not, from the frequency response viewpoint, markedly superior to the zero-order hold strategy.

4.18 Figure 4.24 shows (a) a discrete-time system and (b) a continuous system.

Determine the pulse transfer function $G(z)$ for the discrete-time system, assuming that the samplers operate synchronously at intervals of 0.2 second.

Determine the gain of both systems to steady sinusoidal inputs at frequencies 0, 100 and 10^6 rad/s.

Comment briefly on your results.

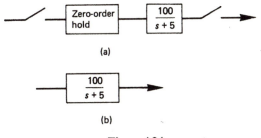

(a)

(b)

Figure 4.24

4.19 Calculate and plot in the z plane the points equivalent to:

(i) $s = j\frac{1}{6}\frac{\omega}{\omega_s}$,   (ii) $s = j\frac{1}{3}\frac{\omega}{\omega_s}$.

Under the four mappings:

(a) $s = \frac{1}{T}\ln z$,

(b) $s = \dfrac{z-1}{T}$,

(c) $s = \dfrac{z-1}{zT}$,

(d) $s = \dfrac{2}{T}\dfrac{z-1}{z+1}$.

Explain with the help of your z plane plot why the mapping (d) is superior to those of (b) and (c) in the discretization of transfer functions $G(s)$, and in stability testing of pulse transfer functions $G(z)$.

4.20 For the system of Figure 4.25, determine the largest value for C that will not cause instability when $\tau = 10$ and $T = 0.1$. Denote this value of C by C_m. The input is a ramp $v(t) = t$. Calculate and sketch the form of the output $y(t)$ when C has the value $C = C_m/2$.

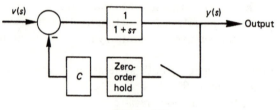

Figure 4.25

4.21 Determine the maximum gain C_1 that ensures stability in the control loop shown in Figure 4.26. Determine also the range of values of the gain to ensure a non-oscillatory closed-loop step response for the system. The sampling interval is one second.

Sketch the root locus for the system.

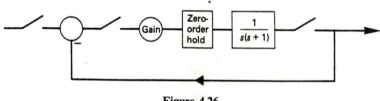

Figure 4.26

4.22 In the system of Figure 4.27, the sampler operates with period T. Use the W transformation to determine the maximum value of C for which the system is stable.

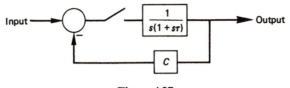

Figure 4.27

4.23 Check the stability of each of the three following transfer functions by:

(a) Jury's test;
(b) W transformation followed by the Hurwitz test;
(c) solution of the cubic equations of the denominator.

$$\text{(i)} \quad G_1(z) = \frac{1}{z^3 + 2.7z^2 + 1.5z + 0.2}.$$

$$\text{(ii)} \quad G_2(z) = \frac{z^2}{z^3 - 1.1z^2 - 0.25z + 0.275}.$$

$$\text{(iii)} \quad G_3(z) = \frac{1}{z^3 - 3.06z^2 + 3.12z - 1.06}.$$

4.24 To obtain practice in dealing with higher-order stability questions, test the transfer function

$$G(z) = \frac{1}{z^6 - z^5 + 0.99z^4 + 0.26z^3 - 0.3225z^2 - 0.0025z + 0.003125}$$

by any method or numerical subroutine of your own choosing.

4.25 A process of transfer function

$$\frac{C}{s(1 + sT_1)}$$

is preceded by a zero-order hold. The process output is sampled every T seconds and is then connected to form a closed loop with unity negative feedback. Transform the system to the z domain and obtain analytic expressions for:

(a) the value of C at which the system root locus intersects the unit circle;
(b) the coordinates of the point(s) where the root locus intersects the unit circle.

4.26 A process has the transfer function $G(s) = 1/(s + c)$. A designer decides to precede $G(s)$ by a compensator of transfer function

$$D(s) = \frac{(b - a)(s + c)}{(s + a)(s + b)}$$

with the aim of obtaining the compensated transfer function

$$G(s)D(s) = \frac{b - a}{(s + a)(s + b)} = H_1(s).$$

In practice, $D(s)$ is realized by its discrete equivalent $D(z)$ followed by a zero-order hold. Compute the transfer function $H_2(z)$ of the implemented combination of $D(z)$, zero-order hold and $G(s)$.

Insert numerical values $a = 5$, $b = 10$, $c = 0.1$, choose a time step for the discrete implementation and compare numerically the step responses of $H_1(z)$ and $H_2(z)$. To what extent are the designer's aims achieved?

5

Digital Control Algorithms

5.1 Introduction

Given a continuous process equipped with adequate measurement and actuating devices and a digital computer closing the control loop (Figure 5.1) how is the computer to be programmed to achieve a particular control objective?

One obvious but fundamental point is that control design proper always begins with a sufficiently accurate mathematical model of the process to be controlled. All the methods given here include the modelling aspect. The structure of the presentation is explained in Section 5.2.

The chapter also covers choice of the sampling interval, quantization problems, robustness and comparisons of performance.

5.2 Design methods

5.2.1 Approaches to digital controller design

We are concerned with the design of digital, discrete-time controllers for continuous-time processes.

There are four basic routes as illustrated in Figure 5.2.

1. Continuous-time modelling to obtain $G(s)$ followed by transformation to $G(z)$ and digital controller design.
2. The all discrete route (digital modelling and digital design).
3. Continuous-time modelling and continuous controller design followed by discretization of the resulting system.
4. Design in the w' domain.

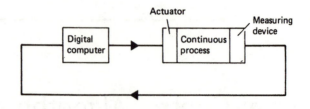

Figure 5.1   A continuous process under closed-loop digital control.

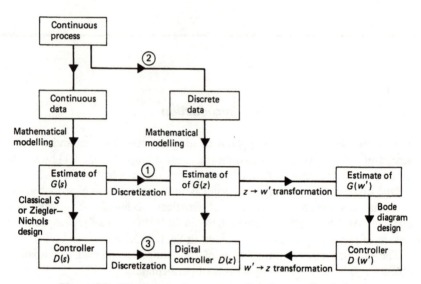

Figure 5.2   The four basic routes to digital controller design.

With the exception of the simple on–off algorithm described in Section 5.3.2, all digital control design procedures start from a mathematical description of the process to be controlled.

Thus the overall control-design task can be represented in two stages:

(a) obtaining a mathematical description of the process to be controlled (modelling);
(b) control design proper.

For a typical industrial problem, the effort required for task (a) (modelling) is often an order of magnitude greater than the effort required for (b) (control design proper).

Any control-design method that requires only a simple process model therefore has a high appeal to those faced with real industrial control problems.

5.2.2 Obtaining a simple continuous-time model approximating the process

As long ago as 1948, Ziegler and Nichols (Z2) suggested that an approximate transfer function could be fitted with little effort to an open-loop step-response curve of the process that is to be controlled.

Figures 5.3 and 5.4 show two different process step responses. If the response is as in Figure 5.3, then the suggested approximating transfer function is

$$G(s) = \frac{Ke^{-sT_1}}{(1 + sT_2)},\tag{5.1}$$

where the coefficients K, T_1, T_2 are obtained by the constructions shown in Figure 5.5 or, more accurately, by curve-fitting techniques.

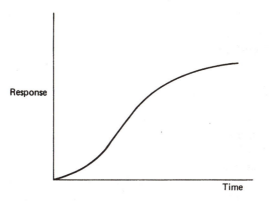

Figure 5.3   The open-loop step response of a typical industrial process.

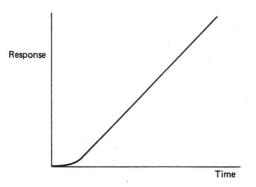

Figure 5.4   The open-loop step response of a process containing an integrator.

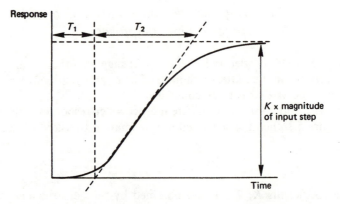

Figure 5.5 Showing how the coefficients in equation (5.1) are
found by simple graphical construction.

If the process step response is as in Figure 5.4, then the suggested
approximating transfer function is

$$G(s) = \frac{Ke^{-sT_1}}{s}. \tag{5.2}$$

Oscillatory open-loop step responses are encountered rather infrequently in
industrial applications, so that these two models suffice for most applications.

It can be seen that in both equations (5.1), (5.2) the order of the system
is, in a sense, modelled by a pseudo dead time that is contained within T_1, in
addition to any true process dead time. The validity of such an approximation
has to be judged in the context of the control-design operation.

5.2.3 Design approach 1: Transformation of G(s) into the z domain followed by digital design of a controller

Figure 5.6 shows a process with transfer function $G(s)$ preceded, as it will always
be in practice, by a hold device (here we have assumed a zero-order hold). The
continuous-time transfer function of the process with zero-order hold is denoted
$G'(s)$.

Let $H(z)$ be the desired transfer function of the closed-loop system: then

$$H(z) = \frac{D(z)G'(z)}{1 + D(z)G'(z)}. \tag{5.3}$$

Thus the required controller characteristic $D(z)$ can be determined from the
equation

$$D(z) = \frac{1}{G'(z)} \frac{H(z)}{1 - H(z)}. \tag{5.4}$$

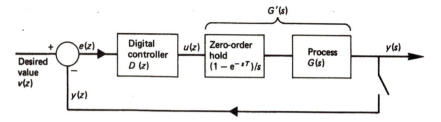

Figure 5.6 A closed-loop in which the transfer function $D(z)$ of the controller is to be synthesized.

Notice that

$$G'(z) = \mathscr{L}\left\{\frac{1 - e^{-sT}}{s} G(s)\right\} = (1 - z^{-1})\mathscr{L}\left\{\frac{1}{s}G(s)\right\}. \qquad (5.5)$$

Substituting for $G(s)$ from equation (5.1) and combining with equation (5.5) yields

$$D(z) = \frac{H(z)}{1 - H(z)} \frac{(z - e^{-T/T_2})}{Kz^{-T_1/T}(1 - e^{-T/T_2})}. \qquad (5.6)$$

GUIDELINES ON THE CHOICE OF $H(z)$

$H(z)$, the desired transfer function for the closed-loop system, must be chosen so that the controller transfer function $D(z)$ is realizable and stable. In practice this involves avoiding attempts to cancel out either process dead time or process zeros that are outside the unit circle in the z plane.

An attempt to cancel out a process dead time of magnitude pT synthesizes a controller with a difference equation of the form

$$u(k) = e(k + p).$$

This equation cannot be used in a control algorithm since it requires knowledge of the future values of e.

An attempt to cancel process zeros that are outside the unit circle synthesizes a controller having poles outside the unit circle. Cancellation can never be perfect. Let the error in attempting to cancel a real zero at $z = a$, $a > 1$, be ε; then the error ε gives rise to a term $\varepsilon/(z - a)$ in the resultant transfer function.

To confirm this, let

$$D(z) = \frac{z - a}{z - (a + \varepsilon)}, \qquad a > 1, \varepsilon > 0$$

and observe that

$$\mathscr{L}\{D(z)\} = (a + \varepsilon)^k - a^k.$$

This expression becomes infinite as k increases, showing that any error in cancellation, however small, will lead to an unusable system.

A further restriction on $H(z)$ results from the fact that the controller $D(z)$ must not have a numerator whose order exceeds that of the denominator. For example, if

$$G'(z) = \frac{1}{(z-1)(z-0.5)} \qquad \text{and} \qquad H(z) = \frac{1}{z},$$

then

$$D(z) = \frac{(z-1)(z-0.5)z^{-1}}{(1-z^{-1})} = z - 0.5.$$

$D(z)$ is not realizable since it requires a difference equation of the form

$$u(k) = e(k+1) - 0.5e(k).$$

This cannot be implemented since it requires a knowledge of future values of the input e. A different choice of $H(z)$ (for instance $H(z) = 1/z^2$) could lead to a physically realizable controller.

Requirements on $H(z)$ to ensure exact steady-state following of a particular type of input can be determined by using the final-value property. Suppose, as usual, that $y(z) = H(z)v(z)$. Then error

$$e(z) = v(z) - y(z),$$

$$v(z) - e(z) = H(z)v(z),$$

$$e(z) = v(z)(1 - H(z)).$$

Assume that $H(z)$ is to be specified so that there is zero steady-state error to a linear ramp input. By the final-value theorem

$$e(t)|_{t \to \infty} = v(z)(z-1)(1 - H(z))|_{z \to 1}$$

$$= \frac{Tz}{(z-1)^2}(z-1)(1 - H(z))|_{z \to 1}.$$

This, for zero error with the ramp input, the condition is

$$\frac{Tz}{(z-1)}(1 - H(z))|_{z \to 1} = 0. \tag{5.7}$$

Alternatively, the desired closed-loop transfer function may be specified with the aid of Figure 5.7 (previously derived in Section 3.7.3) to ensure a particular dynamic behaviour as characterized by ω_n (undamped natural frequency) and ζ (damping factor).

Suppose that the sampling interval T has been fixed and the aim is to obtain a closed-loop system with $\omega_n = 2\pi/(10T)$ and $\zeta = 0.8$. The poles of the closed-loop system must be located at (see Figure 5.7) $0.55 \pm j0.22$.

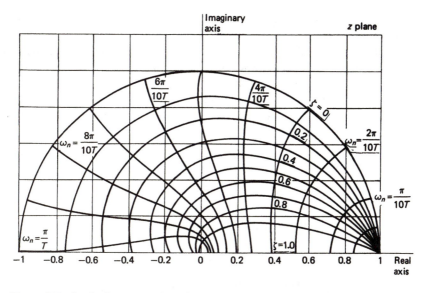

Figure 5.7 Loci of constant damping factor (ζ) and constant undamped natural frequency (ω_n) in the z plane.

The desired closed-loop transfer function is therefore given by

$$H(z) = \frac{N(z)}{(z - 0.55 + j0.22)(z - 0.55 - j0.22)},$$

where the numerator $N(z)$ remains to be chosen.

$N(z)$ needs to be chosen to give a suitable steady-state gain and to add zeros as recommended in Section 4.2.4.

CONVERTING A TYPICAL INDUSTRIAL CONTROL REQUIREMENT INTO A DESIRED
CLOSED-LOOP TRANSFER FUNCTION (GENERAL POINTS AND USEFUL RULES)

A typical industrial control problem might be stated as follows:

> Currently, 4% of the output of a particular process plant has to be discarded because of products failing to meet dimensional tolerance requirements. Suggest an economically viable control scheme that will reduce the percentage of product that has to be discarded.

The steps in the control design are then:

(a) convert the industrial control requirement into an equivalent desired closed-loop transfer function;
(b) obtain a dynamic model of the process that is to be controlled;
(c) from (a) and (b) put forward an economically viable design that takes into account any significant constraints not represented in (a) or (b).

This section is concerned only with point (a), the conversion of an industrial requirement into a desired closed-loop transfer function. (Point (b) is given due consideration in each of the design procedures of this chapter, point (c) represents the engineering reality of each particular implementation and some appreciation may be obtained from the case histories of Chapter 7.)

As an intermediary between the industrial requirement and a desired closed-loop transfer function, we can sometimes use the following performance-related system parameters:

Rise time the time for a step response to rise from 10% to 90% of its final value;

Time for the step response to reach 50% of its final value;

Settling time the time for the step response to settle to within 5% of its final value—smaller settling bands are used by some authors in this definition;

Steady-state error in response to step, ramp or parabolic input functions;

Closed-loop system bandwidth ω_b the frequency at which the closed-loop gain falls 3 dB below its zero-frequency ω_n;

Closed-loop damping factor ζ.

SOME USEFUL INTER-RELATIONS BETWEEN THE PARAMETERS
Let $G(s)$ be a process to be controlled by a controller of transfer function $D(s)$ and let

$$H(s) = \frac{G(s)D(s)}{1 + G(s)D(s)}.$$

The following relations are given without proof; they can be confirmed by reference to many basic texts on continuous control. Truxal (T4) is particularly useful despite its age.

(a) Let $H(s)$ have no finite zeros and be dominated by one pair of complex poles. Then

$$\omega_b = \omega_n[1 - 2\zeta^2 + (2 - 4\zeta^2 + 4\zeta^4)^{1/2}]^{1/2}. \tag{5.8}$$

Some typical values are given in Table 5.1.

Table 5.1

ζ	Relation between ω_b and ω_n
0.5	$\omega_b = 1.27\omega_n$
$1/\sqrt{2}$	$\omega_b = \omega_n$
0.95	$\omega_b = 0.7\omega_n$

(b) Define

$$k_p = \lim_{s \to 0} G(s)D(s)$$

$$k_v = \lim_{s \to 0} sG(s)D(s),$$

$$k_a = \lim_{s \to 0} s^2 G(s)D(s).$$

$$(5.9)$$

Then for stable $H(s)$, the k_p, k_v, k_a relate to the steady-state errors as follows:

steady-state error to step input $= \dfrac{1}{1 + k_p}$;

steady-state error to ramp input $= \dfrac{1}{k_v}$;

steady-state error to parabolic input $= \dfrac{1}{k_a}$.

The three quantities are referred to as the *position, velocity* and *acceleration constants* respectively.

(c) Time to reach 50% of step response $\simeq 1/k_v$. (See Reference T5, equation (1.265).)

(d) $k_v = \dfrac{\omega_n}{2\zeta}$

and in conjunction with equation 5.8, it is found that when $\zeta = 1/\sqrt{2}$, $k_v = 0.79\omega_b$. (See Reference T5, equation (5.43).)

(e) Rise time $\simeq 2.5/\omega_b$. (See Reference T5, equation (1.240).)

Note: There are limitations on the applicability of the relations (c)–(e) which should be understood by following up the references.

(f) For all real processes, $G(s)_{s \to \infty} = 0$ and hence for large values of s

$$H(s) \simeq G(s)D(s).$$

Now an ideal system could be visualized as one satisfying $H(s) = 1$, but this would imply a controller transfer function $D(s)$ that could not be realized. (Try this on a simple example of $G(s)$ to convince yourself.)

(g) System bandwidth ω_b is a measure, in part, of the ability of a system to follow time-varying inputs and to reject time-varying disturbances.

Given a desired value of ω_b, we may using methods to be given later select a suitable sampling interval T. Then either:

1. Choose $H(s)$ based on guidelines given above and discretize to obtain a desired $H(z)$, or

2. Choose ζ, ω_n based on guidelines given above and use Figure 5.7 to synthesize a desired $H(z)$.

Inexperienced engineers tend to design systems under simplistic assumptions where process models are known, time-invariant, linear and noise-free.

In practice a majority of industrial control loops exists mainly to reject disturbances that are not deterministic, and the performance that can be obtained is limited by process nonlinearity, noise and the high capital cost of providing equipment for rapid corrective actions.

Referring back to the industrial control problem cited at the beginning of this section: Equipment dimensions limit force and hence acceleration and hence rate of correction of product dimensions (assuming a mechanical production process). Actuator cost is related to maximum available actuator force and this also sets a limit on maximum rate of correction. Thus, maximum performance is fixed either by a limit on the physical strength of the equipment or by a cost limit on the power of the actuator that moves the equipment—neither limit can be represented in a linear model.

DEAD-BEAT CONTROL

Suppose the response of the closed-loop system to a unit step is required to equal unity at every sampling instant after the application of the unit step (sometimes called *dead-beat* control).

We have

$$v(z) = 1 + z^{-1} + z^{-2} + \cdots = \frac{1}{1 - z^{-1}}, \tag{5.10}$$

$$y(z) = z^{-1} + z^{-2} + \cdots = \frac{z^{-1}}{1 - z^{-1}} \tag{5.11}$$

Then the desired closed-loop model must have the pulse transfer function $H(z) = z^{-1}$.

From equation (5.4), the controller has the transfer function

$$D(z) = \frac{z^{-1}}{G'(z)(1 - z')}. \tag{5.12}$$

Suppose that the process transfer function $G(s)$ has been identified in open-loop by the method of Section 5.2.2 as

$$G(s) = \frac{e^{-sT_1}}{(1 + sT_2)}.$$

Then, allowing for the inclusion of the zero-order hold, the combined transfer function G' is given by an expression of the form

$$G'(z) = \frac{(1 - \alpha)z^{\beta}}{(z - \alpha)}, \tag{5.13}$$

where α, β depend on T_1, T_2 and the magnitude of the sampling interval T.

Notes

(a) The controller will not be physically realizable if the dead time in the process is equal to or greater than the sampling interval.
(b) Large initial values of control may be produced by the algorithm. In a practical implementation this may drive the receiving actuator into saturation.
(c) The response of the controlled system may be highly oscillatory, although equal to unity at the sampling instants.

MODIFICATION TO THE DEAD BEAT ALGORITHM TO OBTAIN A SMOOTHER RESPONSE
(DAHLIN'S MODIFICATION)

To avoid undesirable transient oscillations, the response of the closed-loop system to a unit step can be required to follow a smooth curve.

In continuous time, the required step response can be expressed as

$$y(s) = \frac{e^{-as}}{1 + sp} \frac{1}{s} \tag{5.14}$$

with a, p chosen to give the required response (see Figure 5.8). Let $a = nT$; then

$$y(z) = \frac{(1 - e^{-T/p})z^{-n-1}}{(1 - z^{-1})(1 - e^{-T/p}z^{-1})}. \tag{5.15}$$

This leads to the controller characteristic $D(z)$

$$D(z) = \frac{(1 - e^{-T/p})z^{-n-1}}{1 - e^{-T/p}z^{-1} - (1 - e^{-T/p})z^{-n-1}} \frac{1}{G'(z)}, \tag{5.16}$$

where, as usual, $G'(z)$ denotes the transfer function of the process to be controlled, together with its associated zero-order hold.

Further details of the algorithm can be found in Reference D1.

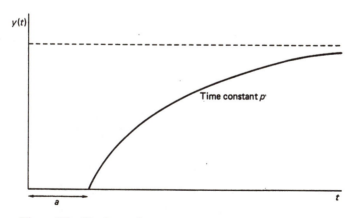

Figure 5.8   The form of response represented by equation (5.14).

Ringing   It sometimes happens that a digital controller, synthesized so as to obtain a particular desired system behavior, has one or more poles close to the $z = -1$ point. The output of such a controller oscillates strongly in a manner that is often considered unsatisfactory even though the process is being controlled as was intended.

In the literature, poles near to $z = -1$ are often referred to as *ringing poles*. A crude but common strategy to remove ringing poles involves replacing them by an equivalent steady state gain. Thus, a controller term $1/(z + 1)$ is replaced by a term

$$\frac{1}{z + 1}\bigg|_{z \to 1} = \frac{1}{2}.$$

KALMAN's ALGORITHM

Kalman's algorithm can be considered as a modified dead-beat algorithm where steps are taken to prevent excessive oscillation of the control signal.

The approach is explained for a second-order process with unity steady-state gain—refer to Figure 5.9. The transfer function of the process with zero-order hold is as usual denoted by $G'(s)$.

A unit step is assumed at the reference input:

$$v(z) = \frac{1}{1 - z^{-1}}.$$

The required response y is chosen as

$$y(z) = \alpha z^{-1} + z^{-2} + z^{-3} + \cdots.$$

The required control u is chosen as

$$u(z) = \beta + \gamma z^{-1} + z^{-2} + z^{-3} + \cdots$$

with α, β and γ as yet unspecified. Here it can be seen how the Kalman approach, by specifying the form of the control signal as well as that of the process

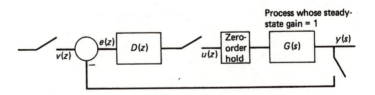

Figure 5.9   The closed-loop system used in the explanation of Kalman's algorithm.

response, is able to prevent excessively oscillatory control actions. Now

$$\frac{y(z)}{v(z)} = (1 - z^{-1})(\alpha z^{-1} + z^{-2} + z^{-3} + \cdots)$$

$$= \alpha z^{-1} + z^{-2} + \cdots - \alpha z^{-2} - z^{-3} - \cdots$$

$$= \alpha z^{-1} + (1 - \alpha)z^{-2} = (\text{say}) \, P(z), \tag{5.17}$$

$$\frac{u(z)}{v(z)} = (1 - z^{-1})(\beta + \gamma z^{-1} + z^{-2} + \cdots)$$

$$= \beta + \gamma z^{-1} + z^{-2} + \cdots$$

$$- \beta z^{-1} - \gamma z^{-2} - z^{-3} - \cdots$$

$$= \beta + (\gamma - \beta)z^{-1} + (1 - \gamma)z^{-2} = (\text{say}) \, Q(z). \tag{5.18}$$

$$G'(z) = \frac{y(z)}{u(z)} = \frac{y(z)}{v(z)} \frac{v(z)}{u(z)} = \frac{P(z)}{Q(z)}$$

and at this point the coefficients α, β, γ that were left undesignated above can be fixed from knowledge of $G'(z)$—see the example below for numerical details.

Finally we use the relation that

$$\frac{y(z)}{v(z)} = \frac{D(z)G'(z)}{1 + D(z)G'(z)}$$

leading to

$$D(z) = \frac{P(z)}{G'(z)(1 - P(z))}$$

$$= \frac{Q(z)}{P(z)} \frac{P(z)}{(1 - P(z))} = \frac{Q(z)}{1 - P(z)} \tag{5.19}$$

and the control design is complete.

Note that the number of intermediate values required in the control sequence $u(z)$ depends on the order of the process that is to be controlled—in general the process response has $n - 1$ intermediate values and the control u has n pre-steady values, where n is the order of the process to be controlled.

Example   **Kalman's algorithm: a second-order process with unity steady-state gain**   Assume that

$$G'(z) = \frac{0.5 + 0.22z}{(z - 0.1)(z - 0.2)}$$

and that, as above

$$\frac{y(z)}{v(z)} = P(z) = \alpha z^{-1} + (1 - \alpha)z^{-2},$$

$$\frac{u(z)}{v(z)} = Q(z) = \beta + (\gamma - \beta)z^{-1} + (1 - \gamma)z^{-2}.$$

We note that the sums of the coefficients of the polynomials $P(z)$ and $Q(z)$ are both unity and that $G'(z)$ may be scaled so that its numerator and denominator coefficients also satisfy the same requirement and the powers of z match. (Because $u(z)$ is chosen having regard to the process order and steady-state gain, this normalizing operation can always be achieved by simply dividing the terms in $G'(z)$ by the sum of the numerator coefficients.)

Normalization produces

$$G'(z) = \frac{0.22z^{-1} + 0.5z^{-2}}{z^{-2}(z^2 - 0.3z + 0.02)}$$

$$= \frac{0.305z^{-1} + 0.695z^{-2}}{1.39 - 0.417z^{-1} + 0.028z^{-2}} = \frac{P(z)}{Q(z)}.$$

Now

$$D(z) = \frac{Q(z)}{1 - P(z)} = \frac{1.39 - 0.417z^{-1} + 0.028z^{-2}}{1 - 0.305z^{-1} + 0.695z^{-2}}.$$

The simplicity of the approach can be seen from the example. The controller coefficients are obtained directly from the (normalized) pulse-transfer function of the process.

NUMERICAL DETERMINATION OF THE COEFFICIENTS IN A CONTROLLER OF SPECIFIED FORM

As earlier in this section, let the process, together with a zero-order hold, be represented by $G'(s)$.

The form of the controller D is supposed chosen, whereas its coefficients $a_1, \ldots, a_n$ remain to be determined. An off-line computation is arranged as shown in Figure 5.10. Some initial values are assumed for the vector $a = (a_1, \ldots, a_n)^T$ of controller parameters.

A test signal is input to the configuration, and the simulated time response is compared with the desired response. A scalar valued criterion $J(a)$ is calculated after some suitable time t_1 has elapsed. A typical expression for $J(a)$ might be

$$J(a) = \int_0^{t_1} (y(t) - y_d(t))^2 \, dt. \tag{5.20}$$

A hill-climbing routine (see Reference L4) is utilized to automatically manipulate

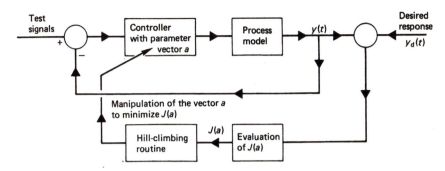

Figure 5.10   Numerical determination of the coefficients in a controller of specified form.

the vector a iteratively to minimize $J(a)$, i.e. to choose the controller parameters that make the simulated system response match the required response as closely as is possible.

Notice that it will normally be the aim of the design exercise to obtain a good response not only at sampling times but also between sampling times. This is why we have shown the expression for $J(a)$ (equation (5.20)) as a continuous integral, although we know that it will be evaluated as a summation. The time steps in the simulation and in the summation must be chosen to be very short compared with the proposed system's sampling time, so that a sufficiently accurate estimation of the behavior between sampling instants can be obtained.

A suitable form of controller for many purposes is given by

$$D(z) = \frac{a_0 + a_1 z^{-1} + a_2 z^{-2}}{b_0 + b_1 z^{-1} + b_2 z^{-2}}. \tag{5.21}$$

THE ROOT-LOCUS APPROACH

An alternative to the numerical search for controller coefficients is to use the root-locus technique. Normally, the technique can be used to determine the value of only one parameter, the loop gain.

If the controller is represented by a gain k in series with a fixed element $F(z)$ and the process together with zero-order hold is represented by $G'(z)$ then (Figure 5.11) the closed-loop transfer function is

$$\frac{kF(z)G'(z)}{1 + kF(z)G'(z)}. \tag{5.22}$$

The root locus consists of a plot of the path of the roots of the equation

$$1 + kF(z)G'(z) = 0 \tag{5.23}$$

as k varies.

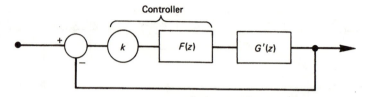

Figure 5.11 The block diagram corresponding to equation (5.22).

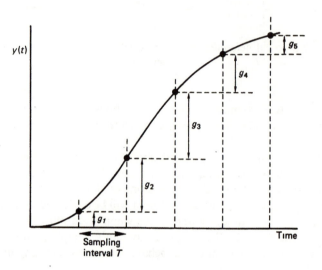

Figure 5.12 Open-loop process step response used as a basis
for the discrete-time model.

In conjunction with lines of constant damping factor, a suitable value of
k can be chosen.

5.2.4 Design approach 2: Digital modelling followed by
digital design

AN ALGORITHM BASED ON DIGITAL APPROXIMATION TO THE PROCESS OPEN-LOOP
STEP RESPONSE

Let the open-loop unit step response of a process be as shown in Figure 5.12.
Define the sequence g_i, $i = 1,\ldots$ as shown in the figure. Clearly, for a
non-oscillatory process whose step response reaches a steady state, the sequence
g_i converges.

Let $G(z)$ be the transfer function of the process; then the response to a unit step is given by

$$y(z) = G(z)\frac{1}{(1 - z^{-1})}. \tag{5.24}$$

But from the diagram

$$y(z) = g_1 z^{-1} + (g_1 + g_2)z^{-2} + \cdots = \sum_{k=1}^{\infty} \sum_{i=1}^{k} g_i z^{-k}, \tag{5.25}$$

so that

$$G(z) = (1 - z^{-1}) \sum_{k=1}^{\infty} \sum_{i=1}^{k} g_i z^{-k}$$

$$= \sum_{i=1}^{\infty} g_i z^{-i}. \tag{5.26}$$

Since we want to avoid reading a large number of values g_i from the diagram, we truncate equation (5.26) to yield

$$G(z) \simeq \sum_{i=1}^{n-1} g_i z^{-i} + \frac{g_b z^{-n}}{1 - pz^{-1}}, \tag{5.27}$$

in which the parameter p governs the rate of decay of the sequence $\{g_i\}, i = n, \ldots$. $G(z)$ can also be written

$$G(z) = \sum_{i=1}^{n} b_i z^{-i}/(1 - pz^{-1}), \tag{5.28}$$

where

$$b_1 = g_1,$$

$$b_i = g_i - pg_{i-1}, \qquad i = 2, \ldots, n.$$

p is chosen so that the correct steady-state value is obtained for y.

Let y_s be the steady-state value of y after the application of a unit step; then

$$p = 1 - \left(g_n \middle/ \left(y_s - \sum_{i=1}^{n-1} g_i\right)\right). \tag{5.29}$$

A so-called *finite-time settling response* will be obtained, provided that the response can be expressed as a finite polynomial in the operator z. This can be achieved by using the following algorithm:

$$u(k) = k_I \sum_{i=0}^{k} (v(i) - y(i)) - \sum_{j=1}^{n} k_j x_j(k), \tag{5.30}$$

where x_j represents the jth element in the state vector of the process and u is the control signal to be applied to the process;

$$\left. \begin{array}{c} k_I = 1 \Big/ \left(\sum\limits_{i=1}^{n} b_i \right), \\[2mm] k_1 = 0, \qquad k_2 = k_3 = \cdots = k_{n-1} = k_I, \\[2mm] k_n = \left(1 + p - \sum\limits_{i=1}^{n-1} g_i \Big/ k_I \right) \Big/ g_n. \end{array} \right\} \qquad (5.31)$$

See Takahashi [T1] for the details of the derivation. The implementation of the algorithm is shown in Figure 5.13.

Often the state vector x is not available to be fed back. Although x may be estimated, this complicates the application and the need for the state vector to be available must be considered a disadvantage.

DIFFERENCE-EQUATION MODELLING

A continuous process with input $u(t)$ and output $y(t)$ can be modelled approximately by the discrete transfer function $G(z)$, where $G(z)u(z) = y(z)$.

Since $G(z)$ is a ratio of polynomials in z, it can be expanded and then arranged as the difference equation

$$y_k = - \sum_{i=1}^{n} a_i y_{k-i} + \sum_{i=1}^{n} b_i u_{k-i}, \qquad (5.32)$$

where the coefficients a_i, b_i depend on the chosen sampling interval T.

Define

$$y = \begin{bmatrix} y_k \\ \vdots \\ y_{k+2n-1} \end{bmatrix}, \qquad u = \begin{bmatrix} u_k \\ \vdots \\ u_{k+2n-1} \end{bmatrix},$$

$$\theta = \begin{bmatrix} a_1 \\ \vdots \\ a_n \\ b_1 \\ \vdots \\ b_n \end{bmatrix},$$

$$\Lambda = (-z^{-1}y, \ldots, -z^{-n}y, z^{-1}u, \ldots, z^{-n}u).$$

then, by sampling $2n$ times, equation (5.32) can form the basis for the vector matrix equation

$$y = \Lambda\theta, \qquad (5.33)$$

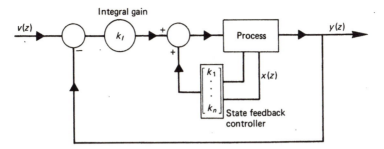

Figure 5.13 The finite settling time controller implemented.

which at least in theory can be rearranged to give identification of the parameter vector θ using

$$\theta = \Lambda^{-1} y. \tag{5.34}$$

The inversion of Λ is easily ill-conditioned since the column vectors in the matrix Λ are often very similar to each other.

The situation can be improved by sampling many more than $2n$ times. It is still true that

$$y = \Lambda\theta, \tag{5.35}$$

although the matrix Λ is now non-square. Equation (5.35) can be manipulated into the form

$$\theta = (\Lambda^T \Lambda)^{-1} \Lambda^T y. \tag{5.36}$$

The estimate for the parameter vector θ can be recognized as a standard result in regression analysis—it is a best least squares estimate.

More manageable in practice would be a recursive version of equation (5.36). This development is left as an exercise (see Exercise 5.24).

This type of difference-equation model is very suitable for use directly in numerical controller design as in Section 5.2.3. Alternatively, the difference equation can be transformed to a transfer function $G(z)$ and the synthesis techniques of Section 5.2.3 used for controller design.

The recursive version of the algorithm has been found useful for on-line modelling as part of adaptive control schemes. Where the model is required to follow a time-varying process, the identification must be made to work through a data 'window', so that the effects of very old data on the estimate θ are minimal. The generic name for a system that determines θ on-line and then proceeds automatically to choose and implement controller parameters is *self-tuning regulator*. See Chapter 12 for a detailed treatment of adaptive and self-tuning systems.

5.2.5 Design approach 3: Continuous modelling, continuous control design, discrete approximation

ZIEGLER AND NICHOLS' METHOD

Many years ago Ziegler and Nichols [Z2] put forward a simple idea of great utility that is still valuable for practical control design. The problem is to control an unknown process P whose open-loop response to a unit step is available. (The approach has already been introduced in Section 5.2.2.) The steps in the argument are essentially as follows:

(a) P is unknown but its step response curve is available.
(b) If P is open-loop stable then it may be approximated by a first-order process in series with a dead-time element; that is,

$$P(s) \simeq \frac{k}{1 + sT_2} e^{-sT_1} = (\text{say}) \ G(s)$$

for some choice of k, T_1, T_2.
(c) The numerical values in $G(s)$ can be found from the response curve of P as shown in Figure 5.14.
(d) Best average settings for a three-term controller can be determined to control the process of transfer function $G(s)$. Ziegler and Nichols produced simple explicit formulae for determination of the controller settings from a knowledge of k, T_1, T_2 in $G(s)$.
(e) The final assumption is that the controller settings will be satisfactory when implemented to control the process P.

The numerical details of the method are as follows.
The suggested three-term continuous controller for the process with transfer

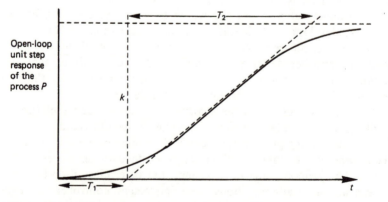

Figure 5.14 Geometrical construction for the determination of model parameters by the Ziegler–Nichols method.

function

$$G(s) = \frac{ke^{-sT_1}}{(1 + sT_2)} \tag{5.37}$$

has the transfer function

$$D(s) = c\left(1 + \frac{1}{T_I s} + T_D s\right) \tag{5.38}$$

in which, by the Ziegler–Nichols rules [Z2]

$$c = \frac{1.2T_2}{kT_1}, \qquad T_I = 2T_1, \qquad T_D = 0.5T_1.$$

For a two-term $P + I$ controller, the rule is

$$D(s) = c\left(1 + \frac{1}{T_I s}\right)$$

with

$$c = \frac{0.9\ T_2}{k\ T_1}.$$

For a proportional only controller, $D(s) = c$ with

$$c = \frac{1}{k}\frac{T_2}{T_1}$$

Using simple forward rectangular approximations to differentiation and integration the discrete-time form of the three-term controller can be derived as

$$u(k) = c\left[e(k) + \frac{T}{T_I}\sum_{i=0}^{k} e(i) + \frac{T_D}{T}(e(k) - e(k-1))\right] + u(0). \tag{5.39}$$

Alternatively, the algorithm can be expressed in the so-called 'incremental form'

$$\Delta u(k) = c\left[e(k) - e(k-1) + \frac{T}{T_1}e(k) + \frac{T_D}{T}(e(k) - 2e(k-1) + e(k-2))\right]. \tag{5.40}$$

Equation (5.40) follows from equation (5.39) by forward rectangular discretization approximation of the form

$$\frac{de(k)}{dt} = \frac{e(k) - e(k-1)}{T_1}, \qquad \frac{d\sum_{i=0}^{k} e(i)}{dt} = \frac{e(k)}{T},$$

$$\frac{d}{dt}(e(k) - e(k-1)) = \frac{e(k) - 2e(k-1) + e(k-2)}{T},$$

$$\frac{du(k)}{dt} = \frac{\Delta u(k)}{T}.$$

At each sampling instant, equation (5.39) supplies an absolute value, whereas equation (5.40) supplies an increment.

A number of practical questions arise such as:

(a) Will there be an unacceptably large disturbance when the algorithm is first switched on?
(b) Will the integral term be driven into saturation?
(c) Will the actuator have completed its movement by the end of each sampling period?

Points such as these decide which of the two algorithms is most suitable for a particular service.

The sampling interval needs to be chosen with care. The problem is discussed in Section 5.5.1.

The one great advantage of the Ziegler–Nichols method just described is its great simplicity! Before dismissing the approach as being too crude, ponder on the limitations of process models, however obtained. Most processes are so complex that any model is really a gross simplification. The largest error in the design chain is likely to arise because the original process data is not sufficiently representative.

Final comment: an important note of realism   Readers should be aware that it is only in the relatively ideal world of servomechanism design that transfer-function type models can be easily derived and that specifications in terms of ζ, ω_n, bandwidth, error constants and the like can realistically be set out. A typical industrial process tends to be poorly defined with, perhaps, at best, a step response available. The specification to be met is not in general easy to translate into servomechanism terms. It is under these typical industrial conditions that the apparently crude route through step-response, Ziegler–Nichols procedure and three-term controller implementation is preferred in practice, although it may appear inferior as a scientific method.

5.2.6 Design approach 4: Transformation of the problem into the w' domain, followed by control design using the Bode diagram

In classical continuous control system design, frequency-response methods of synthesis, using Bode and Nyquist diagrams, are of acknowledged importance. These diagrams are not practicable in the z domain, largely because of the irrationality of the function e^{sT}. However, the irrationality can be avoided by the use of the w domain (or its numerically more convenient variant, the w' domain). In this section, we describe the design of digital controllers by the use of the Bode diagram in the w' domain.

For completeness we first review how the Bode diagram is applied classically to continuous system design in the frequency domain.

THE BODE DIAGRAM AS A DESIGN TOOL

The Bode diagram is a plot of the magnitude and phase angle of a transfer function against frequency. It has the following very useful characteristics.

(a) It can be sketched by very simple techniques to an accuracy sufficient for preliminary system design purposes. In particular, on the logarithmic axes that are usually used, a rational transfer function has a magnitude curve that is representable by straight-line segments.

(b) The closed-loop behavior of a system can be predicted by inspection of the open-loop Bode diagram.

(c) The transfer functions of proposed controllers can be added graphically very easily to check their effects on the closed-loop behavior. Such controllers may be suggested by experience or trial and error, or they may be synthesized with the aid of the Nichols chart or the root-locus diagram.

(d) The magnitude curve of a minimum phase transfer function (no zeros in the right half plane) completely characterizes the closed-loop stability behavior. In particular, the slope with which the magnitude curve crosses the unity gain line governs closed-loop stability.

CONTROL DESIGN USING THE CONVENTIONAL BODE DIAGRAM

The Bode diagram, with its easily sketched curves, is the best graphical tool for design in the frequency domain. In design using the Bode diagram, it is usually the aim to achieve the highest loop gain that is consistent with a sufficient stability margin. This sufficient stability margin is usually stated in terms of *gain margin* and *phase margin*. These two terms are defined in Figure 5.15. A typical stability margin requirement is

$$\text{gain margin} \geqslant 8 \text{ dB}$$

and

$$\text{gain margin} \geqslant 30°.$$

Satisfaction of the two criteria has been found to lead to satisfactory stability margins in many practical applications.

A useful guide to help in achieving both criteria simultaneously is that the gain plot should cross the 0 dB line with a slope of -20 dB per decade at a point that satisfies the phase-margin requirement.

Example The use of the Bode diagram in classical continuous control synthesis. A process has the transfer function

$$G(s) = \frac{1}{s(1 + 0.1s)(1 + 0.2s)}$$

and its Bode diagram is sketched, approximately, in Figure 5.16.

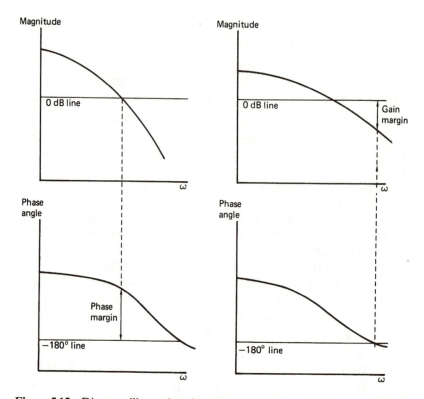

Figure 5.15   Diagram illustrating the meaning of the terms 'gain margin' and 'phase margin'.

A simple proportional controller can then be designed immediately by inspection of the Bode diagram. We see that to produce an 8 dB gain margin, a gain of 14 dB is required in the controller (equal to a proportional linear gain of 5 times). Notice from the Bode diagram that in this case the phase-margin requirement of 30° is satisfied simultaneously.

Of course, we shall usually require more than a simple proportional controller and we may, if we so wish, investigate many possible controllers whose forms are suggested by the form of the process to be controlled. All of the power of the accumulated servomechanism design method may be brought to bear on the problem at this point. However, to keep a hold on reality, recall that either noise or nonlinearity will limit what can be achieved and that neither is being modelled. Thus, it is necessary to pay heed to accepted guidelines on degrees of performance improvement that can be realistically aimed for. One such guideline relates to the use of a lead compensator. Such a compensator has the transfer function

$$D(s) = \frac{\alpha(1 + sT_c)}{1 + \alpha sT_c}$$

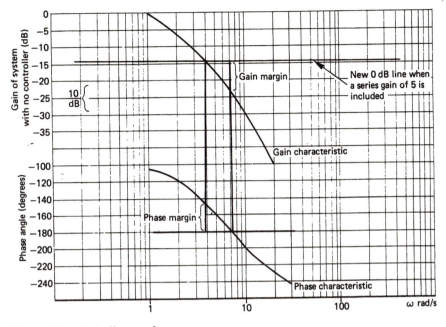

Figure 5.16 Bode diagram for

$$G(s) = \frac{1}{s(1 + 0.1s)(1 + 0.2s)}$$

and conventional wisdom decrees that T_c is set to cancel the dominant pole in the process while α takes on a value of approximately $\alpha = 0.1$. The α term in the numerator is present to allow the transfer function to be realized easily using a simple analog circuit.

Thus we have

$$D(s) = \frac{0.1(1 + 0.2s)}{1 + 0.02s}$$

and

$$G(s)D(s) = \frac{0.1}{s(1 + 0.1s)(1 + 0.02s)}.$$

The Bode diagram for this transfer function is given in Figure 5.17. The phase characteristic reaches the $-180°$ line at a frequency some three times higher than for the uncompensated system. The series gain C is finally fixed to satisfy the phase margin requirements. (The gain margin then comes out at 12 dB.) The controller so obtained has the transfer function

$$CD(s) = \frac{16(1 + 0.2s)}{1 + 0.02s}$$

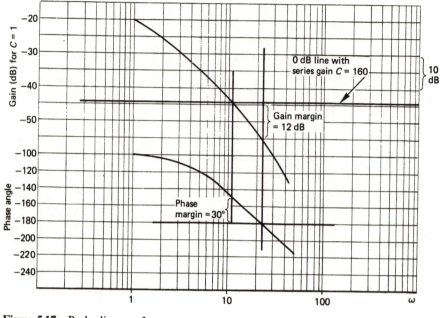

Figure 5.17 Bode diagram for

$$G(s) = \frac{0.1}{s(1 + 0.1s)(1 + 0.02s)}$$

and this can be discretized to produce a digital controller by methods described earlier.

Notice how the velocity constant for the system with phase advance is some 3 times greater than for the system with a series gain only, implying a better ability to follow a ramp input.

A BRIEF REVIEW OF THE CLASSICAL APPROACHES TO COMPENSATOR DESIGN

Phase compensation In phase compensation, dynamic elements are inserted into the control loop to modify the phase angle. The three principal elements are:

The *lead compensator*, with transfer function

$$\frac{1 + sT_c}{1 + \alpha s T_c}, \qquad \alpha < 1$$

(as applied in the example above). The element is used to increase system stability margins and/or to increase system bandwidth.

The *lag compensator*, with transfer function

$$\frac{1 + \alpha s T_c}{1 + s T_c}, \qquad \alpha < 1,$$

is used mainly to reduce steady-state error. The element tends to slow system response.

The *lag–lead compensator*, with transfer function

$$\frac{(1 + s/a)(1 + s/c)}{(1 + s/b)(1 + s/d)}, \qquad \frac{b}{a} = \frac{c}{d} > 1,$$

gives a phase lag at low frequencies and a phase lead at higher frequencies. To a large extent, the element achieves the combined aims of the lead compensator and the lag compensator.

Pole-cancellation compensation A cancellation compensator contains zeros to cancel some or all of the plant poles. The cancelled poles are then replaced by controller poles.

Some specific compensation objectives are:

(a) replacement of a slow pole by a faster pole to increase system speed of response;
(b) replacement of a dominant slow pole by a slower pole to increase system steady-state accuracy;
(c) replacement of a dominant complex pole pair by a different complex pair to modify system transient response.

Dipole addition A dipole is a pole–zero pair, where the pole and zero are placed very close together in the s plane. A dipole affects system steady-state gain without significantly affecting dynamic behavior.

Example of dipole addition A plant of transfer function

$$G(s) = \frac{100}{s^2 + 10s}$$

is considered to have a satisfactory transient behavior when connected in a unity feedback loop. However, the steady-state ramp-following ability of the system is considered unsatisfactory. Insertion of the compensating dipole $(s + 0.1)/(s + 0.02)$ into the loop reduces the steady-state error in following a ramp by a factor of five compared with the uncompensated case, with little effect on the high-frequency behavior of the system.

Checking the closed-loop behavior A control design arrived at by the essentially open-loop Bode techniques of this section meets, by definition, closed-loop gain and phase-stability margins as well as steady-state accuracy requirements. However, it is usually desirable to check that the closed-loop frequency response of the controlled system is satisfactory. To obtain the necessary closed-loop information, a Nichols chart may be used. A Nichols chart is a Cartesian plot of open-loop gain (dB) against open-loop phase angle, overprinted with loci of constant closed-loop gain and of constant closed-loop phase angle. The open-loop gain versus phase plot may be obtained directly from the Bode diagram and the closed-loop frequency response is then generated at the intersections with the overprinted loci.

CONTROL DESIGN USING THE BODE DIAGRAM IN THE w' DOMAIN

Bode diagrams are not a practicable tool in the z domain because the irrationality of the inevitable $e^{j\omega T}$ terms destroys the simple character that is the attraction of the Bode approach.

However, a Bode diagram in the w' domain of a transfer function $G(z)$ that has been w' transformed allows all the usual Bode-oriented design techniques to be used to design a controller. The resulting controller may then be transformed from w' to z domain to be developed into a realizable digital algorithm.

The advantage of using the w' domain plot, compared with the usual s domain approach, is that effects caused by sampling are modelled and taken into account during the design phase rather than being considered at a later, discretization, stage of the design process.

The w and the z domains are linked by the relation

$$w = \frac{z-1}{z+1}. \tag{5.41}$$

Denote the imaginary part of w by ω_w; then ω_w is the analog of frequency ω. The relation between ω_w and ω is given by

$$\omega_w = \tan\left(\frac{\omega T}{2}\right). \tag{5.42}$$

The w' domain   To obtain a closer analogy between pseudo frequency and actual frequency, it is common to modify the w transformation into a scaled version that is often denoted by w'. In this case, the w' and z domains are linked by the relation

$$w' = \frac{2w}{T} = \frac{2}{T}\frac{z-1}{z+1}. \tag{5.43}$$

Let $\omega_{w'}$ be the imaginary part of w'; then the relation between $\omega_{w'}$ and actual frequency ω is given by

$$\omega_{w'} = \frac{2}{T}\tan\frac{\omega T}{2}. \tag{5.44}$$

Table 5.2 Comparison of $\omega_{w'}$ with ω

T	ω	$\omega_{w'}$	
0.1	0.1	0.1	
0.1	0.5	0.5	
0.1	1.0	1.0	
0.1	5	5.11	$\omega_s = 2\pi/T = 62 \text{ rad/s}$
0.1	10	10.9	
0.1	20	31	
0.1	25	60	
0.1	30	282	
1	0.1	0.1	
1	1	1.09	
1	1.5	1.86	$\omega_s = 6.2 \text{ rad/s}$
1	2	3.11	
1	2.5	6	
1	3	28	

Under conditions where

$$\omega < \frac{\omega_s}{3}, \tag{5.45}$$

it can be seen from Table 5.2 that $\omega_{w'} \simeq \omega$, i.e. values of pseudo frequency $\omega_{w'}$ are approximately equal to true frequency. Thus we prefer the w' domain to the w domain because $\omega_{w'}$ better approximates frequency ω than does ω_w.

An example illustrates the approach. Note that the necessary transformations produce unwieldly expressions making hand calculation unattractive for all but the simplest problems.

Example Examine the problem of digitally controlling the process whose transfer function is

$$G(s) = \frac{1}{s^2}$$

with the help of the w' domain Bode plot. Allow for a zero-order hold to precede the process. Use $T = 0.1$ for the sampling interval.

The transfer function $G'(s)$ for the process preceded by zero-order hold is

$$G'(s) = \left(\frac{1 - e^{-sT}}{s}\right)\left(\frac{1}{s^2}\right),$$

leading to

$$G'(z) = \frac{T^2}{2}\frac{(z+1)}{(z-1)^2}.$$

$G'(z)$ is now transformed to the W' domain by setting $z = (2 + Tw')/(2 - Tw')$:

$$G'(w') = \frac{T^2}{2}\left[\left(\frac{2+Tw'}{2-Tw'}+1\right)\Big/\left(\frac{2+Tw'}{2-Tw'}-1\right)^2\right]$$

$$= \frac{1-w'T/2}{(w')^2}.$$

Notation To simplify the expressions that follow we define $v = \omega_{w'}$. To plot the Bode diagram, we wet $w' = jv$ so that v is the analog of true frequency ω. That is,

$$G'(jv) = \frac{1-jvT/2}{(jv)^2} = \frac{jvT/2-1}{v^2}.$$

A Bode plot for this transfer function is given in Figure 5.18. The plot indicates that severe stability problems can be expected in closed loops unless the characteristic is significantly modified by the inclusion of the controller.

Clearly the phase angle must be increased in the region where $v < 2$. A reasonable choice of phase compensator seems to be the lead compensator with break points at $v = 0.3$ and $v = 3$. (A quick calculation shows that the 10 to 1 ratio between break points gives a phase lead of 55° at $v = 0.95$.) Thus, the proposed compensator has the transfer function

$$D(jv) = \frac{1+jv/0.3}{1+jv/3}.$$

A trial Bode sketch using this controller shows that it is necessary to add a series gain of $1/3.16$ (equivalent to -10 dB) to obtain a system with sufficient phase margin. The final Bode plot for the system with the proposed controller is given in Figure 5.19.

The controller has the transfer function

$$D(w') = \frac{1}{3.16}\frac{(1+w'/0.3)}{(1+w'/3)}.$$

Transforming to the z domain yields

$$D(z) = \frac{1}{3.16}\left(\frac{(3T+20)z+3T-20}{(3T+2)z+3T-2}\right).$$

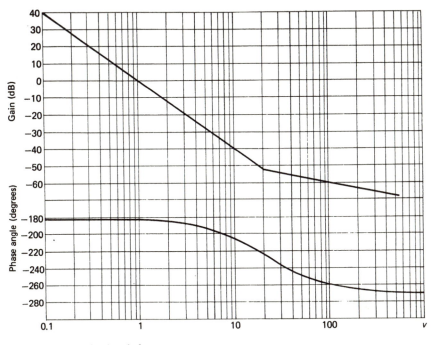

Figure 5.18   Bode sketch for

$$G'(jv) = \frac{jv/(2T) - 1}{v^2}$$

with $T = 0.1$.

Setting $T = 0.1$, we obtain

$$D(z) = 2.8\left(\frac{z - 0.97}{z - 0.74}\right).$$

Notice how the required controller emerges with the effect of sample and hold and sampling interval already taken into account.

Compare the situation with the admittedly simpler approach where the controller is designed using the normal continuous Bode technique followed by discretization. It can be seen that the w' domain approach, in return for somewhat greater effort, allows greater design precision than the continuous approach followed by discretization.

We have pointed out above how v is a reasonable approximation for real frequency ω over a particular range of the Bode diagram. For applications where precise frequency specifications have to be met (and these occur more often in communications than in control applications) a technique known as pre-warping may be used. This consists in pre-distorting the specification of the

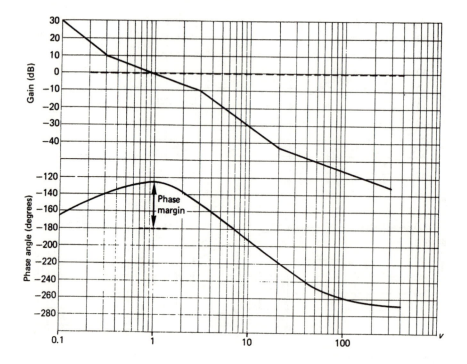

Figure 5.19   Bode sketch for

$$G'(jv)D(jv) = \frac{(jv/(2T) - 1)(1 + jv/0.3)}{3.16v^2(1 + jv/3)}$$

with $T = 0.1$.

problem by an amount equal and opposite to the mismatch between v and ω, so that the original frequency specification is exactly met.

We do not pursue pre-warping here. It can be found as a topic in many texts on digital filtering. Such a text on digital filters constitutes useful background reading since digital filters and digital algorithms have many features in common.

Note finally that in the methods described in this section, essentially open-loop techniques are used to design closed-loop systems. In any practical application, it is necessary to check the closed-loop behavior of the proposed system using Nichol's chart, root-locus or simulation techniques. In particular it will be usual to check the closed-loop bandwidth and the maximum amplification (M factor) of an input sinusoid.

5.3 *Design descriptions*

5.3.1 *Discrete-time versions of feedforward and cascade-control algorithms*

Any continuous-time algorithm can be transformed into a discrete-time version by suitable approximation. We give below the digital versions of two well-tried continuous-time algorithms.

CASCADE-CONTROL ALGORITHM
Some types of process (a reactor surrounded by a water jacket is the best known example) lend themselves well to so-called *cascade control*.

Here an inner controller D_1 controls an intermediate process variable x (the temperature of the water jacket) while the second, outer controller, D_2, controls the main process variable y (the temperature inside the reactor) by operating on the desired value of the inner controller. This can be understood by reference to Figure 5.20. (Standard sampling arrangements are assumed and G_1 contains a zero hold circuit.)

The equations for the system are

$$y(z) = G_2 G_1(z) D_1(z) (D_2(z)(v(z) - y(z)) - x(z)), \qquad (5.46)$$

$$x(z) = G_1(z) D_1(z) (D_2(z)(v(z) - y(z)) - x(z)). \qquad (5.47)$$

The configuration works best when the inner loop responds quickly compared with the outer loop (as in the reactor example). The coefficients for the controllers D_1 and D_2 are best found numerically by methods similar to those given in Section 5.2.3.

COMBINED FEEDFORWARD–FEEDBACK CONTROLLER
Feedback control systems suffer from the defect that correction begins only after an error has been detected. For some process applications, an unacceptably large quantity of off-target material is produced before correction can be obtained by feedback control. An example is the rolling mill of Figure 5.21. It

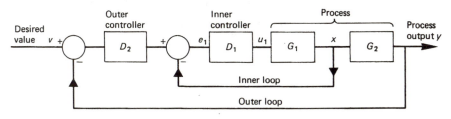

Figure 5.20   The cascade control of the sequential process $G_1 G_2$.

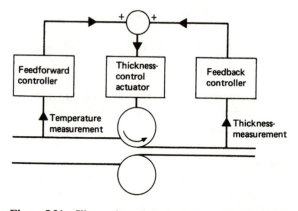

Figure 5.21   Illustration of the operation of feedforward control.

is assumed that temperature deviations in the incoming material cause output thickness deviations. Correction to the thickness cannot begin until the material arrives at the measuring device. If the temperature of the incoming strip is measured before entry to the mill, corrections to the roll gap can be made to be exactly synchronized with the measured temperature disturbance.

Feedback control is still required in almost every case since, inevitably, the model driving the feedforward strategy is subject to error. In particular, processes alter their characteristics slowly with time due to wear, thermal drift, etc., in a manner that is particularly difficult to reflect in a model. Such slow drifts can be compensated by a low-gain feedback loop which, in a sense, calibrates the model used in the feedforward scheme. The resulting feedforward/feedback scheme uses feedforward to correct short-term disturbances and feedback to correct slowly changing errors.

Let w be a measurable process disturbance. The algorithm is

$$u(k) = -\frac{P_1(z^{-1})}{P_2(z^{-1})}e(k) + \frac{P_3(z^{-1})}{P_4(z^{-1})}w(k), \qquad (5.48)$$

where the P_i are polynomials in z^{-1}. A simple lead–lag filter is often sufficient for synchronization. It can be realized through the arrangement

$$\frac{P_3(z^{-1})}{P_4(z^{-1})} = \frac{a_0 + a_1 z^{-1}}{1 + b_1 z^{-1}}. \qquad (5.49)$$

5.3.2 The simplest algorithm: on–off control

The algorithm

$$\left.\begin{array}{ll} u_k = m, & e_k > 0 \\ u_k = -m, & e_k \leqslant 0 \end{array}\right\} \qquad (5.50)$$

may be entirely suitable for applications where long time constants smooth the effects of the rectangular waveforms that will be applied to the plant.

The algorithm is very simple and, most important practically, it requires only an on–off (rather than a continuous) actuator. The cost savings on the actuator are usually the biggest incentive to use the algorithm.

Another advantage of the algorithm is that it can be applied without accurate quantitative knowledge of the dynamics of the process to be controlled.

For temperature control, the algorithm is modified to

$$\left.\begin{aligned} u_k &= m & e_k > 0\\ u_k &= 0, & e_k \leqslant 0 \end{aligned}\right\} \tag{5.51}$$

and for straightforward applications it may be entirely satisfactory.

These systems are nonlinear and may need to be investigated by a suitable nonlinear technique such as the describing-function method [L5].

5.4 Design realizations: realization of algorithms in terms of unit delays and constant multipliers

A control algorithm, represented, let us say, by the transfer function $G(z)$, must be realized in the computer in terms of unit delays and constant multipliers.

A particular transfer function has a number of alternative realizations. Mathematically, the alternative realizations are completely equivalent but, considered as alternative programming approaches, they differ in general in terms of ease of programming, computational efficiency, noise transmission, and sensitivity to parameters errors.

Some specimen realizations are given below. A more comprehensive discussion of realization will be encountered in Chapter 8.

5.4.1 Direct method

Let $G(z) = N(z)/D(z) = y(z)/u(z)$ be the transfer function to be realized; then

$$y(z) = \frac{N(z)}{D(z)} u(z). \tag{5.52}$$

We can write the equation in the form

$$y(z) = N(x)\left(\frac{u(z)}{D(z)}\right) = N(z)q(z) \tag{5.53}$$

where

$$q(z) = \frac{u(z)}{D(z)}. \qquad (5.54)$$

This equation is realized by a configuration of unit delays to yield the intermediate variable q. The realization of equation (5.52) is then completed by noting that $y(k)$ is a linear combination of

$$q(k), q(k+1), \ldots$$

The approach is made clear by an example.

Example Given the third-order difference equation

$$y(k+3) + a_1 y(k+2) + a_2 y(k+1) + a_3 y(k)$$
$$= b_1 u(k+2) + b_2 u(k+1) + b_3 u(k) \qquad (5.55)$$

determine a realization in terms of unit delays.

First we introduce the dummy variable q to allow an intermediate step. q is defined to satisfy the equation

$$q(k+3) + a_1 q(k+2) + a_2 q(k+1) + a_3 q(k) = u(k). \qquad (5.56)$$

The equation is realized by the arrangement shown in Figure 5.22.
Now

$$y(k) = b_1 q(k+2) + b_2 q(k+1) + b_3 q(k) \qquad (5.57)$$

and hence the original equation, (5.55), can be realized by the arrangement of Figure 5.23.

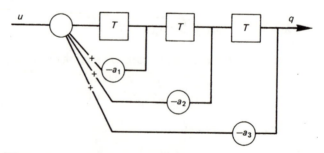

Figure 5.22 Realization of equation (5.56) by the direct method.

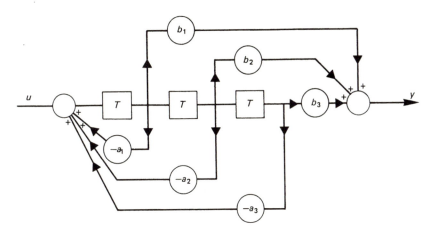

Figure 5.23 Realization of equation (5.55) by the direct method.

5.4.2 Parallel method

The transfer function $G(z)$ is decomposed into partial functions and then is represented as a parallel network. The following example makes the method clear.

Example Let

$$G(z) = \frac{z + 2}{z^2 + 6z + 5} = \frac{z + 2}{(z + 1)(z + 5)}. \tag{5.58}$$

In terms of partial fractions

$$G(z) = \frac{1}{4(z + 1)} + \frac{3}{4(z + 5)} = \frac{y(z)}{u(z)}.$$

The diagram, Figure 5.24, follows immediately.

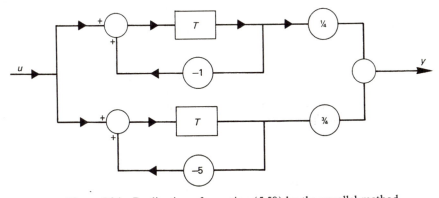

Figure 5.24 Realization of equation (5.58) by the parallel method.

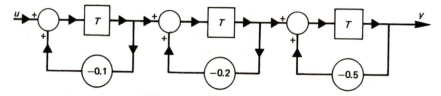

Figure 5.25   Realization of equation (5.59) by the factorization method.

5.4.3 Factorization method

If the transfer function $G(z)$ can conveniently be factorized, then $G(z)$ can be represented by a sequence of elements as the following example makes clear.

Example   Let

$$G(z) = \frac{1}{z^3 + 0.8z^2 + 0.17z + 0.01} \tag{5.59}$$

$$= \frac{1}{(z + 0.1)(z + 0.2)(z + 0.5)}.$$

$G(z)$ can immediately be represented as a sequence of elements as in Figure 5.25.

5.5 Design considerations

5.5.1 Choice of sampling interval

Every time a digital control algorithm is designed, a suitable sampling interval must be chosen. Choosing a long sampling interval reduces both the computational load and the need for rapid analog/digital conversion and hence the hardware cost of the project.

However, as the sampling is increased, a number of potentially degrading effects start to become significant. For a particular application, one of these degrading effects sets the upper limit for the sampling interval. The process dynamics, the type of algorithm, the control requirement and the characteristics of input and noise signals all interact to set the maximum usable value for T.

Below we discuss some of the degrading effects. We then pass on to suggest ways of deciding on the sampling interval as part of the design process.

DEGRADING EFFECTS ON SYSTEM PERFORMANCE AS SAMPLING INTERVAL T IS INCREASED OR MADE EXCESSIVELY SHORT

Figure 5.26 shows the three main degrading effects that may impair system performance as T is increased.

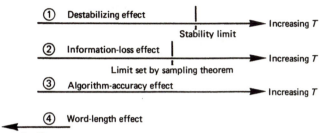

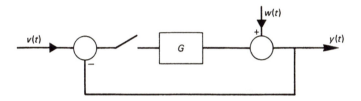

Figure 5.26 Illustration of the main degrading effects of too long or too short a sampling interval.

Figure 5.27 Information loss effect. The sampler must operate fast enough to allow the information in the signals $v(t)$, $w(t)$ to be preserved if the control loop is to operate with zero error.

Destabilizing effect   Because of conversion times and computation times a digital algorithm contains a dead time that is absent from its analog counterpart. Dead time has a very marked destabilizing effect on a closed-loop system due to the phase shift caused.

It is rather easy, using some of the methods already encountered (stability tests, root-locus technique, frequency-response plots) to determine the stability limit as T is increased. Note however that long before the stability limit is reached, a closed-loop system becomes highly oscillatory.

Information-loss effect   *Time varying signals $v(t)$* Suppose, as in Figure 5.27, that a control system receives input signals $v(t)$ whose frequency spectrum is bounded by an upper limit ω_v. For an ideal control system, $y(t) = v(t)$ to make the error zero. In a digital system, this means that the control loop must reconstruct $y(t)$ from samples, and to achieve this the sampling rate must satisfy

$$\omega_S > 2\omega_v. \tag{5.60}$$

Perfect construction would require a knowledge of future input values (not physically possible). To allow for non-ideal reconstruction, a higher sampling rate than indicated by inequality (5.60) has to be used.

Step inputs in $v(t)$ When a system has to be designed to respond to step inputs, the spectrum of $v(t)$ has to be assumed to extend to infinity and the inequality suggests infinitely fast sampling. A practical approach is to argue alternatively in terms of the acceptable delay before the response to the step input begins. A mean delay of $T/2$ and a maximum delay of T will arise, due to sampling, before the response begins.

From a knowledge of the step response when no sampler is present, T can be chosen so that the delay is either negligible or is acceptable for the particular application.

Disturbance inputs $w(t)$ If $v(t)$ is held constant, then the system can be considered as a regulator whose task is to hold the output constant, despite disturbances $w(t)$ entering the system. The closed-loop system has to construct a copy of the signal to be rejected. Let ω_w be the highest frequency that must be rejected; then ω_s must satisfy

$$\omega_s > 2\omega_w. \tag{5.61}$$

Notice that ω_w refers to disturbance signals to which the closed-loop system must respond in its regulatory role. High-frequency disturbances that will be removed by the low-pass open-loop character of the system do not need to be considered. It is necessary to ask questions such as: does the disturbance need to be counteracted? Has the control loop sufficient bandwidth to carry out such counteraction? The highest frequency that must and can be regulated is then designated as ω_w for use in inequality (5.61).

Example Figure 5.28 shows an (idealized) rolling mill without control. Strip of thickness $H(t)$ enters the mill and leaves at thickness $h(t)$. The thickness profile $H(t)$ contains variations that cause corresponding variations in $h(t)$.

In a thickness control system, $h(t)$ is to be kept constant at a given value

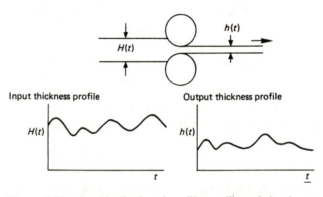

Figure 5.28 An idealized strip-rolling mill and its input thickness profile and output thickness profile.

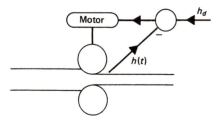

Figure 5.29 The strip-rolling mill with thickness control.

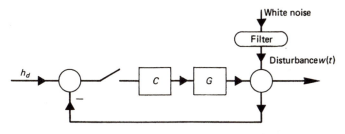

Figure 5.30 Block diagram equivalent of Figure 5.29.

h_d. A measurement of $h(t)$ is compared with h_d and the error is fed to a digital algorithm that drives a motor to adjust the roll gap by moving the top roll in a vertical direction (Figure 5.29).

The control loop can be represented diagrammatically as in Figure 5.30. Several factors act to disturb the system output—the disturbances are lumped into the single signal $w(t)$.

The disturbance signal $w(t)$ can be partly characterized by its variance σ_w^2 and its upper frequency limit ω_w. σ_w^2 can be calculated from the expression

$$\sigma_w^2 = \frac{1}{t} \int_0^t (h(\tau) - \bar{h}(\tau))^2 \, d\tau \tag{5.62}$$

where

$$\bar{h}(t) = \frac{1}{t} \int_0^t h(\tau) \, d\tau \tag{5.63}$$

over some suitable observation time t_* on the uncontrolled system. ω_w can be determined approximately by inspection of records of $h(t)$ or, more accurately, by the use of a spectrum analyzer.

In the design of a continuous-control loop for the rolling mill it will clearly be necessary for the closed-loop bandwidth to exceed ω_w if the disturbances are to be compensated. When a digital controller is introduced, its sampling

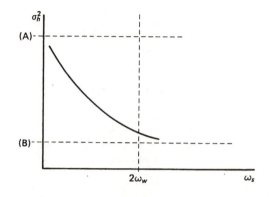

Figure 5.31 The dependence of the variance of
$h(t)$ on sampling frequency ω_S: (A) is the variance
of the signal $h(t)$ when no control is exercised;
(B) is the variance lower limit set by the
constraints in the control loop.

rate must satisfy the sampling theorem, i.e. $\omega_S > 2\omega_w$. If lower sampling rates
are used, the system becomes progressively less able to reject the disturbances.
We show this diagrammatically in Figure 5.31. The particular shape of the
curve depends on how the disturbance signal is constituted—the most difficult
control problem occurs when most of the noise power is concentrated at high
frequencies near to ω_w.

It is clearly a requirement of good engineering that the sampling frequency
should be specified as part of the total specification. Any serious mismatch in
this respect usually implies that expensive over-provision has been made in
respect of computer power or actuation devices.

Notice that the example above has concentrated on steady-state
performance. If the most critical control task is concerned with initial transient
effects, rather than steady-state disturbances, the approach to specifying
sampling rate must be modified accordingly.

System outputs not completely specified by the output $\mathscr{Z}$ transform
Consider the case where a continuous process is manipulated in closed loop
by a discrete-time controller. $\mathscr{Z}$ transform methods will give information on
the response of the continuous system only at intervals T, where T is the period
associated with the discrete-time controller. What can be said about the behavior
of the continuous system between sampling instants? Ideally we need some
theorem analogous to a converse of the sampling theorem.

To illustrate the problem, suppose that the step response of a continuous
system under closed-loop discrete-time control is found by $\mathscr{Z}$ transform analysis
to be as shown in Figure 5.32(a).

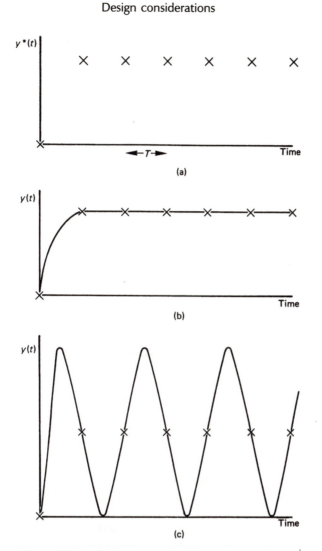

Figure 5.32 (a) $y^*(t)$ as specified by a $\mathscr{L}$ transform; (b) and (c) two possible continuous signals $y(t)$ corresponding. to $y^*(t)$.

The form of the continuous response is not specified uniquely by Figure 5.32(a) unless some constraint can be put on the frequency content. For instance, the actual continuous response could be either as in Figure 5.32(b) or as in Figure 5.32(c). For many applications, the response of Figure 5.32(c) would be considered to be completely unsatisfactory. But how is such behavior to be predicted from $\mathscr{L}$ transform analysis, which does not give any indication of the presence or absence of the oscillations?

To answer the question, we need to know what frequencies will be present in the output of a closed-loop system. Neglecting harmonics that would be generated by nonlinearities, the output $y(t)$ will not contain frequencies higher than ω_v or ω_w due to forcing by control or disturbance inputs. However, each time the zero-order hold at the output of a digital controller takes on a new level the process is subject to a step input.

Thus the output of such a system is a set of connected open-loop step responses. Provided that the open-loop step response cannot by itself be responsible for unacceptable high-frequency inter-sample fluctuations, then satisfaction of the inequality

$$\omega_s > 10\omega_b,$$

where ω_b is the bandwidth of the closed-loop system, usually ensures that the output $\mathscr{L}$ transform completely characterizes the system output.

Attention may also need to be devoted to intermediate signals, such as the controller output, to make sure that they are not performing high-amplitude high-frequency oscillations that are undetected at the output of a low-pass process.

Algorithm accuracy effect   A number of digital-control and digital-filtering algorithms are derived from continuous analog algorithms by a process of discretization (Section 5.2.5). In the transformation of an algorithm from continuous- to discrete-time form, errors arise and the character of the digital algorithm differs from that of its continous counterpart.

In general, these errors occurring during the discretization process become larger as the sampling interval increases.

It is clear that this effect should rarely or never be allowed to dictate a shorter sampling interval than would otherwise have been needed. Rather, an alternative, perhaps purely discrete, design approach should be used to allow a longer sampling interval without the introduction of unacceptable error.

Finally, we note that algorithms designed to work with a long time interval are in general more demanding of accurate information on the process to be controlled. Specifically, an algorithm suited to operating with a long sampling interval usually needs to be designed on the basis of an accurate knowledge of the process parameters. To put this another way, we can expect an algorithm designed for long sampling intervals to be more sensitive to process parameter drift than a similar algorithm designed to work on a shorter sampling interval.

Word-length effect   As the sampling interval T becomes very short, a digital system does not tend to the continuous case because of the finite word length. To visualize this effect, we can imagine that, as a signal is sampled more frequently, so adjacent samples have more similar magnitudes. If the beneficial effects of shorter sampling are to be realized, longer word lengths are needed to resolve the differences between adjacent samples. This effect will be discussed in Section 5.5.2 under the title quantization.

In many cases that we have encountered in practice, the shortening of the sampling interval has brought poles and zeros close to the point $z = 1$, due to the presence of terms of the form $(z - e^{aT})$. As an example, consider the case of the phase advance circuit with transfer function

$$D(z) = \frac{z - a}{z - \alpha a}, \qquad 0 < a < 1, \qquad 0 < \alpha < 1. \tag{5.64}$$

The phase angle ϕ_N produced by the numerator term $z - a$ is given by

$$\phi_N = \tan^{-1} \frac{(\sin \omega T)}{(\cos \omega T - a)},$$

$$\frac{d\phi_N}{da} = \frac{\sin \omega T}{1 - 2a \cos \omega T + a^2}.$$

To achieve a particular angle ϕ_N, we must have

$$a = \cos \omega T - \frac{\sin \omega T}{\tan \phi_N}.$$

We are interested in the effects as T becomes small, hence we can set $\sin \omega T = \omega T$ and $\cos \omega T = 1$ to obtain

$$\frac{d\phi_N}{da} = \frac{\omega T}{1 - 2(1 - \omega T/\tan \phi_N) + (1 - \omega T/\tan \phi_N)^2}$$

$$= \frac{(\tan \phi_N)^2}{\omega T}.$$

The phase angle ϕ_D, contributed by the denominator of $D(z)$, is given by the same type of derivation, with αa substituted for a. The sensitivity of the phase-advance circuit increases without limit as the sampling interval T goes to zero.

Discussion At this point we have come to the conclusion that the sampling frequency must at least satisfy the following inequalities

$$\omega_S > 2\omega_v \qquad (1),$$

$$\omega_S > 2\omega_w \qquad (2),$$

$$\omega_S > 2\omega_b \qquad (3).$$

Because of the far from ideal data reconstructors that are usually used, the inequalities must in practice be satisfied by a large margin if problems are to be avoided.

We note that if inequality (3) is satisfied, then an excessively oscillatory response—not apparent from $\mathscr{Z}$ transform analysis—will automatically be avoided. (The absence of local oscillations at intermediate points in a control loop cannot, of course, be guaranteed, even though inequality (3) is satisfied.)

EMPIRICAL RULES FOR THE SELECTION OF SAMPLING FREQUENCY
Practical experience and simulation results have produced a number of useful approximate rules for the specification of minimum sampling frequencies.

1. Assume that a process has a dominant time constant T_{dom}; then the sampling interval T is required to satisfy

$$T < T_{\text{dom}}/10 \qquad (5.65)$$

   when closed-loop digital control is implemented. This rule is widely applied by practising engineers but it is full of danger under the conditions where high closed-loop performance is forced from a process with a low open-loop performance.

2. Assume that a process has a Ziegler–Nichols open-loop model

$$G(s) = \frac{e^{-sT_1}}{(1 + sT_2)};$$

   then Goff [G3] gives curves implying that T must satisfy

$$T < T_1/4. \qquad (5.66)$$

3. Assume that a process under closed-loop control is required to have
   (a) a settling time T_S, or
   (b) a natural frequency ω_n.
   Then
   (a) $T < T_S/10$,
   (b) $\omega_S > 10\omega_n$,
   where ω_S is the sampling frequency ($\omega_S = 2\pi/T$).

Rules based on experience of particular applications From experience of process industries it has been found in general that sampling intervals of 1 second are short enough for most loops to control flow, pressure, level, and temperature, although certain flow control loops may require sampling intervals of less than 0.5 seconds. Fast-acting electromechanical systems require much shorter sampling intervals, perhaps down to a few milliseconds.

Sideband energy method An alternative approach that has been put forward by Bennett and Linkens [B2] should also be mentioned. They give an inequality that sets an upper bound on the average power of the error caused by the sidebands generated by sampling. The reference goes on to suggest an interactive procedure by which the sampling frequency can be specified for a particular control system subject to known inputs.

Numerical example: Choice of sampling interval   A motor of time constant $\tau_M = 0.5$ s is used in a position-control system. In design A, the gain C is set at $C = 0.6$—this yields $\zeta \simeq 0.9$. In design B, the gain is increased to $C = 12.5$ and tachometer feedback is introduced to maintain the same damping factor $(\zeta = 0.9)$ as in case A. Choose a sampling interval for digital implementation of the designs using the $\omega_S = 10\omega_b$ criterion.

Find the approximate damping factor for the sampled data version of case A through a determination of closed-loop pole locations. Compute and compare the continuous and discrete closed-loop step responses for case A.

Step 1   The open-loop transfer functions
Case A has the open-loop transfer function

$$G_1(s) = \frac{2c}{s^2 + 2s} = \frac{1.2}{s^2 + 2s}.$$

Case B has the open-loop transfer function

$$G_2(s) = \frac{2c}{s^2 + (2 + cp)s} = \frac{25}{s^2 + 9s}$$

(p is the gain of the tachometer loop; here p is set to $p = 0.56$ to give $\zeta = 0.9$).

Step 2   Determination of the closed-loop bandwidths
We have sketched in Figure 5.33 the open-loop Nyquist diagrams for the two cases, A and B, and by superimposing the M circle for $M = 1/\sqrt{2}$, it is apparent that the closed-loop bandwidths are approximately

   for case A, $\omega_b = 0.78$ rad/s,
   for case B, $\omega_b = 3.5$ rad/s.

Step 3   Selection of sampling intervals
If we use the rule that $\omega_S \geqslant 10\omega_b$, then recommended sampling intervals are found to be

   for case A $T = \dfrac{2\pi}{7.8} \simeq 1$ s,

   for case B $T = \dfrac{2\pi}{35} \simeq 0.2$ s.

Step 4   Approximate calculation of ζ for case A
Allowing for the inclusion of a zero-order hold as in Figure 5.34 the open-loop

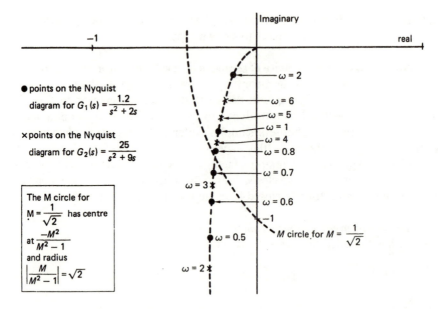

Figure 5.33　Open-loop Nyquist diagrams for the two cases A and B. The M circle for $M = 1/\sqrt{2}$ has center at $-M^2(M^2 - 1) = 1$ and radius $|M/(M^2 - 1)| = \sqrt{2}$.

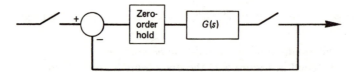

Figure 5.34　The closed loop system analyzed in the example.

transfer function for case A is

$$\mathscr{Z}\left\{\left(\frac{1 - e^{-sT}}{s}\right)\left(\frac{1.2}{s^2 + 2s}\right)\right\}.$$

Now

$$\mathscr{L}^{-1}\left\{\frac{1.2}{s^2(s + 2)}\right\} = \mathscr{L}^{-1}\left\{\frac{0.3}{s + 2} + \frac{0.6 - 0.3s}{s^2}\right\}$$

$$= 0.6t - 0.3 + 0.3e^{-2t}.$$

Hence

$$G(z) = (1 - z^{-1})\left(\sum_{k=0}^{\infty} (0.6kT - 0.3 + 0.3e^{-2kT})z^{-k}\right)$$

$$= (1 - z^{-1})(0.3)\left(\frac{2Tz}{(z - 1)^2} - \frac{z}{z - 1} + \frac{z}{z - e^{-2T}}\right).$$

Setting $T = 1$ as calculated above, the closed-loop transfer function is found to be

$$\frac{G(z)}{1 + G(z)} = \frac{0.26(z + 0.52)}{z^2 - 0.794z + 0.313} \tag{5.67}$$

and the closed-loop poles are at the positions

$$z = 0.397 \pm j0.394.$$

Now the continuous system had

$$\zeta = 0.9, \qquad \omega_n = 1.1,$$

whereas the discrete-time system with $T = 1$ has approximately, from Figure 5.7,

$$\zeta = 0.57, \qquad \omega_n = 3\pi/10 = 0.94.$$

The conversion of a continuous loop to sampled operation has, as usual, increased the system settling time.

Step 5   Determination of step responses
Next we compare the step response of the continuous loop, case A, and its discrete-time counterpart with $T = 1$ second.
   The continuous system has the step response

$$y(t) = 1 - \frac{e^{-\zeta\omega_n t}}{\sqrt{(1 - \zeta^2)}} \sin\left(\sqrt{(1 - \zeta^2)}\omega_n t + \tan^{-1}\frac{\sqrt{(1 - \zeta^2)}}{\zeta}\right).$$

   For the discrete-time implementation of case A, the difference equation representing closed-loop behavior is

$$y(k + 2) = 0.794y(k + 1) - 0.313y(k) + 0.26(u(k + 1) + 0.52u(k))$$

and if $v(k) = 1.313$ for all $k \geqslant 0$, then the difference equation has the same steady-state value as the differential equation.

Solution of the difference equation assuming zero initial conditions
The continuous-time and discrete-time responses are plotted together in Figure 5.35.

Table 5.3

k	0	1	2	3	4	5	6	7	8	9	10
$y(k)$	0	0	0.52	0.954	1.13	1.127	1.06	0.985	0.986	0.994	1.002

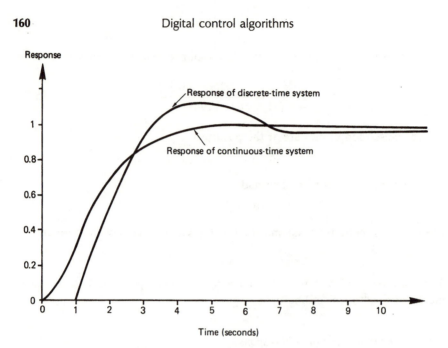

Figure 5.35   Response of the continuous system $2c/[s(s + 2)]$ and of the discrete equivalent.

5.5.2 Quantization effects

Within the closed loop of a control system, a number of quantization operations occur, where numbers have to be approximated in order to fit the finite word length of the digital computer.

The approximations occur, for instance:

At A/D conversion.

After multiplication—when the result must be truncated or rounded before being stored.

When inserting parameters to be used as the constants in an algorithm.

The effects, taken together, are called *quantization effects*. An exact and complete analysis of these effects will rarely be attempted. Rather it will be the aim to obtain an order-of-magnitude estimate of the effects insofar as they influence the behavior of the control loop.

A PRACTICAL VIEWPOINT OF THE MATERIAL THAT IS TO FOLLOW
It is essential to obtain an appreciation of the possible effects of quantization if properly matched control systems are to be designed and if certain types of control-system malfunction are to be understood.

However, in almost every control application, it is the coefficients of the continuous-time process, rather than of any digital device in the loop, that are subject to major uncertainty and drift with time. Thus, every real-world designer builds a large safety margin into every design to allow for inevitable process variations. The inability to place controller poles with perfect precision, due to finite word length of the computer used for implementation, is quite insignificant in the overall design and scarcely needs to be considered.

From an application point of view, we are interested only in those quantization effects that are significant in the context of the overall control design.

The effects that we describe below are therefore those that need to be considered in system design or that may become significant for particular choices of coefficients. It is particularly important to avoid situations where a very small error is greatly amplified to a point where it degrades system behavior. Such effects are discussed and illustrated by examples.

QUANTIZATION REGARDED AS A GENERATOR OF NOISE

Errors occur at the input of a digital controller due to the operation of an analog/digital converter of finite word length. The passage of the errors (and their possible amplification) through the controller is of interest.

It is usual to consider the difference between the exact input signal and the quantized signal from the analog/digital converter as a noise source. In considering the passage of the noise through the controller, three approaches are normally used:

(a) To obtain upper bounds on the errors that can be caused by quantization effects.
(b) To neglect transient effects and to analyze only the steady-state effects.
(c) To treat the quantization errors as noises arising from stochastic sources and to proceed with a statistical analysis.

Once some general points have been established these three approaches will be explained.

QUANTIZATION: SOME GENERAL POINTS

When an m-bit number is to be stored in an n-bit register, $n < m$, two alternative strategies are possible:

(a) *Truncation*—in which the remainder of the representation after the nth digit is ignored.
(b) *Rounding*—in which the nth digit in the representation is chosen to minimize the error between the m-bit number and the number to be held in the n-bit register.

The two strategies are illustrated in Figures 5.36 and 5.37, where it is assumed that the signal to be held to n-bit accuracy is originally continuous.

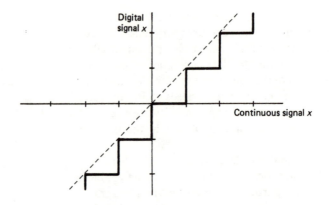

Figure 5.36   Truncation.

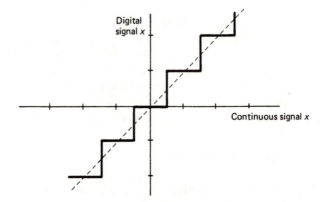

Figure 5.37   Rounding.

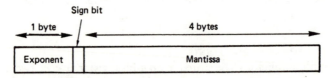

Figure 5.38   Storage of a floating-point number.

In floating-point arithmetic, numbers are typically stored, in an 8-bit microprocessor, as shown in Figure 5.38. In the example illustrated in the figure, an integer of up to

$$2^{32} - 1(\, = 4\,294\,967\,295)$$

can be stored with perfect accuracy. Thus, stored parameters may be held, even

Most significant bit   **Least significant bit**

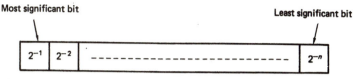

Figure 5.39   Storage of a fixed-point number.

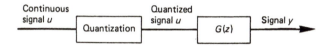

Figure 5.40   The transmission of quantization noise through
a system.

in a modest machine, to at least 9 (decimal) figure accuracy. Clearly, parameters
are never known to such accuracies in the first place and the effect of quantization
on parameters is not expected to be very significant in the execution of control
algorithms. However, in iterative algorithms, the gradual accumulation of errors
may be significant and a very high accuracy of storage of intermediate results
may be required.

We shall assume for purposes of the following analysis that each number
is represented in fixed-point manner in an n-bit binary register as shown in
Figure 5.39.

When truncation is used, the maximum error due to quantization is 2^{-n}.
When rounding is used, the maximum error due to quantization is
$\frac{1}{2} \cdot 2^{-n} = 2^{-n-1}$. We define,

$$\alpha = 2^{-n} \qquad \text{(truncation)},$$

$$\alpha = 2^{-n-1} \qquad \text{(rounding)}.$$

THREE METHODS OF ANALYSIS

Worst case analysis to obtain error bounds   The error $x(z)$ at the output
of the system of Figure 5.40, due to an error input $e(z)$, is given by

$$x(z) = G(z)e(z) \tag{5.68}$$

or, in the time domain,

$$x(k) = \sum_{m=0}^{k} g(m)e(k - m). \tag{5.69}$$

Then, using well-known inequality relations

$$|x(k)| \leqslant \sum_{m=0}^{k} |g(m)||e(k - m)| \tag{5.70}$$

and, since $e \leqslant \alpha$,

$$|x(k)| \leqslant \sum_{m=0}^{k} |g(m)| \alpha. \tag{5.71}$$

Inequality (5.71) may be used to find an upper bound on the error due to quantization.

Steady-state analysis   Assume that a system, as in Figure 5.40, is under study, in which a signal u is quantized and then input to a dynamic element whose output is the signal y.

Using the final-value theorem we can write

$$|y(k)|_{k \to \infty} \leqslant |G(z)|_{z \to 1} \alpha. \tag{5.72}$$

Equation (5.72) may be used to find an upper bound for the effect of quantization on steady-state accuracy.

Stochastic analysis   If we assume that all values of quantization error are equally likely, up to the maximum value of α ($\alpha = 2^{-n}$ for truncation and $\alpha \doteq 2^{-n-1}$ for rounding) then the quantization noise can be considered to be a random variable uniformly distributed from 0 to α or $-\alpha$ to α. (The assumption of randomness is only completely justifiable where u itself is random. However, even where u is deterministic, then, provided that n is large, the assumption can be expected to be reasonably justified for many practical cases.)

With truncation, the mean value of the error caused by quantization is $-\alpha/2$. Equivalently we can say that the *expected value* of the error

$$\mathscr{E}\{e\} = -\alpha/2. \tag{5.73}$$

With rounding, the mean value of the error is zero, i.e.

$$\mathscr{E}\{e\} = 0. \tag{5.74}$$

For both truncation and rounding,

$$\mathscr{E}\{e^2\} = \frac{1}{q} \int e^2 \, de, \tag{5.75}$$

the integral to be taken over the width q of the quantization band. That is,

$$\mathscr{E}\{e^2\} = \frac{1}{q} \left. \frac{e^3}{3} \right|_{-q/2}^{q/2} = \frac{1}{q} \left(\frac{q^3}{24} + \frac{q^3}{24} \right) = \frac{q^2}{12}; \tag{5.76}$$

i.e. the variance of the quantization error e is $q^2/12$ for both truncation and rounding.

Let

$$x(z) = G(z)e(z)$$

be the equation that we wish to study: that is, we wish to study the transmission of quantization error $e(z)$ through a dynamic element of transfer function $G(z)$.

Calculation of the mean value of the transmitted error The mean value of the transmitted error x can be most easily obtained in the time domain. For instance if $G(z) = 1/(z - 0.5)$, then in the time domain

$$x(k + 1) = 0.5x(k) + e(k).$$

In the steady state $\mathscr{E}\{x(k + 1)\} = \mathscr{E}\{x(k)\}$ and we can write

$$\mathscr{E}\{x(k)\} = 2\mathscr{E}\{e(k)\}.$$

Alternatively the final-value theorem may be used in conjunction with $G(z)$ to yield the same result.

Calculation of the variance of the transmitted error As usual, let

$$G(z) = \sum_{k=0}^{\infty} g(k)z^{-k};$$

then

$$x(k) = \sum_{m=0}^{k} g(m)e(k - m). \tag{5.77}$$

Equation (5.77) gives $x(k)$ by convolution.

$$|x(k)|^2 = \sum_{m=0}^{k} |g(m)|^2|e(k - m)|^2 \tag{5.78}$$

(this equation is valid because e is a random signal) and if we let $k \to \infty$ and insert the expected value of (error)^2 we obtain

$$\mathscr{E}\{x^2\} = \sum_{m=0}^{\infty} |g(m)|^2\mathscr{E}\{e^2\}. \tag{5.79}$$

Equation (5.79) assumes that e and x have zero mean values. In case they have nonzero means, this can be taken into account during calculation. Note that for the right-hand side of equation (5.79) to exist, the summation must converge and this it will do provided that $G(z)$ is an asymptotically stable transfer function.

Example A system of transfer function

$$G(z) = \frac{1}{z - e^{CT}} = \frac{1}{z - a}, \tag{5.80}$$

where $a = e^{CT}$, receives an input signal u whose exact z transform is $u(z)$, but because of the operation of a finite word length analog/digital converter, the signal that is fed to the algorithm is not $u(z)$ but $\hat{u}(z)$, where

$$\hat{u}(z) = u(z) + e(z),$$

where e is the quantization error of the analog/digital converter.

Let $y(z)$ be the noise-free output of the system and $\hat{y}(z)$ be the output of the system with noise. Then if we define the error at the output by $x(z) = \hat{y}(z) - y(z)$, we obtain by linearity that $x(z)/e(z) = G(z)$, where G is the same as in the formulation of the example, equation (5.80).

Worst-case analysis of the example We start with equation (5.80) and obtain a deterministic upper bound on the output error. From equation (5.71),

$$|x(k)|_{\max} = \sum_{k=0}^{\infty} |h(k)|\alpha.$$

In this case, the impulse response is

$$h(k) = a^k$$

and

$$\sum_{k=0}^{\infty} a^k = \frac{1}{1-a},$$

and therefore the upper bound on the output error is obtained as

$$|x(k)|_{\max} = \frac{\alpha}{1-a} = \frac{\alpha}{1-e^{CT}}. \tag{5.81}$$

Stead-state analysis of the example Here we use the final-value theorem to obtain the worst-case steady state using equation (5.80). Let x_s be the steady-state error due to the quantization error $e(z)$; then for the case in the example,

$$|x_s| \leqslant |G(z)|_{z=1}\alpha = \left| \frac{\alpha}{1-e^{CT}} \right|. \tag{5.82}$$

Stochastic analysis of the example Given the expected mean and variance of the input error signal, we need to calculate the corresponding mean and variance of error at the system output.

For rounding, we are assuming that the situation is unbiased, resulting in

$$\mathscr{E}\{e(k)\} = 0 \qquad \text{(rounding)}.$$

For truncation, the error is always negative, being (we assume) uniformly spread over the interval -2^{-n} to zero. Thus we make the assumption

$$\mathscr{E}\{e(k)\} = -2^{-n-1} \qquad \text{(truncation)},$$

or we can agree to write

$$\bar{e} = 0 \qquad \text{(rounding)},$$

$$\bar{e} = -2^{-n-1} \qquad \text{(truncation)}.$$

Substituting values

$$\bar{x} = a\bar{x} + \bar{e},$$

$$\bar{x} = \frac{\bar{e}}{1 - a}.$$

The variance of x is the same for either rounding or truncation. It is calculated as follows.

The value $x(k)$ is first expressed as a summation of impulse responses

$$x(k) = \sum_{m=0}^{k} g(m)e(k - m).$$

In this example, we obtain

$$x(k) = \sum_{m=0}^{k} a^m e(k - m)$$

and the variance of $x(k)$ is then obtained from

$$\sigma_{x(k)}^2 = \sum_{\substack{m=0 \\ k \to \infty}}^{k} (a^m)^2 \sigma_e^2. \tag{5.83}$$

The variance of the error at the input is given by equation (5.76) as

$$\sigma_e^2 = \mathscr{E}\{e(k)^2\} = \frac{q^2}{12}. \tag{5.84}$$

After the control algorithm (5.80) is first put into operation, it can be seen from equation (5.83) that the variance of output noise $\sigma_{x(k)}^2$ rises as k increases. If $a = 1$ then $\sigma_{x(k)}^2 \to \infty$ as $k \to \infty$; that is, the noise in the output increases without limit. If $a > 1$, the algorithm (5.80) is unstable, but we note additionally that it would produce ever-increasing output noise.

For a usable algorithm, $a < 1$, and under such a case, $\sigma_{x(k)}^2$ tends to a steady value as k increases. We call this steady value σ_x^2 and calculate it by

$$\sigma_x^2 = \left(\sum_{m=0}^{\infty} a^{2m} \right)(\sigma_e^2).$$

$$= \frac{1}{(1 - a^2)} \frac{q^2}{12} = \frac{q^2}{12(1 - e^{2CT})}. \tag{5.85}$$

Comparison of estimates for the example   Suppose that the algorithm $G(z)$

of equation (5.80)

$$G(z) = \frac{1}{z - e^{CT}}$$

is operated with $C = -1$ and with rounding after A/D conversion with n working bits. Then the upper bound on quantization error in the output is obtained from equation (5.81) as

$$|x(k)|_{max} = \frac{\alpha}{1 - e^{-T}},$$

where $\alpha = 2^{-n-1}$. That is,

$$|x(k)|_{max} = \frac{2^{-n-1}}{1 - e^{-T}}. \qquad (5.86)$$

Using equation (5.79), the variance of x is found to be

$$\sigma_x^2 = \frac{q^2}{12(1 - e^{2CT})} = \frac{2^{-n}}{12(1 - e^{-T})}. \qquad (5.87)$$

NUMERICAL VALUES FOR THE QUANTIZATION ERROR IN THE SYSTEM OUTPUT
In Table 5.4 we have listed $|x(k)|_{max}$ and $\mathscr{E}\{x^2\}$ as a function of T for three different word lengths. We have also calculated the signal-to-noise ratio in the output signal, as follows:

(a) We have assumed that the mean value $\bar{u}$ of the input signal u is half the maximum value allowed by the A/D converter.
(b) We have assumed that $\bar{y} = \bar{u}$, giving a usefully pessimistic view of the effects of noise.
(c) Under this condition, the worst-case signal-to-noise ratio is given by

$$S_{min} = \frac{\bar{y}}{|x(k)|_{max}}. \qquad (5.88)$$

(Note how S_{min} is dependent on point (a). It will be doubled if u reaches the maximum value allowed by the A/D converter and correspondingly reduced as $\bar{u}$ is reduced towards zero.)

We can understand this by recalling that the magnitude of the quantization error x is fixed while the signal $\bar{y}$ depends on the input signal.

If we decide that a signal-to-noise ratio of 20:1 is desirable, we can suggest a minimum word length, n_{min}, for each sampling interval; for instance, additional computation yields Table 5.5. These data are plotted in Figure 5.41.

The graph is a useful guide to matching sampling time with word

Table 5.4   (*Note*: I have assumed $\bar{y} = 0.0100000$ (binary), $\bar{y} = 0.25$ (decimal))

Number of working bits in A/D converter n	Sampling interval T	σ_x^2	$\|x(k)\|_{max}$	Signal-to-noise ratio S_{min}
7	10	6.5×10^{-4}	0.004	64
	1	0.001	0.006	40
	0.1	0.007	0.04	6.09
	0.01	0.065	0.39	0.63
	0.001	0.65	3.9	0.063
12	1	3.2×10^{-5}	1.93×10^{-4}	1300
	0.1	0.0002	0.001	194
	0.01	0.002	0.01	20
	0.001	0.02	0.122	2
	0.0001	0.203	1.22	0.2
15	1	4×10^{-6}	2.4×10^{5}	10 360
	0.1	2.6×10^{-5}	1.6×10^{4}	1560
	0.01	2.5×10^{-4}	0.0015	163
	0.001	0.0025	0.015	16.4
	0.0001	0.025	0.15	1.63
	0.00001	0.25	1.53	0.16

Table 5.5   The 'signal-to-noise ratio' column in Table 5.4 allows us to suggest the following inequalities matching word length with sampling interval.

T	n_{min}	Resulting minimum signal-to-noise ratio
10	6	32
5	6	32
2	6	28
1	6	20.2
0.5	7	25
0.1	9	24
0.05	10	25
0.01	12	20
0.005	13	20
0.001	(15	16)
0.001	16	33
0.0005	17	33
0.0001	19	26

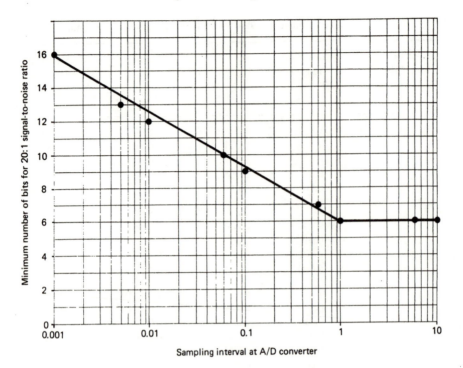

Figure 5.41 Graph relating word length with sampling interval to guarantee a 20:1 signal-to-noise ratio in the example.

length provided that the following points are borne in mind:

(a) A 20:1 signal ratio is assumed adequate and this is calculated by assuming that the signal to be A/D converted has half the maximum allowed magnitude while the noise achieves its upper limit.
(b) The graph is based on a first-order algorithm of a particular type. Practical algorithms are likely to be much more complex with possibility of increased noise magnification.
(c) Noise from other sources, e.g. from truncation or rounding after multiplication, has not been considered.

The first point (pessimistic) tends to compensate for the other two (optimistic) points.

An example to illustrate the importance of scaling at the A/D converter An A/D converter with eight working bits (it has an additional bit for sign indication) feeds into a system of transfer function

$$G(z) = \frac{y(z)}{u(z)} = \frac{10(z-1)}{z}. \tag{5.89}$$

It receives an input sequence $\{u(k)\}$ defined in column 1 in Table 5.6 and should produce an output sequence $\{y(k)_{ideal}\}$, which is listed in column 2 of Table 5.6. Column 3 lists the binary equivalents of the input signals received by an A/D converter whose most significant bit represents 8 volts. Column 4 lists $\{y(k)_A\}$ as produced by the algorithm G from the truncated input of the A/D converter.

Continuation of the example—improvement of the accuracy by improved scaling   We use an analog bridge circuit to subtract a constant 8 volts from the input signal and adjust the algorithm to allow for this. The ideal output $y(k)_{ideal}$ remains the same as in Table 5.6 but we repeat it here for convenience. Table 5.7 shows the performance of the same A/D converter with the same algorithm but with the scaled input signal. The most significant bit of the A/D converter now represents 1 volt, and $y(k)_B$ indicates the system output.

Note that various alternative methods of treating negative binary numbers exist but the point of this example is unaffected by any changes in such representation. The results of the example are plotted in Figure 5.42

EXAMPLES OF QUANTIZATION EFFECTS IN RECURSIVE ALGORITHMS
Quantization effects may cause large errors in recursive algorithms. The problem may be visualized with the aid of Figure 5.43. The top diagram illustrates the progress of the solution of the equation

$$y(k) = 0.8y(k-1) \tag{5.90}$$

from an initial condition $y(0) = 7$ towards the correct solution of $y(k) = 0$ as $k \to \infty$.

The lower diagram shows the same equation being solved with quantization bands of width $q = 0.5$ centered at 0, 0.5, 1, 1.5, etc., with rounding of numbers to the center of each band after each iteration of the solution. As the diagram shows, the solution reaches $y(k) = 1$ for $k = 8$ and it remains there for all further iterations.

We next investigate in detail the performance of three simple recursive algorithms under different word-length operation. We then derive some approximate guidelines on how to avoid the types of problem seen in the examples.

Investigation of the algorithm $D(z) = 1/(z - e^{-0.672T})$ with $T = 0.1$, fixed-point arithmetic with 5 working bits and truncation   The algorithm is equivalent to the difference equation

$$y(k) = 0.935y(k-1) + u(k-1). \tag{5.91}$$

To investigate numerically we set $y(0) = 7.5$, $u(k) = 1$ for $k \geqslant 0$. Table 5.8 shows how the algorithm proceeds.

Digital control algorithms

Table 5.6

k	1	2	3										4	
	Input signal u (volts)	$y(k)_{ideal}$	Binary equivalent of u										Decimal equivalent of truncated u	$y(k)_A$
			2^3	2^2	2^1	2^0	2^{-1}	2^{-2}	2^{-3}	2^{-4}	(bits lost by truncation)			
0	7.0		0	1	1	1	0	0	0	0	0 0		7.0	
1	7.4	4.0	0	1	1	1	0	1	1	0	0 1		7.375	3.75
2	7.7	3.0	0	1	1	1	1	0	1	1	0 0		7.6875	3.125
3	7.9	2.0	0	1	1	1	1	1	1	0	0 1		7.875	1.875
4	8.0	1.0	1	0	0	0	0	0	0	0	0 0		8	1.25
5	8.1	1.0	1	0	0	0	0	0	0	1	1 0		8.0625	0.625
6	8.3	2.0	1	0	0	0	0	1	0	0	1 1		8.25	1.875
7	8.6	3.0	1	0	0	0	1	0	0	1	1 0		8.5625	3.125
8	9.0	4.0	1	0	0	1	0	0	0	0	0 0		9.0	4.375

Table 5.7

k	Input signal $(u-8)$ volts	$y(k)_{ideal}$	Binary equivalent of $(u-8)$											Decimal equivalent of truncated u	$y(k)_B$
0	-1		1	0	0	0	0	0	0	0	0	0	0	-1	4.063
1	-0.6	4	0	1	0	0	1	0	1	1	1	1	0	-0.59375	2.968
2	-0.3	3	0	0	1	0	0	1	0	0	1	1	0	-0.296875	2.031
3	-0.1	2	0	0	0	1	0	1	1	1	1	0	0	-0.09375	0.938
4	0	1	0	0	0	0	0	0	0	0	0	0	0	0	0.938
5	0.1	1													
			2^0							2^{-7}					
										bits lost					
6	0.3	2					by truncation								2.031
7	0.6	3				Results repeat by symmetry									2.968
8	1	4													4.063

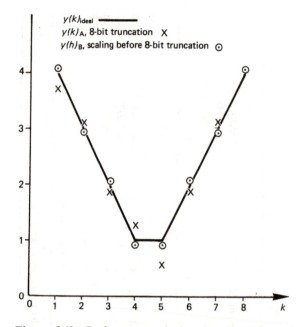

Figure 5.42 Performance of the algorithm $G(z) = 10(z-1)/z$ with infinite word length, with truncation and with scaling followed by truncation.

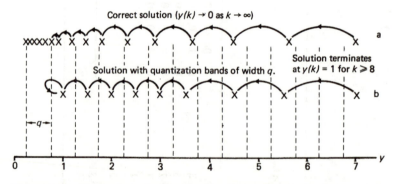

Figure 5.43 Illustration of the progress of the numerical solution of the equation $y(k) = 0.8y(k-1)$, (a) with infinite word length; (b) with quantization bands of $q = 0.5$ and rounding.

Investigation of the algorithm $D(z) = 1/(z - e^{-0.14728T})$ with $T = 0.1$, fixed-point arithmetic with 7 working bits and truncation The algorithm is equivalent to the equation

$$y(k) = 1.0148y(k-1) + u(k-1). \tag{5.92}$$

Table 5.8

	5-bit working with truncation (binary and decimal values are given)		Reference solution with long word length	
k	$y(k-1)$	$y(k)$	$y(k-1)$	$y(k)$
0	111.1 0000 = 7.5	1000.0000 = 8.0	7.5	8.0125
1	1000.0000 = 8	1000.0111 = 8	8.0125	8.49
2	1000. = 8	1000. = 8	8.49	8.93
3	1000. = 8	1000. = 8	8.93	9.35
			9.35	9.75

Solution remains at $y(k) = 8$ for all k

Solution moves monotonically to $y(k)_{k \to \infty} = 15.385$

To investigate numerically we set $y(0) = 128$, $u(k) = 0$ for all k. Table 5.9 shows how the algorithm proceeds.

Investigation of the algorithm $D(z) = 1/(z + e^{-T})$ with $T = 10^{-5}$, fixed-point arithmetic with 16 working bits and rounding The algorithm is equivalent to the equation

$$y(k) = -0.99999y(k-1) + u(k-1) \qquad (5.93)$$

We set $y(0) = 1.5000305$ and $u(k) = 0$ for all k. Table 5.10 shows how the algorithm proceeds.

This example shows very nicely how a short sampling interval needs to be accompanied by an appropriately long word length.

To give a further quantitative feeling for the magnitude of the effects that we have demonstrated in the three examples above, we have considered the algorithms

$$D_1(z) = \frac{1}{z \pm a}, \qquad D_2(z) = \frac{1}{z \pm e^T},$$

and have calculated approximate values of $a_{max}(a < 1)$ and T_{min} as a function of working word length. The results are presented in Table 5.11.

Table 5.9

	7-bit working with truncation		Reference solution with long word length	
k	$y(k-1)$	$y(k)$	$y(k-1)$	$y(k)$
0	1000000 0.0 = 128	1000000 1.11 = 128	128	129.9
1	128	128		
2	128	128		

Reference solution side:

Solution moves monotonically to $y(k) = \infty$ as k increases

7-bit side:

Solution remains at $y(k) = 128$ for all k

Table 5.10

	16-bit working with rounding	
k	$y(k)$	$y(k+1)$
	1.100000000000001 (binary) = 1.5000305	$y(k+1)_{\text{exact}} = -1.5000155$ Now all values between -1.5000153 and -1.5000458 are rounded to -1.5000305 hence $y(k+1)_{\text{rounded}} = -1.5000305$ and the solution oscillates between the values ± 1.5000305

The exact solution oscillates towards $y(k)_{k \to \infty} = 0$.

Observation of these limits should prevent the undesirable behavior demonstrated in the previous examples. (Values outside the limits quoted may fail to take the solution out of a quantization band, resulting in either an incorrect steady solution or an incorrect oscillatory solution.)

QUANTIZATION EFFECTS CAUSED BY MULTIPLICATION
Every time two n-bit numbers are multiplied together in fixed point arithmetic, the result is a number of $2n$ bits. The rounding or truncation of the products

Table 5.11

Working word length (bits)	a_{max} in $D_1(z)$	T_{min} in $D_2(z)$
4	0.89	0.11
5	0.94	0.06
6	0.97	0.03
7	0.985	0.015
8	0.992	0.008
12	0.9995	5×10^{-4}
16	0.99997	3×10^{-5}

of multiplication to fit a word of less than $2n$ bits is a source of quantization error. We note the following points:

(a) A typical algorithm may have quite a large number of multiplications in its realization. (A linear controller cannot have signal $\times$ signal multiplications so we are mostly concerned with coefficient $\times$ signal multiplications.)

   To avoid errors growing large due to accumulating effects, it is therefore necessary to use a larger word length for the multiplication register than is necessary for the A/D converter.

(b) A control designer is able to choose between alternative realizations of any particular algorithm. Each realization has multipliers at different points in the configuration and the quantization noise in the output of the algorithm is a combination of the separate quantization noises.

(c) In general, noise is simplified by transmission through the building blocks that form an algorithm. Hence a parallel realization of a particular algorithm is usually superior to a sequential realization, in terms of quantization noise produced by multiplication. (Note that the transmission of A/D conversion noise through an algorithm is not realization dependent.)

(d) Floating-point arithmetic, as usually implemented in a microprocessor, achieves an accuracy sufficient to allow quantization error to be neglected. However, floating-point operations take considerably longer than fixed-point operations.

(e) Hardware multipliers, by achieving good accuracy at high speed, should be considered for control applications.

Example   Comparison of three realizations of a transfer function in respect of quantization errors caused by multiplication   The transfer function to be realized is

$$G(z) = \frac{2z}{(z-1)(z-0.4)}. \qquad (5.94)$$

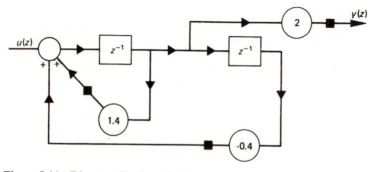

Figure 5.44   Direct realization of

$$G(z) = \frac{2z}{(z-1)(z-0.4)}$$

(the points marked ■ indicate locations where rounding after multiplication takes place).

Direct realization   Let $y(z) = G(z)u(z)$; then

$$(z^2 - 1.4z + 0.4)y(z) = 2zu(z)$$

or

$$y(k+2) = 1.4y(k+1) - 0.4y(k) + 2u(k+1). \qquad (5.95)$$

Figure 5.44 shows the resulting realization diagram. The points marked by solid squares indicate locations where rounding after multiplication takes place.

Parallel method   Taking partial functions, we obtain

$$G(z) = \frac{10}{3(z-1)} - \frac{4}{3(z-0.4)}. \qquad (5.96)$$

The parallel realization then follows as shown in Figure 5.45.

Factorization method

$$G(z) = \frac{2}{(z-1)} \frac{z}{(z-0.4)} \qquad (5.97)$$

leading to the sequential realization shown in Figure 5.46.

Each of the realizations was simulated with a constant input and rounding to two decimal places at each multiplication point. The results together with an exact solution are given in Table 5.12.

Comments on the results obtained   The results show that the parallel realization suffered most from the errors caused by rounding after multiplication, under the conditions simulated.

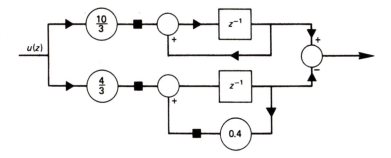

Figure 5.45   Parallel realization of

$$G(z) = \frac{2z}{(z-1)(z-0.4)}$$

(the points marked ■ indicate locations where rounding after multiplication
takes place).

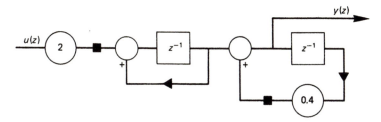

Figure 5.46   Sequential realization of

$$G(z) = \frac{2z}{(z-1)(z-0.4)}$$

(the points marked ■ indicate locations where rounding after multiplication
takes place).

It is not possible to draw general conclusions from the example. In fact
the parallel method of implementation is often preferred on the grounds that
it is expected to perform well in regard to amplification of quantization noise
caused by multipliers.

The lesson is rather that each proposed realization should be checked
through and in critical cases simulated to ensure that a satisfactory accuracy
will be obtained.

LIMITING CYCLING BEHAVIOR DUE TO QUANTIZATION EFFECTS
A system with no input may behave like a stable oscillator in the presence of

Digital control algorithms

Table 5.12 With rounding two decimal places after each
multiplication

Accurate solution	Direct method	Parallel method	Factorization method
0	0	0	0
2	2	2	2
4.8	4.8	4.8	4.8
7.92	8	7.9	7.9
11.168	11.2	11.1	11.2
14.467	14.4	14.4	14.5
17.787	17.8	17.7	17.8
21.115	21.2	21	21.1
24.446	24.4	24.3	24.4
27.778	27.8	27.6	27.8
31.111	31.2	30.9	31.1
34.445	34.4	34.2	34.4
37.778	37.8	37.5	37.8
41.111	41.2	40.8	41.1
44.444	44.4	44.1	44.4
47.777	47.8	47.4	47.8
51.111	51.2	50.7	51.1
54.444	54.4	54	54.4
57.777	57.8	57.3	57.8
61.111	61.2	60.6	61.1
64.444	64.4	63.9	64.4
67.777	67.8	67.2	67.8
71.111	71.2	70.5	71.1
74.444	74.4	73.8	74.4
77.777	77.8	77.1	77.8
81.111	81.2	80.4	81.1
84.444	84.4	83.7	84.4
87.777	87.8	87	87.8
91.111	91.2	90.3	91.1
94.444	94.4	93.6	94.4
97.777	97.8	96.9	97.8
101.111	101.2	100.2	101.1

certain types of nonlinearity. As an illustration, consider the equation

$$\ddot{y} + f(y)\dot{y} + y = 0$$

where $\hspace{10cm}$ (5.98)

$$f(y) < 0, \qquad |y| < 1,$$
$$f(y) > 0, \qquad |y| > 1.$$

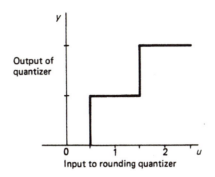

Figure 5.47   The input/output characteristic of a rounding quantizer.

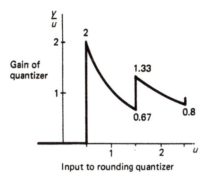

Figure 5.48   The gain (defined as
output/input) of a rounding quantizer
as a function of input.

Intuitively, the system has negative damping for small y and oscillations can
be expected to grow. Analogously, the system has positive damping for large
y and therefore large oscillations tend to be attenuated. Such a system does in
fact oscillate stably and it is then said to be 'limit cycling'. (More rigorously,
nonlinear systems theory defines a limit cycle as a unique closed trajectory in
the phase plane. Around one cycle of such a closed trajectory, there is no net
loss or gain of energy.)

The amplitude-dependent gain introduced by quantization is illustrated in
Figures 5.47 and 5.48. Figure 5.47 simply repeats for convenience the
input–output characteristic of a rounding quantizer. Figure 5.48 plots quantizer
gain, defined as output/input against input. It is easy to visualize how an
element with such a gain characteristic may cause oscillation in a closed-loop
system. Suppose that $y/u > 1$ creates an unstable loop; then when u satisfies
$0.5 < u < 1$, instability occurs and signal levels increase. As soon as u satisfies

$1 < u < 1.5$, the condition $y/u > 1$ no longer holds and signal levels decrease. The closed-loop system then settles into a permanently oscillatory state.

5.5.3 Interaction between quantization effects and sampling rate effects

The type of interaction that is possible between quantization effects and sampling-rate effects is illustrated by an example.

Figure 5.49 shows the action of a differentiation algorithm using the same quantization levels but with two different sampling intervals, T_1, T_2. The interaction may be important in practice since the output with high amplitude spikes may saturate plant actuators causing loss of algorithm equivalence, apart from other possible disagreeable consequences. (The output from a continuous differentiator would have been constant for a ramp input.)

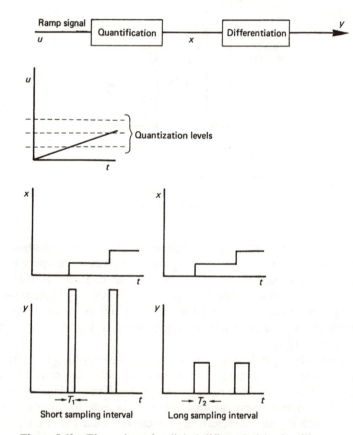

Figure 5.49   The action of a digital differentiation algorithm on a quantized ramp signal.

This interaction is also important in considerations of system behavior as the sampling interval T tends to zero. In order for the system behavior to approach that of a continuous system, it is necessary for the number of bits by which numbers are represented to approach infinity. A practical aspect of this effect is that if the benefits of very rapid sampling rate to be obtained, it will have to be accompanied by a commensurate increase in word length.

The quantitative interplay between sampling rate and word length was deliberately brought out in some of the examples in Section 5.5.2. In particular, Figure 5.41 plots the interaction for one example. It would be useful to scan through the results of the examples in Section 5.5.2 to form an understanding of the nature and extent of the interaction.

It will be seen that there are two main effects:

(a) Numerical ill-conditioning of algorithms containing terms of the form $z - e^{CT}$ in the denominator. As $T \to 0$, the pole corresponding to this term approaches unity and a long word length is needed to obtain good accuracy under such a condition.

(b) As the sampling interval becomes very short, a variable whose rate of change is small will appear constant over several consecutive samples unless a long word length is used. An algorithm that uses the difference between consecutive samples to calculate an output is particularly affected when too short a sampling instant is used. It produces zero output at some sampling instants with a sudden spike of output when a change of the least significant bit occurs.

5.5.4 Effects of noise

A control system designed around assumptions of an exactly known process with noise-free signals may perform badly in practice when the process is not exactly as modelled and noise signals are acting. A usable control system has to be robust—able to retain its characteristics largely unchanged despite changes in the operational environment.

The performance of a control loop may be degraded by the effects of noise. Proper care must be devoted to the specification of filters having regard to the estimated frequency spectrum of the noise, the type of algorithm, the sampling rate and the process dynamics.

It can be assumed that every process model is incorrect to a greater or lesser extent. Thus, an algorithm designed around a particular model may not perform well on the actual process.

It is therefore necessary to test a proposed control strategy to ensure that its performance is satisfactory under conditions of noise and/or process parameter variations. Such testing is most easily carried out by simulation.

In general, a digital control algorithm operating with a long sampling interval is more sensitive to disturbance by process parameter drift than the same type of algorithm operating with a short sampling interval. In some cases, the upper limit for the sampling interval is dedicated by this consideration.

If the main task of a control loop is to counteract the effects of random disturbances, it may be best to design a controller on stochastic principles rather than relying on a deterministic controller to perform well in a statistical sense.

Stochastic systems are really outside the scope of this book, but we are able below to indicate in a simple way one possible approach to the design of a controller on stochastic principles.

DESIGN OF A CONTROLLER FOR REJECTION OF DISTURBANCES THAT ARE DESCRIBED STATISTICALLY

Many control systems have the task of keeping some quantity constant despite the influence of stochastic disturbances. Examples are: ships' autopilots operating in the presence of waves, radio telescope control in the presence of wind gusts and many process control systems in the presence of raw material variations. It is often satisfactory to design a control system on deterministic principles even though the main task of the system will be to offset stochastic disturbances. However, another approach is possible—that of designing the controller on statistical principles. This is a large field of work and we can do no more here than to introduce some of the elementary concepts.

Consider Figure 5.50. A system of transfer function $G(z)$ has its output corrupted by an unwanted disturbance signal $w(z)$. The function $G(z)$ is assumed known and the statistical properties of the signal $w(z)$ are assumed to be known and to be time invariant.

The problem is to design a controller $D(z)$ so that $y(z)$ is kept as close to zero as possible despite the disturbance signal.

Because we are used to working with transfer functions, we invent (in practice this means deriving numerically from records of the signal $w(z)$) a transfer function $N(z)$ that will generate the signal $w(z)$ from idealized white noise $p(z)$.

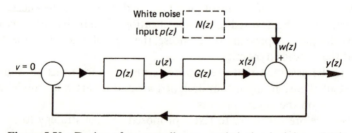

Figure 5.50 Design of a controller on statistical principles. $w(z)$ is a noise signal that is only known statistically. $D(z)$ is to be designed to counteract the disturbing effects of the signal.

p is assumed to have the properties usually ascribed to white noise, i.e.

$$\mathscr{E}\{p(t)\} = 0, \qquad \mathscr{E}\{p(t)^2\} = 1. \tag{5.99}$$

$\{\mathscr{E}\}$ indicates *expected value* and here the expressions indicate that $p(z)$ has zero mean value and unity mean square value.

We assume that some degree of prediction of $w(z)$ is possible and hence we intend to fix each value of control input $u(k)$ so as to minimize the disturbance that will be caused by the future signal $w(k+1)$. This one-step-ahead prediction is intended to give the process $G(z)$ time to respond.

A cost function of the form

$$J = \mathscr{E}\{y^2(k+1)\} \tag{5.100}$$

will be minimized by choice of the control signal $u(k)$ and this will lead to a suggested controller $D(z)$.

Since $y(z) = G(z)u(z) + N(z)p(z)$, an expression for $y^2(k+1)$ can be obtained for inclusion in the equation for J. The statistical properties of the signal p are substituted and J is differentiated with respect to $u(k)$ to find the optimum value for $u(k)$. A closed form for the controller transfer function $D(z)$ is then obtained using the relation

$$D(z) = \frac{-u(z)}{y(z)}.$$

Numerical example   Let

$$N(z) = \frac{z - 0.1}{z - 0.2}, \qquad G(z) = \frac{1}{z - 0.2}$$

in Figure 5.50. A controller $D(z)$ is to be designed to minimize $J = \mathscr{E}\{y^2(k+1)\}$ as outlined above.

From Figure 5.50,

$$zy(z) = 0.2y(z) + u(z) + (z - 0.1)p(z),$$

$$J = (0.2y(k) + u(k) - 0.1p(k))^2$$
$$+ 2(0.2y(k) + u(k) - 0.1p(k))\mathscr{E}\{p(k+1)\} + \mathscr{E}\{p(k+1)^2\}$$
$$= (0.2y(k) + u(k) - 0.1p(k))^2 + 1,$$

minimizing J by choice of $u(k)$ is achieved by setting

$$2(0.2y(k) + u(k) - 0.1p(k)) = 0$$

or

$$2y(z) + 10u(z) = p(z).$$

Now

$$y(z) = G(z)u(z) + N(z)p(z)$$

and the required controller

$$D(z) = -\frac{u(z)}{y(z)}.$$

From these relations

$$2y(z) + 10u(z) = \frac{y(z)}{N(z)} - \frac{G(z)u(z)}{N(z)}.$$

Hence

$$D(z) = \frac{2 - [1/N(z)]}{10 + [G(z)/N(z)]} = \frac{2 - [(z - 0.2)/(z - 0.1)]}{10 + [1/(z - 0.1)]} = 0.1.$$

(In this particular case, the required control is achieved through a simple constant multiplier.)

5.5.5 Comparison of algorithms

Hard information on the relative advantages of different algorithms is very scarce. The author's industrial experience leads to the following broad conclusions, which should not be taken as axioms.

(a) PID algorithms perform surprisingly well in practice. They are robust and difficult to improve on significantly on a day-to-day basis unless very considerable effort is expended.
(b) Feedforward and cascade algorithms perform well in those situations for which they were designed (as described earlier in this chapter).
(c) More complex algorithms tend to lose their supposed advantages once process parameters drift or noise begins to affect the measurements.
(d) In general, the use of a long sampling interval greatly increases the sensitivity of a control loop to process parameter drift. Complex algorithms tend to use relatively long sampling intervals and here lies one explanation for the comment under (c).

As a general comment, academic sources tend to give an over-optimistic view of what can be expected from advanced algorithms. When reading such source documents, answers to the following questions should be sought:

(a) What type of process model is needed? How is it to be obtained?
(b) How accuracy must the model be? How much variation can be tolerated as the process changes? (A typical process will change its parameters by 20% over even a nominally fixed type of operation—for instance, such a

simple device as a d.c. machine driving a gear train has been found to vary its parameters by up to 40% as the oil temperature in the gear train rises during operation. Many processes change by even greater margins as they are subjected to different operating requirements.)

(c) Is there mention of at least one verifiable implementation where, even on a temporary pilot scale, the algorithm has been shown to work properly?

(Simulation results may be reliable but they should be regarded with skepticism and inspected critically to see exactly what they really show.)

5.6 Oscillations in control loops

Unwanted oscillations in a newly implemented control loop are sometimes particularly difficult to eliminate, because of the variety of sources from which they might originate.

Consider as an example a position-control system in which a linear motion is required to follow a ramp signal of desired position. Suppose that when the system is commissioned, the load, instead of following the required ramp, moves in a series of discontinuous jumps. What are the possible sources of such behavior? Informed reasoning suggests the following partial list of possible causes:

(a) Too long a sampling interval in conjunction with a particular choice of control algorithm.
(b) Too large a quantization step.
(c) Limit cycle behavior superimposed on the underlying ramp movement. (Note that a limit cycle may occur in the control loop or, locally, within a control algorithm.)
(d) Unexpected noise sources feeding into the system.
(e) Stick–slip motion due to static friction in the linear motion mechanism.

Trouble-shooting will necessarily involve diagnostic tests. For instance: Can the actuator be made to move slowly and smoothly in open loop?

Bearing in mind that sustained oscillators are associated with non-linearities, one obvious strategy is to inject sinusoidal signals into the system and to track the points where they become nonsinusoidal.

Exercises

5.1 It is required to design a digital control algorithm for a single-input/single-output process. The process characteristics are unknown but input–output data are available.

Discuss critically the possible ways forward towards design of a finished algorithm. Which features in the process might necessitate corresponding special features in the algorithm?

5.2 Show that a closed-loop system whose transfer function is $H(z)$ has zero steady-state error for a step input provided that $H(z)|_{z=1} = 1$.

5.3 A process G described by a difference equation

$$y(k + 1) + 0.5y(k) = u(k)$$

in a preceded by a compensator $D(z)$. The open-loop poles of $G(z)D(z)$ are to be at $z = \pm j/2$. Determine the difference equation for the compensator D to achieve this. Sketch the step response of $G(z)$ alone and of $G(z)D(z)$.

The system $G(z)D(z)$ is put into closed-loop form with unity feedback. Determine the pulse transfer function of the closed-loop system and plot the poles in the z plane.

5.4 (a) Show that a necessary condition for a system to have a finite settling time, after the application of a step input, is that all the system poles should be at the origin of the z plane.

(b) An electric motor of transfer function $K/[s(1 + 0.5s)]$ is preceded by a sampler and zero-order hold and a digital controller D. A sampling interval of 0.1 s is to be used. Determine the z transfer function $D(z)$ of the controller such that the closed-loop step response of the resulting system settles in finite time with the output following one sampling interval behind the reference step input.

5.5 The plant shown in Figure 5.51 is known to have the open-loop transfer function

$$G(z) = \frac{0.2(z + 0.8)}{(z - 1)(z - 0.6)}$$

Design a forward path compensator $D(z)$ to ensure that the closed-loop system is stable and has zero steady-state position error to ramp input signal. Sketch the following error in response to the ramp signal. Finally, check whether the step response of the system is reasonable.

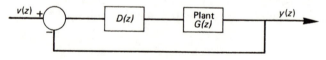

Figure 5.51

5.6 For the system shown in Figure 5.52,

$$G(s) = \frac{1}{s(1 + s)},$$

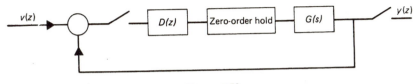

Figure 5.52

$T = 1$ s. Design the controller $D(z)$ so that the response to a unit step at $v(t)$ is

$$y(k) = 0, 0.5, 1, \ldots.$$

5.7 A digital controller of the form

$$D(z) = \frac{u(z)}{e(z)} = \frac{a_0 + a_1 z^{-1} + \cdots + a_m z^{-m}}{b_0 + b_1 z^{-1} + \cdots + b_n z^{-n}},$$

with the coefficients to be determined numerically, was suggested in equation (5.21). Show that the controller may be realized in the time-domain by the relation

$$u^*(t) = \frac{1}{b_0}\left(\sum_{k=0}^{m} a_k e^*(t - kT) - \sum_{k=0}^{n} b_k u^*(t - kT)\right)$$

5.8 Derive an expression for a controller $D(z)$ that will achieve dead-beat control of a process of transfer function $G(z)$.

Explain briefly why Dahlin and Kalman were motivated to modify the basic algorithm and go on to derive the Dahlin and Kalman variants of the algorithm.

5.9 A process was found to have the open-loop unit-step response shown in Table 5.13. The response reaches the value 5.0 for large t.

Table 5.13

Time (seconds)	0	1	2	3	4	5	6	7	8	9
Response	0	1	1.5	2.05	2.475	2.85	3.17	3.44	3.67	3.86

Time (seconds)	10	11	12	13	14	15	16	17	18
	4.03	4.18	4.30	4.40	4.50	4.56	4.63	4.68	4.73

A digital controller is to be designed that will bring the closed-loop rise time to about 3 s. Design the controller, making and stating further assumptions as necessary. (*Hint:* The data above may be generated by a second-order difference approximation to the process.)

5.10 A process with transfer function

$$G(s) = \frac{ke^{-sT_1}}{(1 + sT_2)}$$

is preceded by a zero-order hold and a digital controller $D(z)$. The complete closed-loop system is required to have the transfer function $H(z)$. Derive an expression for the controller $D(z)$.

Suppose next that the closed-loop system is required to have a dead-beat response to a step input. Derive an expression for the required controller characteristic.

5.11 A process has the transfer function Ke^{-sT_d}/s. When put into closed loop with unity feedback, the system oscillates at 0.01 Hz. Calculate T_d.

Suggest settings for a three-term controller to control the process. Comment on the quality of control you would expect to obtain.

5.12 A system has the open-loop response to a unit step shown in Figure 5.53. The system is to be controlled by a proportional P + I integral controller. Fit an approximate transfer function to the open-loop response and use Ziegler–Nichols rules to determine the numerical values in a continuous-time P + I controller.

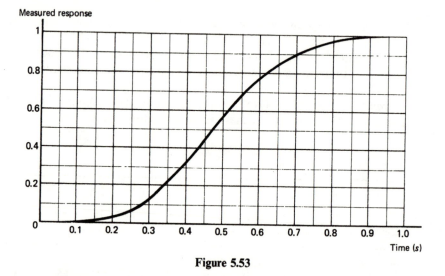

Figure 5.53

Determine a suitable sampling rate for a digital control loop based on the criterion of closed-loop bandwidth. Using this sampling rate and discretization based on trapezoidal approximation, develop a difference equation to represent the discrete version of the necessary controller.

5.13 A process has a unit step response as shown in Figure 5.54. The process is to be controlled by a proportional-only controller of suitable gain C.

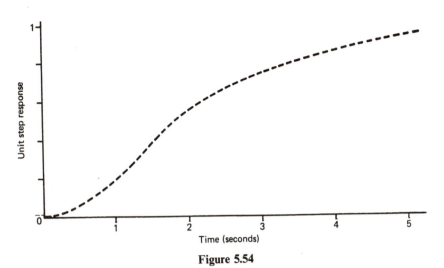

Figure 5.54

From the step-response graph, fit an approximate transfer function to the process of the form $e^{-sT_1}/(1 + sT_2)$, by determining numerical values for T_1 and T_2. Use the empirical rule $C = T_2/T_1$ to determine a suitable value for C.

The correct transfer function of the process can now be revealed as:

$$G(s) = \frac{2}{(s + 0.5)(s^2 + 2s + 4)}.$$

Investigate and comment on the behavior of this process in closed loop under proportional-only control, using the value for C that you found above.

To assist in this:

(a) find the gain margin of the system;
(b) plot the positions of the closed-loop poles as accurately as you can.

5.14 A process of transfer function $1/(1 + 5s)$ is to be followed by a measuring sensor of transfer function $G_m(s)$. Two possible sensors are available:

(a) with $G_m(s) = e^{-0.1s}$ (b) with $G_m(s) = \dfrac{1}{(1 + s)}.$

A proportional-only controller $G_D(s)$ is to be incorporated in the loop as shown in Figure 5.55. For the two cases (a) and (b), calculate the best

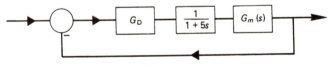

Figure 5.55

(Ziegler–Nichols) setting for G_D and comment on the relative merits of the two measuring devices using the criterion of closed-loop bandwidth.

5.15 Explain why, in the digital control of a continuous process, it is important to choose a sampling interval that matches the application. Describe briefly all the approaches that are available for determination of a suitable sampling interval.

A process has the unit step response shown in Table 5.14. Use Ziegler–Nichols rules to calculate the coefficients of a P + I controller.

Table 5.14

Time t (seconds)	0	1	2	3	4	5	6	7	8	9	10	11
Response $y(t)$	0	0.02	0.05	0.1	0.17	0.4	0.7	0.81	0.9	0.94	0.96	0.98

$$(y(t) \to 1 \text{ as } t \to \infty)$$

Decide on a suitable sampling interval T, using one of the methods described earlier in your solution. Discretize the controller by the trapezoidal method and present the controller as a difference equation using the chosen sampling interval. Comment on any deficiencies in this approach to digital controller design.

5.16 A unity feedback discrete-time system has a forward z-transfer function given by

$$G(z) = \frac{K(z + 0.716)}{(z - 1)(z - 0.368)}, \qquad K = 0.736.$$

Sketch the Bode plots in the w-plane and hence determine the phase and gain margins of the system and the respective crossover frequencies in radians per second.

Assume a sampling period of one second.

5.17 Explain the operation of cascade and feedforward control algorithms and outline the conditions under which it might be advantageous to employ them. Sketch block diagrams showing how the algorithms may be applied to continuous-time processes.

5.18 (a) Sketch realizations of the algorithm

$$D(z) = \frac{(z + 1)}{(z - 1)(z^2 - 1.1z + 0.3)}$$

using the three approaches: direct, parallel and factorization.

(b) Investigate the applicability of the three approaches to the cases where the denominator of $D(z)$ has (i) repeated roots; (ii) complex roots.

5.19 (a) Sketch the parallel realization of the second-order digital controller

$$D(z) = \frac{3 - 2.2z^{-1}}{1 - 1.4z^{-1} + 0.4z^{-2}}.$$

(b) $D(z)$ (given above) is implemented on a microprocessor which uses fixed-point number representation and truncates arithmetic results. Input to the controller is quantized by an A/D converter, and an algorithm corresponding to the parallel realization is used to compute the output.

Determine the minimum word lengths for the A/D converter and microprocessor (assuming that the same word length will be used for both) such that the total quantization error at the output is no worse than 1%.

5.20 Confirm the derivations in the text of the three approaches (worst case, steady-state, stochastic) to estimation of quantization noise at the output of an element of transfer function $G(z)$. An algorithm

$$G(z) = \frac{z(2z - 1)}{z^2 - z + 0.24}$$

is to be implemented by parallel programming in a 16-bit microprocessor where fixed-point multiplication is followed by truncation. Quantization noise in the algorithm output must not exceed 0.2%.

Allowing for quantization noise sources at the multipliers and at the A/D converter, find the minimum number of bits needed in the A/D converter.

5.21 In control design, great care is taken to sample inputs fast enough to properly characterize all the useful information that they carry. Why then is it considered admissible to use such a crude device as the zero hold for signal reconstruction within a control loop?

In control design, can the sampling theorem be invoked to guarantee that no hidden inter-sample oscillations are occurring anywhere in the closed-loop?

5.22 Explain the motivation for using a self-tuning regulator. Describe the principle of its operation, and list briefly its possible disadvantages.

5.23 Explain the principle of operation of a digital self-tuning regulator. Outline the steps required in the design of such a regulator, mentioning points of practical importance in an application.

A process in open loop has input $u(t)$ and output $y(t)$ recorded at one second intervals as shown in Table 5.15. The data are noise free. Determine the numerical coefficients in a second-order difference equation representing the process. Indicate how a digital controller could be designed to achieve dead-beat step response of the process.

If the data in the table had been known to be noisy, how would your approach have been modified?

Table 5.15

Time (seconds)	$y(t)$	$u(t)$	Time (seconds)	$y(t)$	$u(t)$
0	0	1	10	1.875	1
1	0	1	11	1.9375	1
2	0	1	12	2	1
3	1	1	13	2.3125	1
4	2	1	14	2.3125	1
5	2.5	1	15	2.015625	1
6	2.5	1	16	1	1
7	2.25	1	17	1.992	1
8	2	1	18	1.992	1
9	1.875	1			

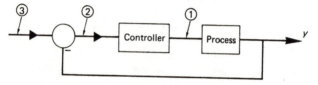

Figure 5.56

In a self-tuning application, y is to be measured as shown in Figure 5.56, whereas the points 1–3 are available as alternative sources of the input signal for the modelling operation. Discuss the relative merits of taking the input signal from each of these three points.

5.24 Derive a recursive version of the algorithm (5.36) of the form

$$\theta_k = \theta_{k-1} + (\text{a correction term}),$$

where θ_k is the best estimate of θ after k data pairs have been accumulated.
You will need to use the following matrix identity:

$$(\Lambda^T\Lambda + \lambda\lambda^T)^{-1} = (\Lambda^T\Lambda)^{-1} - \frac{(\Lambda^T\Lambda)^{-1}\lambda\lambda^T(\Lambda^T\Lambda)^{-1}}{1 + \lambda^T(\Lambda^T\Lambda)^{-1}\lambda},$$

where Λ is of dimension $p \times 2n$, λ is of dimension $2n \times 1$.

5.25 Figure 5.57 shows a control system that is subject to disturbance by the signal w. The characteristics of w are:

$$\mathscr{E}\{w(k)\} = 0, \qquad \mathscr{E}\{w(k)^2\} = 1$$

(where as usual $\mathscr{E}$ indicates expected value).
Determine the controller $D(z)$ that will minimize the cost function

$$J = \mathscr{E}\{y^2(k+1)\}$$

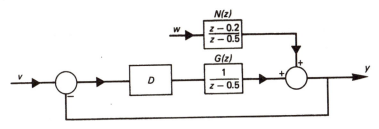

Figure 5.57

under the condition that $v(k) = 0$ for all k. Compute the unit step response of the resulting closed-loop system, assuming zero disturbance inputs. Comment on the type of closed-loop behavior that the statistical approach to control design has produced.

5.26 A continuous sytem of transfer function

$$G(s) = \frac{1.2}{s + 2s}$$

is connected in closed loop in series with a sampler and zero-order hold as shown in Figure 5.34 (repeated here as Figure 5.58).

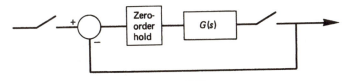

Figure 5.58

Calculate the closed-loop step response as follows:

(a) Knowing that $y(t) = 0$ and $dy/dt = 0$ at $t = 0$ and that $e(t) = 1 - y(0)$ over the interval $0 \leqslant t < 1$, calculate $y(t)$, dy/dt at $t = 1$ by solving the differential equation corresponding with $G(s)$.

(b) Knowing $y(t)$ and dy/dt at $t = 1$ and that $e(t) = 1 - y(1)$ over the interval $1 \leqslant t < 2$, calculate $y(t)$, dy/dt at $t = 2$.

(c) Continue as above until $t = 10$ and plot the results.

(d) Compare the plot with that of Figure 5.35, where the step response is plotted as calculated by the conventional $\mathscr{Z}$ transform method.

(e) Comment on any differences found between the calculated step responses.

5.27 Simple control algorithms can be set up by little more than tuning rules, whereas more ambitious algorithms tend to require access to reasonably accurate process models. In real applications it is rare to find a process whose behavior is sufficiently time-invariant and quantitatively understood

Table 5.16

Application	Control aim
1. Large steerable radio telescope	Angular position is to be controlled
2. Robot arm	Angular position is to be controlled
3. Liquid effluent neutralization	Factory discharge must not be acidic
4. Boiler drum level control	Water level to be maintained constant
5. Fermentation control	Fermenter temperature to be controlled so as to maximize product formation
6. Iron making in blast furnace	Air blast temperature to be controlled to give correct value of silicon in molten iron
7. Rolling of hot steel strip	Roll gap to be controlled to give correct thickness of outgoing strip
8. Automatic ship steering	Rudder position to be controlled to ensure ship follows a given course
9. Temperature control in a concert hall	Temperature to be maintained constant

in the way that the ambitious algorithms require. To appreciate this important point, examine the list in Table 5.16 of typical control applications and, for each, discuss briefly the modelling problem.

5.28 A designer of industrial control systems must frequently convert requirements posed in general commercial terms (e.g. improve product consistency) into forms that are directly applicable in control design (e.g. achieve a particular closed-loop bandwidth).

For the following applications, succest usable control criteria equivalent to the original requirement:

(a) A mass of inertia J has to be rotated from an arbitrary rest position to any other angular rest position within $180°$ in not longer than T_1 seconds.

(b) A sterrable aerial must not be moved out of alignment by more than $e°$ by wind forces.

(c) Cans are to be filled with 1 kg of product to ensure negligible possibility of underweight but with minimum 'give-away'.

(d) The thickness correction scheme for a flat product that is made in batches must satisfy the requirements:

    (i) Product to be within thickness specification within l meters of the start of each batch.

    (ii) Product to be kept within thickness specification during the batch despite the effect of random disturbances.

(e) An endless strip of plate glass of width w meters moves forward at a speed of s meters/second. It is to be cut into lengths of $l \pm e$ meters by a diamond cutter that must be arranged to traverse the strip at right angles using two servomechanisms.

5.29 In Section 5.2.6, a controller

$$D(w') = \frac{1}{3.16} \frac{1 + w'/0.3}{1 + w'/3}$$

was designed by Bode techniques for control of a process of transfer function $G(s) = 1/s^2$.

Determine approximate values of M (the maximum amplification of a sinusoidal input signal) and of bandwidth for the closed-loop system.

6

Hardware Systems for Implementation

6.1 Introduction

Figure 6.1 shows a schematic diagram of a control loop containing a microprocessor. The dotted line encloses the components that would normally be located together and referred to as the microcomputer.

This chapter treats the practical aspects of specifying and interconnecting the elements in such a control loop to ensure their compatible operation.

6.2 Design of input circuits

In a typical control scheme, most measurements will be made by analog transducers. Proper attention must be paid to the associated analog circuitry—in some types of application, considerable effort will have to be devoted to this part of the system.

Some of the questions that need to be considered are the following:

Do the signals need to be amplified, buffered or isolated before being transmitted?

What measures, such as special grounding, shielding or analog filtering are needed to reduce interference?

Do the transducers need local excitation, open-circuit detection or some other application-dependent consideration?

It is clear the the most difficult situation will occur where low-amplitude, wide-bandwidth signals are to be transmitted over long distances through a noisy environment, with a high accuracy requirement.

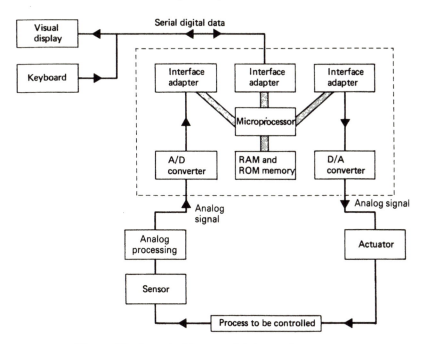

Figure 6.1   A control loop containing a microprocessor.

6.2.1 Some typical measurement inputs

An industrial control scheme may include, typically, fifty or more information-bearing signals arising from a variety of different devices at different geographical locations. Signal levels may vary widely and transmission distances of several hundred meters through electrically noisy environments are common.

Bringing reliably to the computer the information from this wide range of devices is an important operation.

A few examples will illustrate the problem.

MEASUREMENT OF SHAFT ANGULAR POSITION
Where a synchro is used to measure angular position, the signals to the computer consist of three voltages

$$v_1 = v \sin \theta,$$

$$v_2 = v \sin (\theta + 2\pi/3),$$

$$v_3 = v \sin (\theta + 4\pi/3),$$

where θ is the shaft angular position relative to some datum point.

The signals are typically at a level of 10 volts, and if the distance is not too far, they may be transmitted directly from the synchro to an A/D converter

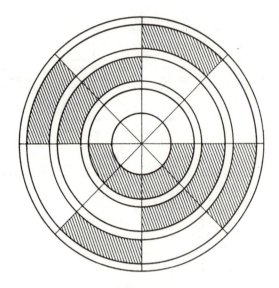

Figure 6.2   Schematic diagram showing the pattern
of a digital encoder.

at the computer interface. Within the computer a simple program can calculate
θ from the three signals.

DIGITAL ENCODERS FOR MEASUREMENT OF SHAFT ANGULAR POSITION
A digital encoder produces a digital signal directly, by the use of a specially
patterned disk mounted on the shaft whose position is to be measured.

The disk pattern is made up of opaque and transparent regions, so designed
that as the disk rotates, its position can be identified by noting the sequence
of opaque or transparent regions along a radius. As a simple example, the disk
shown in Figure 6.2 has eight different radial identifications—they correspond
to binary numbers as can be seen from the figure. It only remains to read the
pattern along a radius by a light source and a bank of photoelectrically driven
gates, and an optical shaft encoder results. The type described here is an absolute
encoder.

A number of different codes, each with its own particular merits, have been
established for the disk pattern. For instance, the so-called Gray code is designed
to ensure that, as the disk rotates, each successive binary number that is generated
differs in only one bit from its neighbors.

Such an arrangement is desirable, since, when several bits need to change
simultaneously, in practice one bit is bound to change before the others. The
Gray code therefore helps the encoder designer, who does not have perfect
radial alignment of either disk pattern or optical system.

Interfacing of digital encoders to a computer is straightforward. They can be connected directly to a parallel input port and read by the program whenever desired.

Optical encoders that measure relative position are also used. As the disk rotates, it generates a stream of pulses that go to an up–down counter whose status gives the angular position. Relative encoders offer high resolution but, like all incremental devices, they have the disadvantage that, under fault conditions, the reference may be lost.

For further information on encoders, Reference W2 should be consulted.

MEASUREMENT OF TEMPERATURE BY THERMOCOUPLE

Industrial furnaces operating at temperatures of less than 1000°C typically use so-called base metal (i.e. nonplatinum) thermocouples for temperature measurement.

A typical base-metal thermocouple produces an output of about 4 mV per 100°C and the relation is nonlinear. The thermocouple calibration holds only if the cold junction of the thermocouple is held at 0°C. This is usually possible only in a laboratory environment and 'cold-junction compensation' needs to be undertaken inside or outside the computer.

Thermocouple signals need to be carried along special 'compensating cables' if disturbance effects due to additional junctions at intermediate temperatures are to be avoided.

Thermocouples usually fail by becoming open-circuited. Such failure must not be interpreted by the computer as indicating low temperature since this would lead to disaster in an automatic control loop. It is essential to have some method of thermocouple break protection in the system.

For transmission of thermocouple signals over any significant distance, they are usually transformed to current signals. A very common type of voltage-to-current converter uses a standard current range of 4 to 20 mA, where 4 mA represents the minimum and 20 mA the maximum value of the signal to be transmitted. At the computer, the current is simply passed through a suitable resistor to provide an input voltage signal.

MEASUREMENT OF TEMPERATURE BY RESISTANCE THERMOMETER

Temperatures in the range $-100°C$ to $+300°C$ are often measured by resistance thermometer. A resistance thermometer is a reel of platinum or other wire whose temperature coefficient of resistance is accurately known. To measure the resistance, the reel needs to be supplied with a current, and usually this is achieved by making the reel one arm of a bridge. Three signals carrying the resistance information at a low voltage level need to be converted to a single signal representing temperature. If the signal has to be transmitted over a significant distance, then the same remarks apply as for the thermocouple signal—a more robust signal such as a 4–20 mA current signal needs to be used for the transmission.

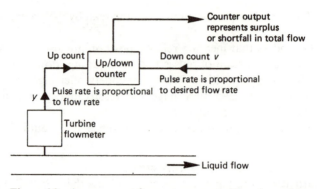

Figure 6.3   Arrangement for measurement of flow by turbine
flow meter.

MEASUREMENT OF FLOW BY TURBINE FLOWMETER

Many liquid blending processes depend on turbine flowmeters for measurement.
A turbine flowmeter produces a sequence of impulses at a rate proportional to
flow rate.

Figure 6.3 illustrates a typical application. A desired flow rate is represented
by a sequence of impulses v and the measured flow rate by a sequence of
impulses y. Both v and y are input to the same counter with y adding to the
total and v subtracting. In a control system, the flow rate y is adjusted to keep
the counter at zero—in this way a correct volume of liquid may be dispensed.

COUNTERS

Mention has been made of counters in the sections on turbine flowmeters and
on digital encoders. A counter receives a stream of pulses, and stores the current
total in a register, which then generates the digital word representing the count.
In outline, the electronic configuration consists of wave shaper, clocked gate,
counter and register (Figure 6.4).

DISCUSSION

This selection of measurement devices has been described to emphasize the
variety of types and levels of signals arising. Where a sensor is located near to
the computer, it is often possible to undertake special computation in a
purpose-built application-oriented input card.

Low-level analog signals need to be made robust enough for transmission
to the computer. Conversion to current signals is the most straightforward
approach. A current signal can be sent over several kilometers since it is not
affected by voltage drop and is more immune to noise than a voltage signal. If
many analog signals arise remotely near to one point, they may be multiplexed,
A/D converted and then transmitted serially along a pair of wires to the
computer.

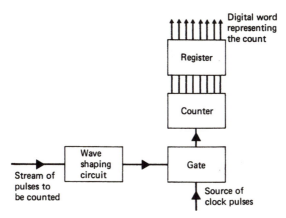

Figure 6.4 Arrangement for counting pulses.

The computer is also required to detect the status (on or off) of switches, such as limit switches or auto–manual change-over switches. Somewhat confusingly, signals arising from simple on–off switches tend to be called digital inputs in the commercial literature. Such signals can be considered as Boolean variables—they are input directly to the computer.

6.3 Actuators

In small-scale applications, such as in the control of laboratory furnaces, it is usual for the actuator to be the most expensive element in the control loop. Again, in a small-scale liquid-level control scheme, the remotely driven flow control valve is often the most expensive item in the loop. Such systems therefore need to be specified with careful consideration of the actuators if a cost-effective design is to be produced. Actuator selection is very application-specific and it therefore cannot be covered here. We go on instead to consider two topics, related to actuators, that are of general applicability.

6.3.1 Choice of control configuration

Consider for concreteness an electrically driven valve that is required to control the rate of flow in a pipe.

Two alternative strategies are possible for controlling the valve:

(a) To close the control loop directly.
(b) To use an additional loop to position the valve.

Figure 6.5 illustrates the two possibilities.

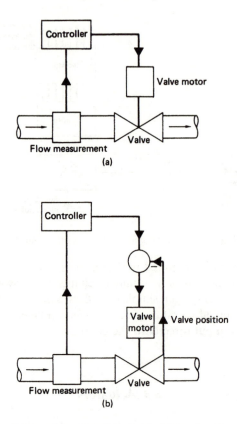

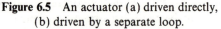

Figure 6.5 An actuator (a) driven directly,
(b) driven by a separate loop.

The choice between the two alternatives is dictated by several factors, such as the dynamics of the process to be controlled, the chosen algorithm and the presence or absence of significant hysteresis in the actuator.

Each type of actuator tends to have its own particular type of non-ideal behavior. Below we consider one such type of behavior—velocity limiting.

6.3.2 Velocity limiting in actuators

An ideal actuator moves at a velocity proportional to the signal that it receives. A real actuator, such as a flow control valve, has a velocity limit. For such an actuator, a large movement takes longer than a small movement.

This amplitude-dependent effect introduces nonlinearity and time delay into the control loop. The valve motor must be specified so that the effects are kept to an acceptable level.

(Other types of nonlinearity such as hysteresis and backlash may be encountered in actuators and may lead to unsatisfactory behavior of the control loop. In such cases, it is usually preferable to include the offending actuator within a local position control loop, as in Figure 6.5(b), to help to decouple the actuator characteristic from the remainder of the system.)

6.4 Analog interfacing

Most processes that we are called upon to control operate in continuous time. This implies that we are dealing, largely, with an analog world to which we must interface to/from the digital computer through which we seek to influence the process.

6.4.1 Control aspects of A/D and D/A conversion

We are interested in the characteristics of these devices in so far as these affect the control loops into which they are connected.

Considering first analog-to-digital (A/D) conversion, assume that the signal $f(t)$ is to be discretized. At time kT the signal $f(t)$ is connected to the A/D converter. Two questions arise:

How long does the conversion take?

How accurate is the conversion?

The two questions may be expected to be interrelated.

Considering digital-to-analog (D/A) conversion, three questions naturally arise. These are:

How long does conversion take?

How accurate is the initial conversion?

Is the output of the D/A converter subject to significant drift between sampling intervals?

We consider D/A conversion first, since every A/D converter necessarily contains a D/A converter.

6.4.2 Digital-to-analog conversion (D/A conversion)

A digital-to-analog converter operates as shown in Figure 6.6. A parallel digital word is converted by a logic and switching network into an equivalent resistance from which an analog voltage is derived. The final amplifier shown in the

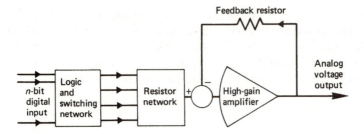

Figure 6.6   Outline of the operation of a D/A converter.

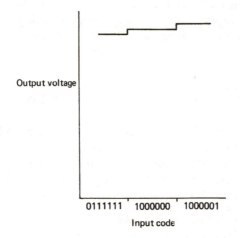

Figure 6.7   The input–output behavior of
an ideal D/A converter.

diagram prevents electrical loading—i.e. it provides appropriate impedance conversion.

The settling time of a D/A converter is determined largely by the characteristics of the buffering amplifier. In some cases, the amplifier is omitted and then the settling time depends on the characteristics of the output circuit.

The output of a D/A converter may contain unwanted transients, sometimes called glitches, due to imperfectly matched switches. Imagine that the digital word 0111111 is being converted and that the word then increases by one unit to 1000000. Ideally the converter output should be as shown in Figure 6.7. However, in practice the switches that control the resistor network may not be perfectly synchronized. Suppose that the six switches that represent the six least significant bits open before the switch that represents the seventh digit has closed. The voltage from the converter then contains a major glitch as shown in Figure 6.8. Notice that major glitches occur only when there are major changes in the binary code—the change from 1000000 to 1000001, for instance,

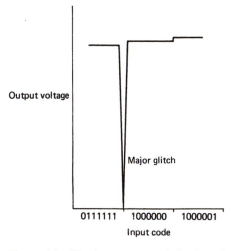

Figure 6.8 The input–output behavior of a non-ideal D/A converter showing the occurrence of glitches.

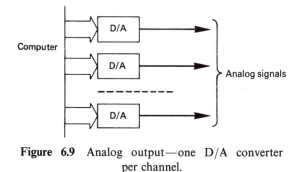

Figure 6.9 Analog output—one D/A converter per channel.

does not generate a glitch. The simplest way to remove glitches is probably to follow the D/A converter by a sample-and-hold device.

6.4.3 Analog output arrangements

Figure 6.9 and 6.10 show two alternative configurations by which multiple analog outputs may be produced. The first alternative (Figure 6.9), in which each channel has its own D/A converter, is faster to respond and less prone to drift than the system of Figure 6.10.

When a group of analog outputs needs to be located some distance, perhaps several kilometers, from the control computer, the configuration of Figure 6.11

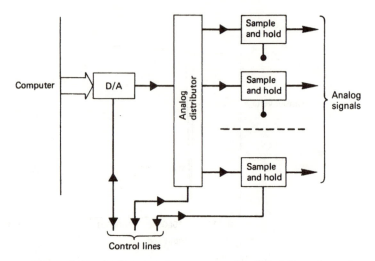

Figure 6.10   Analog output—one sample and hold per channel.

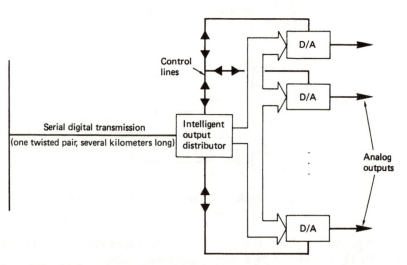

Figure 6.11   Analog output using serial digital transmission to a group of actuators.

may offer a cost-effective solution. The system depends on an auto serial/parallel distributor that can drive a number of D/A converters.

6.4.4 Analog-to-digital conversion (A/D conversion)

An ideal A/D converter would take in a continuous signal $f(t)$ and give out the corresponding signal $f^*(t)$. This section considers practical (non-ideal) A/D conversion.

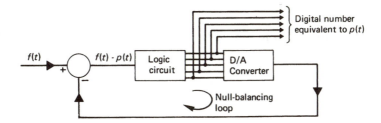

Figure 6.12   A null-balancing form of A/D converter.

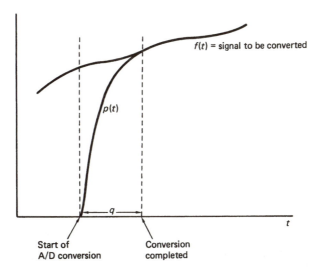

Figure 6.13   The time characteristics of the A/D converter.

A number of alternative electronic techniques exist for A/D conversion. All the techniques have certain features in common, and here we base our presentation around what can be considered as a null-balancing technique.

Assume that a signal $f(t)$ is to be A/D converted. A signal $p(t)$, produced by a D/A converter, is subtracted from $f(t)$ and a counter varies the digital number in the D/A converter until the inequality $|f(t) - p(t)| < \delta$ is satisfied, where δ depends on the number of bits in the digital word—see Figures 6.12 and 6.13. At this point, conversion is complete and the digital number can be read off.

Even with these brief and rather simplistic descriptions of A/D conversion, it becomes obvious that a high accuracy requirement will be in direct conflict with a high speed requirement.

A varying analog signal is applied to an A/D converter. The D/A converter
with the A/D converter can change at a maximum rate of one least significant
bit per clock pulse. This sets a limit on the rate of change of the analog voltage
that can be tracked by the A/D converter. This limit is called the maximum
slewing rate.

Let

v_{max} = the maximum range of the analog input voltage,

f_c = the clock frequency,

n = the number of bits used in the converter.

Then

$$\text{maximum slewing rate} = \frac{v_{max} f_c}{2^n}. \tag{6.1}$$

Assume that a sinusoidal voltage $a \sin \omega t$ is to be converted. Then

$$\frac{d(a \sin \omega t)}{dt} = a\omega \cos \omega t$$

and the maximum rate of change

$$\left.\frac{d(a \sin \omega t)}{dt}\right|_{max\,t} = a\omega. \tag{6.2}$$

The maximum frequency, ω_{max}, that can be allowed in the analog signal without
exceeding the slewing rate of the converter is then given by equating (6.1) and
(6.2) to yield

$$a\omega_{max} = \frac{V_{max} f_c}{2^n}.$$

But $a = V_{max}/2$, hence

$$\omega_{max} = \frac{f_c}{2^{n-1}}. \tag{6.3}$$

6.4.5 Sample-and-hold devices

An ideal sample and (zero-order) hold device takes in an analog signal $f(t)$
and outputs a signal $g(t)$ that is equal to $f(t)$ at each sampling instant and is
constant between sampling instants (Figure 6.14).

Figure 6.15 shows the primitive hardware outline of an actual device. The
sampling switch closes for a finite time q, to allow the capacitor to charge.

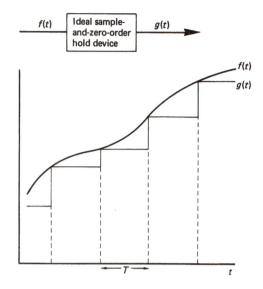

Figure 6.14   The time behavior of an ideal sample-and-hold device.

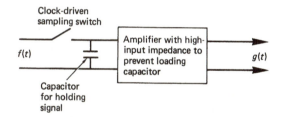

Figure 6.15   Outline of the operation of a sample-and-hold device.

When the sampling switch opens, a transient occurs in the signal $g(t)$. Finally, despite the presence of the amplifier, the signal $g(t)$ does not remain perfectly constant between sampling instants. Figure 6.16 illustrates these effects.

For rapidly changing signals or for rapid sampling (small values of T), the value of q and the length of the switching transient may be critical.

For slow sampling (long values of T), the non-ideal holding due to capacitor discharge is important.

6.4.6 Preceding an A/D converter by a sample-and-hold device

Where a rapidly changing signal is to be A/D converted, an improvement may often be obtained by inserting a sample-and-hold device before the A/D

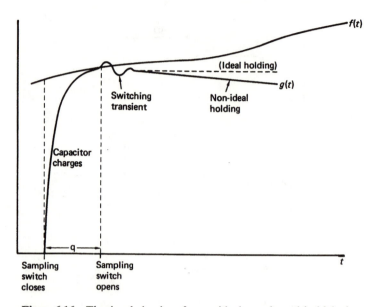

Figure 6.16   The time behavior of a non-ideal sample-and-hold device.

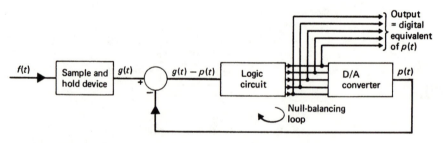

Figure 6.17   An A/D converter preceded by a sample-and-hold device.

converter. Figures 6.17 and 6.18 show the configuration and the time response respectively.

6.4.7 Input multiplexing

A multiplexer is a device for scanning across a number of analog signals and time-sharing them sequentially into a single analog output channel.

The switching is usually performed by JFET or CMOS transistors although mechanical reed relays may still be preferred for some applications.

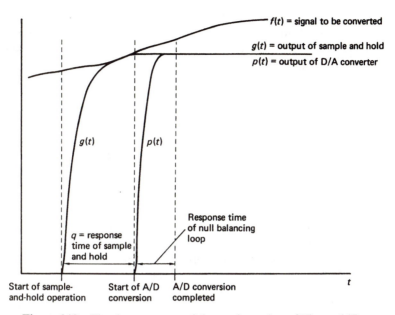

Figure 6.18 The time response of the configuration of Figure 6.17.

The speed of a multiplexer depends on:

(a) the speed of the switch (typical switching times are:
  JFET transistor 10^{-7} s
  Reed relay 10^{-3} s);
(b) the setting time of the circuit fed by the multiplexer.

Let the time constant of the circuit be τ seconds and let the following A/D converter have n bits; then a time T_s must be allowed to elapse before the multiplexer output is A/D converted, where

$$T_s \geqslant \tau \ln(2^n - 1). \tag{6.4}$$

Proof n-bit working gives an accuracy of 1 part in $2^n - 1$.

t seconds after the application of a unit step, the multiplexer circuit has the voltage

$$1 - e^{-t/\tau}.$$

We require that

$$\frac{-t}{\tau} < -\ln(2^n - 1),$$

and the inequality follows.

Explanatory comment   Each time the multiplexer switches, a transient occurs in the signal that is passed on to be A/D converted. Satisfaction of the given inequality guarantees that the transient has died away to a magnitude that cannot cause an error in the digital conversion, even in the worst case, in which the multiplexer switching is between signals at opposite ends of the conversion range.

A typical value for τ might be $\tau = 10^{-6}$ s. This leads to a necessary waiting time of at least

$$5.5 \times 10^{-6}\,\text{s} \quad \text{(8-bit working)}$$

or

$$11 + 10^{-6}\,\text{s} \quad \text{(16-bit working)}.$$

To these times must be added, according to the circumstances, A/D conversion times, computation times, and so on.

When deciding whether the above times are suitable for a particular application, it is necessary to take into account:

(a) The number of signals to be multiplexed, and
(b) the importance of simultaneity. (In some applications it is vital that, so far as possible, the signals are all scanned at the same instant of time—i.e. with minimum time skew.)

(Often a single A/D converter is used in conjunction with a multiplexing switch to scan cyclically a number of analog signals (Figure 6.19). It will often be the intention to scan all the signals at one instant of time, so far as that is possible, and to repeat that process every T seconds, where T is an appropriate sampling interval.)

Notice two points:

(a) If the response time of the A/D converter is significant, compared with T,

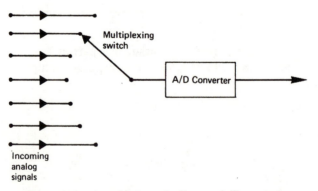

Figure 6.19   A multiplexer feeding an A/D converter.

the scan is skewed. The effect may lead to complications in the interpretation of the data.

(b) The analog signals received by the multiplexer have, in general, quite different values. A worst-case approach must be used so that the converter always has time to settle. (When only a single analog signal is being converted, the maximum rate of change of the signal determines the length of time that needs to be allowed for conversion.)

Simple, order-of-magnitude calculations often suffice to determine whether a particular A/D converter can be used for a particular duty.

Example m analog signals $v_1, \ldots, v_m$ with overall upper frequency limit ω_{max} are to be scanned every T seconds by a multiplexer-fed A/D converter that produces an n-bit signal. Determine approximate relations between m, ω_{max}, T and n for accurate conversion.

Denote by v_{max} the largest signal that can be encoded by the converter. Denote by $(\Delta v_i)_{max}$ the maximum change in the signal v_i over the time period of duration q, where q is as defined in Figure 6.13.

Each individual signal v must not vary significantly over the time period q if it is to be sampled successfully by the A/D converter. This leads to the inequality

$$\frac{(\Delta v_i)_{max}}{v_{max}} < 2^{-n}, \tag{6.5}$$

which must be satisfied by each of the signals.

Assuming sinusoidal signals of maximum frequency ω_{max},

$$(\Delta v_i)_{max} \leqslant q\omega_{max} \frac{v_{max}}{2} \cos \omega_{max} t \bigg|_{max_t} = \frac{q\omega_{max}}{2} v_{max}. \tag{6.6}$$

From (6.5) and (6.6) we obtain the inequality

$$\omega_{max} q < 2^{1-n}, \tag{6.7}$$

which individual signals must satisfy. (Figures 6.20 and 6.21 illustrate the form of this relation.)

Reference C6, Section 8.2 expands this topic.

Assume that the total conversion time for one signal is Γ; then, in order to satisfy the requirement for simultaneous scanning, it is necessary that

$$m\Gamma \ll T. \tag{6.8}$$

In a typical case we might require that

$$m\Gamma \leqslant T/100. \tag{6.9}$$

Finally, we point out that data arriving at a computer through the A/D converter must be transferred by the computer, either to store or to be the subject of

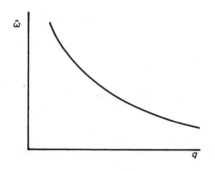

Figure 6.20 The relation between the time period q and the maximum frequency $\hat{\omega}$.

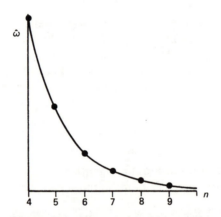

Figure 6.21 The relation between the number of bits in an A/D converter and the maximum frequency $\hat{\omega}$ of the signal to be converted.

further computation. The computer must have sufficient speed in this operation to at least match the speed of incoming data. Sometimes, short but very rapid bursts of data can be transferred to RAM to be dealt with later, once the data stream has ended. In a large scheme, the matching of multiplexing rates, A/D acquisition and conversion times, sampling periods T and computer 'front-end' speeds will repay careful attention.

6.4.8 Choice of multiplexer according to application

The choice of a multiplexer for a particular application involves the familiar compromise between speed and accuracy—if both high speed and high accuracy

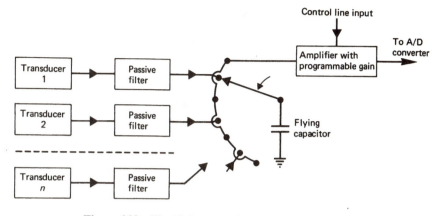

Figure 6.22 The 'flying capacitor' multiplexing circuit.

are needed, the required device will inevitably be relatively expensive. Here we describe three alternative approaches: The flying capacitor method that has been found adequate for many industrial applications; the analog multiplexer that is required for the most exacting applications; so-called digital multiplexing, which is not multiplexing in the sense usually understood. The choice between the last two approaches can only be made by preparing comparative cost and performance budgets for the envisaged application.

The 'flying-capacitor' input circuit (Figure 6.22) provides a speed and accuracy suitable for many industrial applications. The transducer signals are first filtered by passive filters before combined multiplexing and isolation by the flying capacitor device. This, as is shown diagramatically in the figure, charges its capacitor during a connection period and then switches to make the capacitor voltage available at the input of the programmable amplifier. The gain of the amplifier changes in synchronism with the switching of the capacitor to provide appropriate gains for each channel.

For high speed, high accuracy requirements, the system represented in Figure 6.23 would typically be recommended. Each channel uses high-quality dedicated amplifiers and filters before time sharing by an analog multiplexer followed by A/D conversion.

An alternative arrangement (Figure 6.24) is to use one A/D converter per channel and to address the set of converters sequentially by program. Such an arrangement might reasonably be referred to as digital multiplexing.

6.4.9 Multiplexer timing

Figure 6.25 shows a configuration consisting of an analog multiplexer, sample-and-hold device, A/D converter and a PIA chip leading into a microcomputer

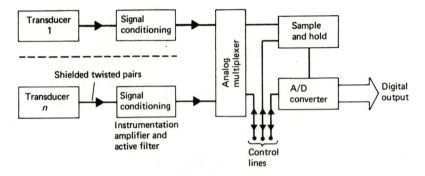

Figure 6.23 A multiplexer and associated circuits for high-accuracy/high-speed applications.

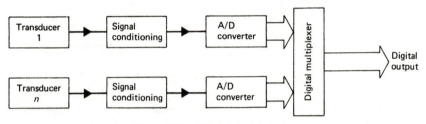

Figure 6.24 Digital multiplexing.

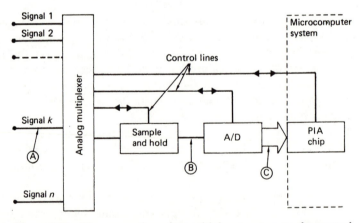

Figure 6.25 The connection of a multiplexer system to the control lines of a microcomputer.

system. Figure 6.26 is an associated timing diagram showing how the analog signal (A) is transmitted through the system via the intermediate signal (B) to become the digital signal (C).

The table within Figure 6.26 lists some typical times, to allow an order of

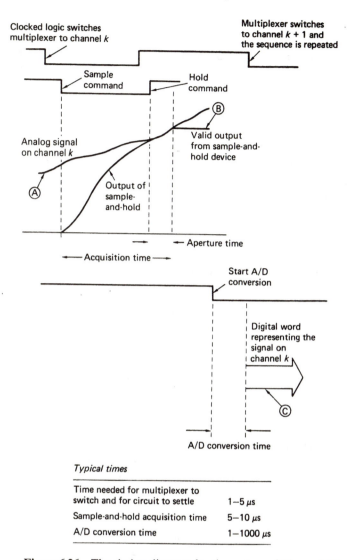

Figure 6.26 The timing diagram for the system of Figure 6.25.

magnitude feeling to be obtained. The times depend on device cost, accuracy required and, in the case of settling times, the capacitance of the circuit into which the device feeds.

The logic to control the data transfer is most easily derived from a programmable interface chip.

An alternative is to use a memory-mapped hybrid input system that contains its own analog instrumentation amplifier and can be connected directly to the address and data bus of a microprocessor.

6.4.10 Typical specifications of data-conversion devices

In order to allow an appreciation of the relative speeds of the most common interfacing devices, Tables 6.1–6.3 show response data for a range of available equipment.

In general there is a trade-off between accuracy and speed, with costs rising steeply when both high accuracy and high speed are required simultaneously.

A/D converters   Cost depends on word length, speed and accuracy. For a 12-bit converter, Figure 6.27 illustrates the approximate relation between speed and cost.

Table 6.1   A/D converters—typical conversion times

Type of converter	8 bits	10 bits	12 bits
Low-cost	10 μs	40 μs	60 μs
General-purpose	5 μs	20 μs	20 μs
Fast	1 μs	2 μs	5 μs
Ultra-fast	0.01 μs	0.1 μs	1 μs

Table 6.2   D/A converters—typical settling times

Type of converter	8 bits	10 bits	12 bits
General-purpose	1 μs	2 μs	4 μs
Fast	0.03 μs	0.25 μs	1 μs
Ultra-fast	0.01 μs	0.02 μs	0.04 μs

Table 6.3   Sample-and-hold devices—typical devices

Type of device	Accuracy	Acquisition time
Low-cost	0.1%	2 μs
	0.01%	5 μs
General-purpose	0.1%	5 ns
Ultra-fast	0.01%	0.1 ns

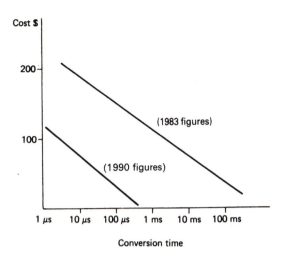

Figure 6.27 Order of magnitude cost for an A/D converter (12 bits) as a function of conversion time (1990 figures with 1983 figures shown for comparison).

6.5 Digital interfacing

6.5.1 Introduction

Digital interfacing is concerned with the technology of digital data transfer between devices. Clearly it is a very large topic indeed and in this section we limit the presentation to aspects in which control engineers need to take a detailed interest.

The chief data transfers that are needed in a control application are:

(a) transfer of process measurement data from an A/D converter to a microprocessor system;
(b) transfer of actuator commands from a microprocessor system to a D/A converter;
(c) transfer of data between a microprocessor system and peripheral devices, such as keyboards, visual display units, tape drives, printers, graph plotters and other computing devices such as larger supervisory computers.

The data transfers are usually achieved by sequential (serial) transfer when the distances are large and by parallel transfer when devices are close together and rapid transfer is required. In the list above (a) and (b) are usually achieved by parallel transfer, serial transmission being reserved for special cases involving relatively long distances. The peripherals in (c) are serviced by a mixture of serial and parallel transfers.

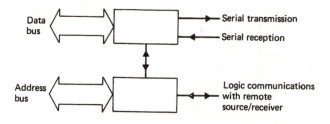

Figure 6.28 Outline of an asynchronous communications
interface adapter (ACIA).

6.5.2 Serial interfacing

Since a microcomputer configuration operates internally by parallel data
transfer, it is necessary to use a serial-to-parallel device to interface a serial line
to the system. Such a serial-to-parallel converter may consist of a register that
is filled, one bit at a time, at a rate dictated by the system clock, by incoming
serial data. When the register is full, it is connected to the system data bus.

Serial data transfer is facilitated by the use of special serial/parallel chips
that contain the logic necessary for organizing the operation. A common device
is the ACIA (Asynchronous Communications Interface Adaptor), sometimes
referred to as UART (Universal Asynchronous Receive and Transmit) device.
A typical ACIA has two serial connections for input and output respectively.
It has logic connections with the remote data source/sender and with the address
bus of the computer and it connects via a buffer register to the computer data
bus (Figure 6.28).

6.5.3 Parallel interfacing

Most of the parallel interfacing required in simple control applications is
achieved through the use of PIA (Peripheral Interface Adaptor) chips. A PIA
is programmable in so far as manipulation of particular bits in its control
register alters the operating configuration.

A PIA is the natural interfacing device to interpose between an A/D or
D/A converter and a microprocessor system. A typical configuration is shown
in Figure 6.29. Much of the interfacing effort expended by control engineers is
devoted to the proper connection of PIA devices which, interposed as described,
are directly in the control loop.

Because of the importance of the PIA chip, we describe below in outline
how the connection to an A/D converter is realized. Then, in Appendix A, we
describe in detail the operation of one particular PIA chip. The incursion into
computer technology represented by Appendix A will, it is hoped, help readers
to strengthen links with their existing knowledge of digital hardware, and will

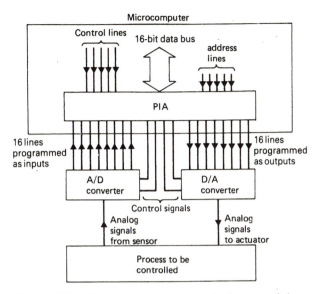

Figure 6.29 A microcomputer in a basic control loop showing the function of a PIA chip.

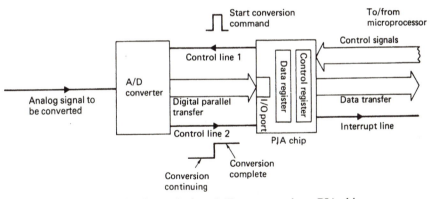

Figure 6.30 Control of an A/D converter by a PIA chip.

give some indication of the type of work that is involved for the control engineer when he undertakes the rather ill-defined activity called interfacing.

CONNECTING AN A/D CONVERTER TO THE PIA CHIP
Figure 6.30 shows a PIA chip with control register CR and data register DR. Its connection to the microprocessor is through the address bus, the data bus and an interrupt line. Its connection to the A/D converter is through a parallel port and two control lines.

A command from the microprocessor along the address bus fixes the configuration of the PIA so that, in particular, the parallel port is designated as an input port.

The operation of the arrangement is then as follows:

(a) The PIA sends a 'start-conversion' signal on control line 1 to the A/D converter.
(b) When conversion is complete, the A/D converter sends a 'conversion completed' signal along control line 2 to the PIA.
(c) The PIA reads the data from the converter into its data register, sets a particular bit in its control register and sends an interrupt signal to the microprocessor.
(d) The microprocessor checks bits in the control registers of all PIAs to see which has raised the interrupt. It then transfers the data from the PIA and clears the bit in the control register.
(e) When the time is reached for a new input signal to be obtained the microprocessor initiates the procedure again.

(Note that a D/A converter can operate autonomously without any control signals. At all times, it simply converts the digital word that it receives. The PIA chip is required to keep constant, between updating instants, the digital word that is to be D/A converted.)

6.6 Microprocessors for control applications

We are concerned in this chapter with relatively simple problems that can be solved by the interconnection of a few computing devices and peripherals. Larger problems that need a network of interconnected computers for their solution are considered in Chapter 10.

Cost, speed and memory size are perhaps the most obvious parameters with which to characterize a computer. According to the application, other parameters need to be considered, such as:

Availability of particular high-level languages.

I/O capability.

Ease and cost of interfacing.

Interrupt facilities.

Availability of software support, cross-assemblers or development systems.

Possibilities for expansion to give, for example, CPU–CPU interconnection for concurrent programming.

The profile of a system to be purchased should, according to one point of view, match the profile of the task to which it is to be applied. An alternative viewpoint is to set great store by modularity and standardization and to seek a standard solution to match a range of problems. The best solution would be a combination of tailoring for individual need using modular building blocks to achieve a degree of standardization. The task of the specifier of systems is made difficult by the rapid rate of change in available equipment—a piece of equipment chosen as a standard building block is likely to be rendered rapidly obsolescent by new products.

Exercises

6.1 Use the literature to survey the availability of measurement devices for two of the following:

   force (include load cells and strain gauges)

   torque

   pressure

   linear and angular position

   angular velocity

   liquid level.

Give consideration to the type of signals produced by the sensors and the need for input circuitry between the sensors and a microcomputer.

6.2 A knowledge of some of the practical aspects of measurement and actuation for common control loops should be obtained by gathering information from the literature on either:

   temperature control—to include consideration of:
      types of measurement devices, their principle of operation, their physical form, range of applicability and need for special input circuitry;
      type of actuators for regulation of the flow of oil, gas, coal or electricity to the heating process;
      the inter-relation of temperature control with combustion control and furnace pressure control; or

   flow control—to include consideration of:
      types of measurement device, their principle of operation, their physical form, their range of applicability;
      noise problems caused by pulsating flow;
      the question of whether to use a subsidiary position control loop for the flow actuator.

Summarize your findings concisely in a brief report.

6.3 Outline the main configurations used for connecting groups of sensors or actuators with a microcomputer system.

6.4 Sketch a block diagram for a configuration for the control of a process by microcomputer, allowing for:

6 analog low-level signals that need to be sampled simultaneously;

1 voltage-free contact closure input signal;

1 shaft encoder digital input signal;

4 analog-driven actuators.

Show the necessary data converters and their outline connection to ACIA, PIA, etc., peripheral chips. Indicate how the timing requirements will be achieved.

If you are familiar with a particular microprocessor family, specify the actual devices that are to be interconnected, in terms of that family.

6.5 Confirm the inequality

$$\omega_{max} q < 2^{1-n}$$

of Section 6.4.7, relating:

maximum frequency of input signal (ω_{max});

number of bits working (n);

A/D conversion time, as defined in Figure 6.13 (q);

for a multiplexer-fed A/D converter.

6.6 Produce sketch graphs showing how the signal-to-noise ratio of a quantized sinusoid varies as a function of:

(a) word length of the quantizer;
(b) $\hat{v}/a$:

where $\hat{v}$ is the maximum permissible voltage to the quantizer and a is the actual amplitude of the sinusoid.

6.7 State the two common digital versions ((1) absolute and (2) incremental) of the PID algorithm. Comment briefly on the relative advantages and disadvantages of the two versions.

An analog ramp signal $u = t/1000$ is input to an analog–digital converter sampling 1000 times per second. The converter represents a signal of 2 volts by a 7-bit number. The converted signal is input to the digital PID algorithm (version (1)) with a derivative time of 4 s. Show that the derivative term will output a sequence of pulses. Find their amplitude and sketch the waveform.

From the viewpoint of process control, does this sequence of pulses have the same effect as the analog derivative it is supposed to approximate? Explain.

Assuming the existence of high-amplitude pulses to be a disadvantage, can their amplitude be reduced by shortening the sampling interval? Explain.

Sketch the waveform of the output from the derivative term of the incremental version of the algorithm.

6.8 Sketch a block diagram showing the principle of operation of an A/D converter—include interconnection with a microcomputer in your diagram.

Under what conditions is it advantageous to precede an A/D converter by a sample and hold?

6.9 Explain the meaning of the terms:

serial data transfer

parallel data transfer

asynchronous operation

handshake

interrupt signal

direct memory access

glitch

tristate.

6.10 What are the principal parameters, from a control systems viewpoint, that characterize:

(a) a D/A converter
(b) an A/D converter
(c) a sample and hold
(d) a multiplexer?

6.11 Explain the procedures by which real-time programs may be produced, tested and implemented in a personal computer. In what departments is a PC solution weak when a real-time industrial system is to be designed and implemented.

6.12 (Mini-project)

Report on the relative merits of personal computers and VMEbus systems as real-time industrial controllers. For each approach consider: program development, robustness in an industrial location, (= industrial credibility), expandability, networking capability.

6.13 Figure 6.31 shows the annotated front panel of a small industrial controller made by the Jumo Company.

Using only the controller specification given in the figure, invent and sketch two widely differing applications that would use the full power of this controller.

Discuss the advantages and limitations of using such an off-the-shelf approach in the applications you envisaged.

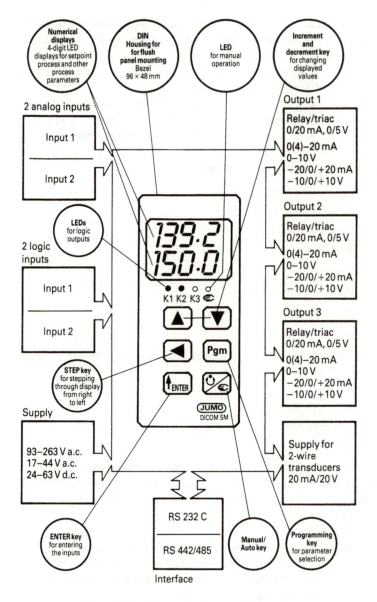

Figure 6.31 Universal Compact Controller.

Tutorial Case Histories

7.1 Introduction

The aim of this chapter is to demonstrate the application of some of the techniques of the earlier chapters in simple situations.

Case history A deliberately uses low-cost ready-made plug-compatible devices programmed in high-level language to achieve temperature control to emphasize the ease with which a digital control loop can be built up.

Case history B puts forward a modular 16-bit solution to one of the most important problems occurring in a metal rolling mill.

7.2 Outline of problem A: Temperature control achieved by a low-cost ready-made configuration, programmed in the RAPID BASIC language

Figure 7.1 shows an electrical oven provided with temperature measurement by a thermocouple and having a remotely controlled, continuously variable power input. The task is to implement a microcomputer, in the position shown by the dotted box, to provide temperature control of the oven.

The functions within the control loop can be broken down as follows:

(a) Sampling of the temperature measurement signal at an appropriate rate.
(b) Transfer of the measurement signal into the computer, followed by conversion and validation.
(c) Comparison of the measured temperature with a stored desired temperature to form an error signal.
(d) Operation on the error signal by an appropriate algorithm to form an output signal.

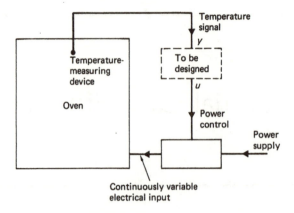

Figure 7.1   Outline of the oven-control problem.

(e) Adjustment of the level of the output signal and transfer through the interface to the power control unit. The power to the oven will be held constant over each sampling interval.

7.3 Computation and interfacing

The computer used in this application is a type 286 IBM compatible personal computer fitted with easily purchasable I/O devices that plug in with no modification or special software requirement. The program, in the RAPID BASIC language, controls the interface through 'call' functions—see below.

A very wide range of interface boards is available for the input/output task. It is recommended to choose one that comes complete with a good utility software package that contains a driver program. This allows all interface functions to be accessed by simple 'call' functions.

The thermocouple produces a small voltage that needs to be amplified before being input to the A/D converter. Thermocouple break protection and cold junction compensation (Section 6.2.1) must also be provided as part of the analog input circuitry. Many interface cards provide this along with other I/O capability.

Figure 7.2 is annotated to give full details of the devices in the control loop.

The power to the oven is controlled through a silicon-controlled rectifier whose input is an analog signal from the D/A converter.

7.4 Control of data transfer

Every T seconds, an instruction within the program of the PC initiates an A/D conversion that takes up to 200 μs to complete. The program allows time for the conversion, making allowance for the worst case, and then reads the A/D

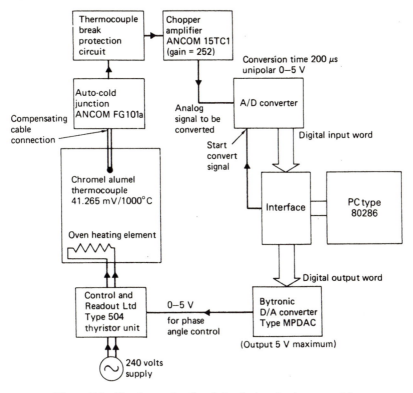

Figure 7.2   Hardware details of the devices in the control loop.

output. The control algorithm then operates on the temperature, as represented by the input word and sends out the result to the register of the D/A convertor. The control action is held constant at a value corresponding to the word in the D/A register until the next update occurs.

7.5 Identification of the oven

Figures 7.3–7.5 show the response of the oven temperature to step changes in the input voltage to the element.

Because the cooling of the oven depends on the degree of insulation whereas the heating depends on the power and location of the heating element, the response curves for cooling and heating differ significantly.

Further, the presence of nonlinearities means that input steps of different magnitudes applied at different starting temperatures produce somewhat different response curves. Table 7.1 summarizes the results of the open-loop

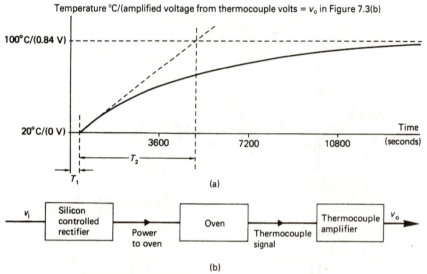

(a)

(b)

Figure 7.3   Open-loop response of the oven heating.

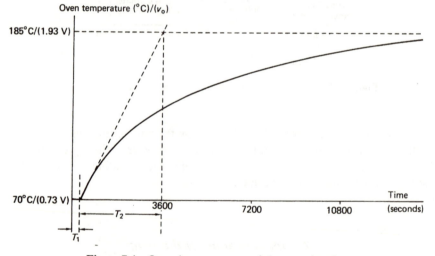

Figure 7.4   Open-loop response of the oven heating.

response tests. The curve of Figure 7.4 is chosen as a reasonable compromise among the curves obtained during experimentation.

Using the Ziegler–Nichols method (Section 5.2.5) the oven was found to have the transfer function

$$G(s) = \frac{1.63\,e^{-270s}}{1 + 3480s}.$$

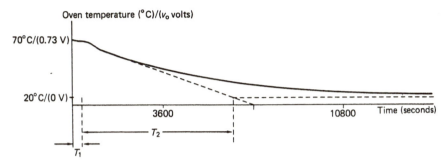

Figure 7.5 Open-loop response of the oven cooling.

Table 7.1 Summary of open-loop response tests

Initial input voltage v_i	Final input voltage v_i	Initial output voltage v_o	Final output voltage v_o	Process gain	Time constant T_2 (seconds)	Dead time T_1 (seconds)	Response curve
0	0.28	0	0.84	3	4800	300	Figure 7.3
0.15	0.89	0.73	1.93	1.63	3480	270	Figure 7.4
0.15	0	0.73	0	4.9	6600	180	Figure 7.5

7.6 Control design

7.6.1 Design of a continuous controller followed by discretization

Here we follow design route number 3 of Chapter 5. First we calculate the coefficients of a three-term (PID) continuous controller using the Ziegler–Nichols rules and discretize by the trapezoidal method to determine the discrete-time equation that will be programmed as an algorithm. To choose the sampling interval T, we calculate the appropriate closed-loop bandwidth ω_b of the continuous system and use the rule (Section 5.5.1) that sampling frequency ω_S should satisfy the inequality

$$\omega_S \geqslant 10\omega_b.$$

CONTINUOUS CONTROLLER DESIGN
The 'compromise model' of the oven has the transfer function

$$G(s) = \frac{1.63\,e^{-270s}}{1 + 3480s}, \tag{7.1}$$

for which the Ziegler–Nichols rules suggest a controller of transfer function

$$D(s) = \frac{1.2}{1.63}\frac{3480}{270}\left(1 + \frac{1}{2 \times 270s} + \frac{270s}{2}\right)$$

$$D(s) = 9.5\left(1 + \frac{0.00185}{s} + 135s\right). \tag{7.2}$$

DISCRETIZATION
For trapezoidal discretization, we set

$$s = \frac{2}{T}\left(\frac{z-1}{z+1}\right)$$

to yield an expression for the digital controller

$$D(z) = 9.5\left[1 + \frac{T(z+1)}{1081(z-1)} + \frac{270}{T}\left(\frac{z-1}{z+1}\right)\right] = \frac{u(z)}{e(z)}$$

$$(z^2 - 1)u(z) = 9.5\left[(z^2 - 1) + \frac{T}{1081}(z+1)^2 + \frac{270}{T}(z-1)^2\right]e(z)$$

$$(1 - z^{-2})u(z) = 9.5\left[(1 - z^{-2}) + \frac{T}{1081}\frac{(z+1)^2}{z^2} + \frac{270}{T}\frac{(z-1)^2}{z^2}\right]e(z)$$

$$u(z) = \frac{u(z)}{z^2} + 9.5\left[(1 - z^{-2}) + \frac{T}{1081}\frac{(z+1)^2}{z^2} + \frac{270}{T}\frac{(z-1)^2}{z^2}\right]e(z)$$

$$u(k) = u(k-2) + 9.5\left[(1 - z^{-2}) + \frac{T}{1081}(1 + 2z^{-1} + z^{-2})\right.$$
$$\left. + \frac{270}{T}(1 - 2z^{-1} + z^{-2})\right]e(z)$$

$$u(k) = u(k-2) + 9.5\left[\left(1 + \frac{T}{1081} + \frac{270}{T}\right)e(k) + \left(\frac{2T}{1081} - \frac{540}{T}\right)e(k-1)\right.$$
$$\left. + \left(-1 + \frac{T}{1081} + \frac{270}{T}\right)e(k-2)\right]. \tag{7.3}$$

7.6.2 Addition of a term to compensate for oven heat losses

At this point we refer to the physics of the problem. Suppose that the desired temperature of the oven is set at some temperature θ_d and that the actual temperature θ of the oven satisfies $\theta = \theta_d$ and that, under this condition, we put the algorithm equation (7.3), into operation with $u(0) = 0$. Then for zero error the algorithm produces zero output. However, to keep the temperature θ equal to θ_d, an input is needed to the oven, sufficient to offset the heat loss. Thus, we need to add a term, say α, preferably dependent on θ_d, to the right-hand side of equation (7.3), which therefore becomes

$$u(k) = u(k-2) + 9.5\left[\left(1 + \frac{T}{1081} + \frac{270}{T}\right)e(k) + \left(\frac{2T}{1081} - \frac{540}{T}\right)e(k-1)\right.$$

$$\left. + \left(-1 + \frac{T}{1081} + \frac{270}{T}\right)e(k-2) + \alpha\theta_d\right]. \tag{7.4}$$

(The term $\alpha(\theta_d)$ is equivalent to the term u_0 in equation (5.39).)

7.7 Choice of sampling interval

When we come to calculate the closed-loop bandwidth of the process with its continuous controller, we find that the ideal differentiator in equation (7.2) causes difficulties—apparently the closed-loop bandwidth is finite!

In an implementation of equation (7.2) by analog techniques, the term $135s$ would need to be approximated by a physically realizable form. For our calculation of closed-loop bandwidth we also could approximate the term and then proceed.

However, we choose an easier way forward. Knowing that the determination of sampling interval can only be very approximate, we calculate the closed-loop bandwidth of the process with $P + I$ controller and use that to specify the sampling interval.

The open-loop response of the oven together with its recommended continuous $P + I$ controller is

$$G(s)D(s) = \left(\frac{1.63\,e^{-270s}}{1 + 3480s}\right)\left(7.1 + \frac{7.1}{891s}\right),$$

and the closed-loop transfer function is

$$\frac{G(s)D(s)}{1 + G(s)D(s)} = \frac{1}{[(1 + 3480s)(891s)/(1.63\,e^{-270s})(6326s + 7.1)] + 1}.$$

The closed-loop bandwidth ω_b was determined numerically as $\omega_b = 0.0042$ rad/s. The suggested sampling frequency is then given as

$$\omega_s = 10\omega_b = 0.042 \text{ rad/s}$$

or

$$T = 2\pi/\omega_s = 150 \text{ s}. \tag{7.5}$$

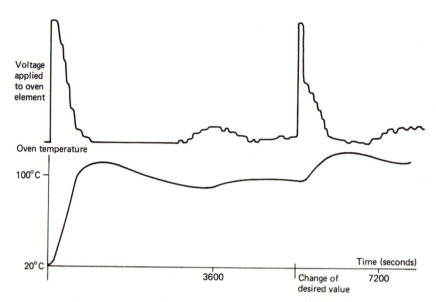

Figure 7.6 Closed-loop control of the oven using a sampling interval of 60 seconds: data from a plant trial.

Table 7.2 Variation of settling time with sampling period: result from a plant trial

Sampling period	Settling time following 20°C to 100°C step in desired temperature (seconds)	Settling time following 100°C to 120°C step in desired temperature (seconds)
10	4 500	2 400
30	5 400	3 600
60	5 400	6 300
150	5 400	7 800
180	5 400	8 500
240	13 500	System did not settle

This is a very long sampling interval compared with the recommendation of equation (5.66), so for the initial implementation we decided to use the value $T = 60$ s. Figure 7.6 shows the performance of the oven under closed-loop control with this sampling interval.

As a practical investigation of the sensitivity of the control loop to variations in T, we varied T over a wide range. Table 7.2 shows how the settling time was affected. Figure 7.7 shows the time–temperature curve for $T = 240$ s. It is clear from these results that the sampling rate may be varied over quite a

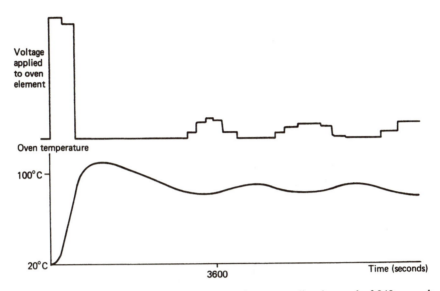

Figure 7.7   Closed-loop control of the oven using a sampling interval of 240 seconds: data from a plant trial.

large range in this application without significantly affecting the closed-loop performance.

7.8 Input quantization

The A/D converter uses 8 bits to represent 5 V. In this application, this corresponds to a resolution of 0.55°C, i.e. the oven temperature can change by a maximum of 0.55°C without changing the digital word in the A/D converter. The calculation is as follows:

> The A/D converter represents a 5 V signal by 8 bits. Two amplifiers of gains 252 and 3.4 are interposed between the thermocouple and the A/D converter.
>
> Voltage to A/D converter = 0.0354 V/°C, volts per least significant bit
>
> (LSB) at the A/D converter = 5/256 = 0.0195 V/LSB.
>
> Therefore, °C/LSB = 0.0195/0.0354 = 0.55°C/LSB.

Note that if the desired temperature is fixed then input quantization leads directly to an equivalent output quantization. The closed-loop system cannot in general find a steady state and must necessarily oscillate between two temperatures above and below the steady temperature that would have obtained in the continuous case.

7.9 Alternative approach: design of a temperature controller to achieve a particular closed-loop behavior

As described in Chapter 5, a number of alternative methods are available for the design of controllers and some of these are suggested as exercises at the end of this chapter.

Here we illustrate briefly the design of a finite settling time controller. In general, as will be found from the exercises, the choice of a short sampling interval when designing a digital controller leads to a high-order controller which is easily numerically ill-conditioned.

We therefore chose in this part of the project a long sampling interval, $T = 300$ s, and a desired closed-loop transfer function $H(z)$ to give finite-time settling.

Recall that the oven has the transfer function

$$G(s) = \frac{1.63\,e^{-270s}}{(1 + 3480s)}.$$

Now, bearing in mind the variation of model parameters found during identification of the oven, the controller can only be approximate and it can be designed around the approximated model

$$G(s) = \frac{1.63\,e^{-300s}}{1 + 3480s}.$$

Allowing for the zero-order hold that precedes the oven and, as usual, denoting the transfer function of the combination of zero-order hold and oven by $G'(s)$,

$$G'(s) = \left(\frac{1 - e^{-sT}}{s}\right) \frac{1.63\,e^{-300s}}{(1 + 3480s)},$$

$$G'(z) = \frac{1.63z^{-1}(1 - e^{-300/3480})}{(z - e^{-300/3480})} = \frac{z^{-1}(0.135)}{z - 0.917}. \tag{7.6}$$

The controller is found from the equation

$$D(z) = \frac{H(z)}{G'(z)(1 - H(z))}. \tag{7.7}$$

We choose $H(z) = z^{-2}$ to ensure that the controller will be physically realizable:

$$D(z) = \frac{z^2}{[z^{-1}(0.135)/(z - 0.917)](1 - z^{-2})}$$

$$= \frac{z^2 - 0.917z}{0.135(z^2 - 1)},$$

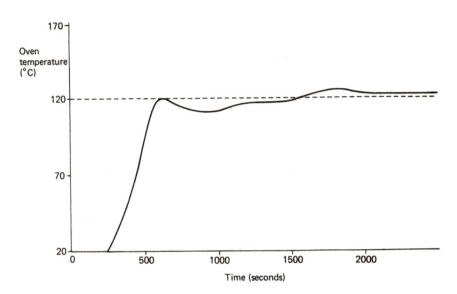

Figure 7.8 Simulated performance of the oven using the algorithm (7.8) with sampling interval $T = 300$ seconds.

leading to the difference equation

$$u(k) = u(k-2) + 7.4e(k) - 6.7e(k-1). \tag{7.8}$$

Figure 7.8 shows the simulated closed-loop performance of the oven with this control algorithm. The performance can be seen to be rather poor—investigation will show that the performance cannot be improved by shortening the sampling interval. The reason is that the synthesis takes the dead-time in the model literally and updates its output infrequently. In this application, the dead-time in the process model is fictitious and a better model for use in digital design would be (say) a third-order transfer function, fitted to the response curves of Figures 7.3–7.5.

7.10 Comments on the temperature/control application

The case history leads to some conclusions of general interest:

(a) The approach to control design through Ziegler–Nichols identification and controller specification followed by discretization led to a controller that worked very well in practice.
(b) The sampling interval calculated on the basis of closed-loop bandwidth proved to be a good choice in practice. In this application, the performance proved to be rather insensitive to changes in sampling interval.

(c) Limitations were encountered in the z domain design approach, mainly caused by the fictitious dead-time in the Ziegler–Nichols model. At the end of this chapter a mini-project is put forward to investigate further the design of a controller in the z domain.

7.11 Outline of problem B: Design of a thickness and flatness control system for metal rolling

Steel or aluminum strip is required to be of a specified thickness and to be flat. We discuss the design of a microprocessor-based controller for this task.

7.12 Thickness measurement and control

In steel or aluminum flat rolling, material of thickness H is entered into a pair of rollers having initial unloaded gap q (Figure 7.9).

Entry generates a force F, so that the outgoing strip has thickness

$$h = q + F/M, \qquad (7.9)$$

where M is a stiffness coefficient (see Figure 7.10).

For thickness control, q and f are measured by synchro and load cell respectively and equation (7.9) is used to generate an estimate of h, which cannot be otherwise measured without a significant time delay. (A measurement of h by nucleonic gauge is made downstream and is used to 'calibrate' equation (7.9).)

The thickness control actuator is an electric motor that moves the upper roll vertically to modify q and hence to modify h.

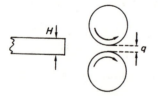

Figure 7.9 Steel strip of thickness H about to enter a rool gap of dimension q.

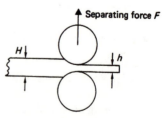

Figure 7.10 Steel strip during rolling.

7.13 Flatness measurement

Poor strip flatness is caused by non-uniform elongation across the strip width and flatness is monitored by measuring and comparing tensions at points across the strip width. (Points on the strip width where the elongation has been greatest have lowest tension—see Figure 7.11.)

In this application we measure tension at five points, sited symmetrically across the strip width. Let the five tension measurements be denoted s_{-2}, s_{-1}, s_0, s_1, s_2 as in Figure 7.12. We fit a polynomial of the form

$$s(\omega) = a_0 + a_1\omega + a_2\omega^2$$

through these five points, using a least-squares criterion.

The parameters a_1, a_2 are the measures of flatness with $a_1 = a_2 = 0$ implying perfect flatness. The parameter a_0 is not of interest since it is merely a measure of the mean tension applied during rolling.

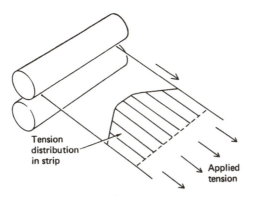

Figure 7.11   Non-uniform elongation. Tension is lowest where elongation is greatest.

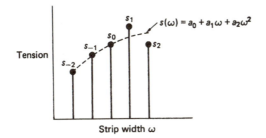

Figure 7.12   Tensions measured at points across the strip width. $s(\omega)$ is a polynomial drawn as a best fit to the five points.

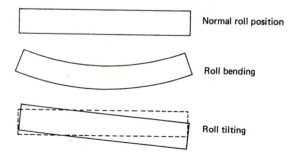

Figure 7.13 Flatness actuators operate through roll
bending and roll tilting as shown.

For our digital system, we need values of $a_1(kT)$, $a_2(kT)$ for input to the control system and ideally we need the five values s_{-2}, s_{-1}, s_0, s_1, s_2 all simultaneously at time (kT), to allow $a_1(kT)$, $a_2(kT)$ to be determined.

7.14 Flatness control

Roll bending through fast-acting hydraulic cylinders achieves largely parabolic modification to the roll gap and largely operates on the measured flatness parameter a_2. Roll tilting through electric actuators largely operates on the measured flatness parameter a_1 (Figure 7.13).

7.15 Control-system specification

The system is to control h, a_1, a_2 in the strip to remain equal to operator-specified desired values h_d, a_{1d}, a_{2d}. An accuracy of 0.1% is to be aimed for. Closed-loop system bandwidths are to be 6 rad/s or higher. The strip outgoing velocity is 5 m/s.

7.16 Choice of sampling interval T for the control loops

Using the criterion that sampling frequency ω_s should satisfy $\omega_s \geqslant 10\omega_b$ leads to the requirement $2\pi/T \geqslant 60$ or $T \leqslant 2\pi/60$. We choose $T = 0.1$ s.

7.17 Choice of input-signal scanning arrangement

Before a control action can be decided upon, the system must be in possession of up-to-date measurements of the parameters h, a_1, a_2. These parameters are calculated from the seven input signals, q, F, s_{-2}, s_{-1}, s_0, s_1, s_2, which must therefore be scanned at a much faster rate than every 0.1 s.

Let us first consider the five signals representing strip tensions. Ideally they should be sampled simultaneously, for otherwise they will represent a diagonal scan across the moving strip. At least, we must ensure that the a_1, a_2 parameters cannot change significantly while the s_i signals are being scanned. Let T_s be the time taken to scan the 5 tension signals; then we require that the a_1, a_2 parameters do not change significantly over that period.

From the closed-loop bandwidth requirement,

$$\left.\frac{da_i}{dt}\right|_{\text{max}\,t} = 6 \text{ units/s}, \, i = 1, 2$$

(where $a_{i\,\text{max}}$ has been normalized to unity). This follows since

$$\left.\frac{d \sin \omega t}{dt}\right|_{\text{max}\,i} = \omega \cos \omega t|_{\text{max}\,t} = \omega.$$

Let Δa_i be the change in parameter a_i while the 5 tension meters are scanned—then we require that

$$\left|\frac{\Delta a_i}{a_{i\,\text{max}}}\right| < 0.001.$$

Since $a_{i\,\text{max}} = 1$ this implies that

$$\left.\frac{da_i}{dt}\right|_{\text{max}} \times T_s < 0.001,$$

or

$$T_s < \frac{0.001}{6} = 0.00017$$

therefore the maximum time for connection to each of the s_i signals is $T_s/5 = 34 \, \mu s$.

Bearing in mind the approximate nature of the bandwidth criterion and the need to allow time for sample acquisition and for switching of input channels, we shall choose an A/D converter with a conversion time of about 25 μs.

The input scan therefore requires a total time of $7 \times 25 \, \mu s$ + small overheads for acquisition and switching. Computation time and output time complete the cyclic time budget. The total time for the three activities can be kept negligibly small compared with $T = 0.1$ s, so that the control algorithms

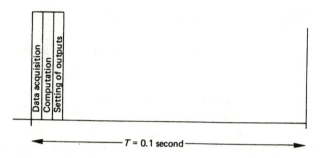

Figure 7.14 Computation timing diagram.

may fix their outputs $u(k)$ depending on current inputs $e(k)$ rather than on previous inputs only. The timing diagram of Figure 7.14 summarizes the situation.

7.18 Choice of word lengths

A 0.1% accuracy requirement implies at least 10-bit working. Because we know that there may be an amplification of input quantization noise we choose a 12-bit A/D converter with 16-bit computation and memory followed by 12-bit D/A conversion.

7.19 Mathematical models of the process

The actuators plus the equipment that they drive constitute the dynamics of the process. The models are

$$\frac{h(s)}{u_3(s)} = \frac{k_3}{s(1+s)} = G_3(s), \tag{7.10}$$

$$\frac{a_1(s)}{u_1(s)} = \frac{k_1}{s(1+s)} = G_1(s), \tag{7.11}$$

$$\frac{a_2(s)}{u_2(s)} = \frac{k_2}{s(1+0.1s)} = G_2(s). \tag{7.12}$$

7.20 Control-algorithm design

A closed-loop process with undamped natural frequency ω_n has the transfer function

$$\frac{\omega_n^2}{\omega_n^2 + 2\zeta\omega_n s + s^2} = H(s) \tag{7.13}$$

(where ζ is the damping factor).

Setting $s = j\omega$ leads to

$$H(j\omega) = \frac{\omega_n^2}{(\omega_n^2 - \omega^2) + j2\zeta\omega_n\omega}. \tag{7.14}$$

We choose a damping factor $\zeta = 1/\sqrt{2}$, and using this value we obtain from Table 5.1 that $\omega_b = \omega_n$.

In our case, ω_b is given as $\omega_b = 6$ rad/s, so we require

$$\omega_n = 6 \text{ rad/s}.$$

To use the z plane loci, Figure 5.7, we need to express ω_n in terms of $\pi/10T$, i.e. to choose p such that

$$\frac{p\pi}{10T} \simeq 6.$$

Since $T = 0.1$, this leads to

$$p = \frac{6}{\pi} = 1.91.$$

We choose $p = 2$, leading to a somewhat higher bandwidth than required and on the appropriate ζ locus, $\zeta = 0.707$, we find that the necessary z-plane poles are $z = 0.57 \pm j0.28$, and thus

$$H(z) = \frac{N(z)}{(z - 0.57 + j0.28)(z - 0.57 - j0.28)} = \frac{N(z)}{z^2 - 1.14z + 0.403}, \tag{7.15}$$

with the numerator to be fixed after further consideration.

7.21 Fixing of the numerator N(z)

We know that the behavior of $H(z)$ is very strongly influenced by the location of numerator zeros. We set

$$N(z) = (z - c_1)(z - c_2)$$

and investigate the possibilities.

The time behavior of the closed-loop system is described by the equation

$$y(k) = 1.14y(k - 1) - 0.403y(k - 2) + v(k)$$
$$- (c_1 + c_2)v(k - 1) + c_1c_2v(k - 2). \tag{7.16}$$

We can see immediately that such a time behavior is impossible to achieve in practice since (for instance) when a step is applied to v, the output y is required to follow perfectly with no delay. To allow the control algorithm to be physically realizable, we must drop one of the numerator zeros and set

$$N(z) = z - c_1,$$

corresponding to the equation

$$y(k) = 1.14y(k-1) - 0.403y(k-2) + v(k-1) - c_1 v(k-2). \quad (7.17)$$

Using guidelines established in Section 4.2.4, we set $c_1 = -1$.
Thus the desired closed-loop transfer function is

$$H(z) = \frac{z+1}{z^2 - 1.14z + 0.403}, \quad (7.18)$$

and this is the same desired transfer function for the three loops to control a_1, a_2 and h.

7.22 Calculation of the control algorithms

The three processes, with their zero-order holds, have the continuous transfer functions

$$G_1'(s) = \left(\frac{1 - e^{-sT}}{s}\right) \frac{k_1}{s(1+s)}, \quad (7.19)$$

$$G_2'(s) = \left(\frac{1 - e^{-sT}}{s}\right) \frac{k_2}{s(1+0.1s)}, \quad (7.20)$$

$$G_3'(s) = \left(\frac{1 - e^{-sT}}{s}\right) \frac{k_3}{s(1+s)}. \quad (7.21)$$

Since the processes are so similar, the calculation will be shown only for algorithms related to equation (7.19).
$\mathscr{L}$ transforming equation (7.19) yields

$$G_1'(z) = (1 - z^{-1})\mathscr{L}\left\{\frac{k_1}{s^2(1+s)}\right\}$$

$$= \frac{k_1[(T - 1 + e^{-T})z + 1 - e^{-T} - Te^{-T}]}{(z-1)(z - e^{-T})}.$$

The constants have values

$$k_1 = 10, \qquad k_2 = 2, \qquad k_3 = 0.5,$$

giving

$$G_1'(z) = \frac{10[(-0.9 + 0.9048)z + 1 - 0.9048 - 0.9048T]}{(z-1)(z - 0.9048)}$$

$$= \frac{1}{20.83}\left(\frac{z + 0.983}{z^2 - 1.9048z + 0.9048}\right). \quad (7.22)$$

Let $D_1(z)$ be the transfer function of the controller for loop 1; then we must have

$$H_1(z) = \frac{G_1'(z)D_1(z)}{G_1'(z)D_1(z) + 1} \qquad (7.23)$$

or

$$D_1(z) = \frac{H_1(z)}{G_1'(z)(1 - H_1(z))}.$$

Substituting values leads to

$$D_1(z) = \frac{(z + 1)/(z^2 - 1.14z + 0.403)}{\dfrac{1}{20.83}\left(\dfrac{z + 0.983}{z^2 - 1.9048z + 0.9048}\right)\left(1 - \dfrac{z + 1}{z^2 - 1.14z + 0.403}\right)}$$

$$= \frac{20.83(z + 1)(z^2 - 1.9048z + 0.9048)}{(z + 0.983)(z^2 - 1.14z + 0.403 - z - 1)}$$

$$= \frac{20.83(z^3 - 0.9048z^2 - z + 0.9048)}{z^3 - 1.16z^2 - 2.7z - 0.59}. \qquad (7.24)$$

Similar calculations are carried out to obtain $D_2(z)$ and $D_3(z)$ in analogous fashion.

ALGORITHM FOR CALIBRATION OF EQUATION (7.9) BY NUCLEONIC GAUGE

The nucleonic thickness gauge is not used directly in the thickness-control loop because of the unacceptably long transport delay that would then be introduced into the loop. (For mechanical reasons, the gauge must be mounted some 2 m downstream so the delay is of the order of $\frac{2}{5} = 0.4$ s.)

Equation (7.9) allows a near instantaneous estimate of h to be made. However, equation (7.9) is subject to errors caused by long-term effects as the roll diameters change due to thermal effects and wear. The nucleonic gauge is therefore used to calibrate the equation to overcome these drift effects.

Let h_m be the measured thickness from the nucleonic gauge and, as before, let h be the thickness calculated using equation (7.9).

A correction term is calculated every n samples as follows:

$$\text{correction term} = \frac{1}{n}\sum_{k=1}^{n}(h_m(k + p) - h(k)), \qquad (7.25)$$

and at the end of n steps the correction term is used to adjust h to obtain a corrected value h' as

$$h'(h) = h(k) + \text{correction term}. \qquad (7.26)$$

Meanwhile a new correction term is calculated and the procedure continues cyclically. For this application $n = 100$.

In equation (7.25), p is set to approximate the offset in time between roll gap and nucleonic gauge (here $p = 4$) so that the values that are compared relate to the same point along the strip.

7.23 Realization of the control algorithms

Using a 16-bit machine, quantization noise amplification is not a factor to influence the decision (parallel programming minimizes such amplification) so we have chosen the method of Section 5.4.1. We illustrate the realization only for $D_1(z)$—the two other controllers are realized in the same way.

As an equation in z the required controller is

$$(z^3 - 1.16z^2 - 2.7z - 0.59)u(z) = 20.83(z^3 - 0.9048z^2 - z + 0.9048)e(z).$$

As in Section 5.4.1, let

$$\frac{u(z)}{e(z)} = \frac{N(z)}{D(z)} \qquad \text{and} \qquad u(z) = N(z)q(z),$$

where

$$q(z) = \frac{e(z)}{D(z)}.$$

Here

$$\frac{u(z)}{e(z)} = \frac{20.83(z^3 - 0.9048z^2 - z + 0.9048)}{(z^3 - 1.16z^2 - 2.7z - 0.59)}.$$

Figure 7.15 illustrates the realization of the algorithm.

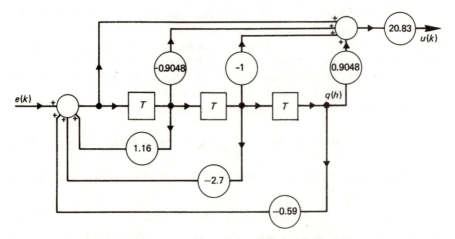

Figure 7.15   Realization of the algorithm for control of the process.

7.24 *Implementation*

The implementation is by rack-mounted modules housed in and powered from a VME standard chassis. Each module is a VME board (double-size Euroboard). The boards connect immediately through the VME bus. Referring to Figure 7.16, module 1 scans the process signals at the necessary rate on receipt of an interrupt that is originated by the real-time clock in module 4. The operation is coordinated by the monoboard microprocessor, module 2, which sends control outputs to the plane via module 5.

Man–machine interaction is through the VDU/keyboard and printer. Data storage is provided on floppy disk via module 8. A serial data link to a VAX machine allows data transfer to take place.

7.25 *Program development*

It is important to understand that the configuration as shown in Figure 7.16 is designed to work with developed and tested software. The configuration cannot handle development and testing of the programs that are needed. Such

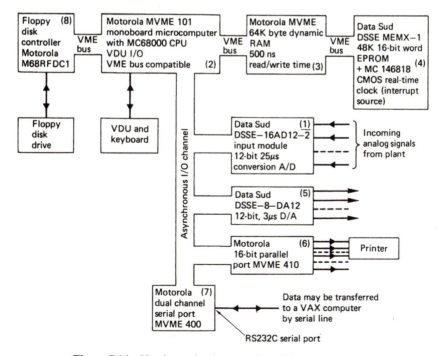

Figure 7.16   Hardware implementation of the control system.

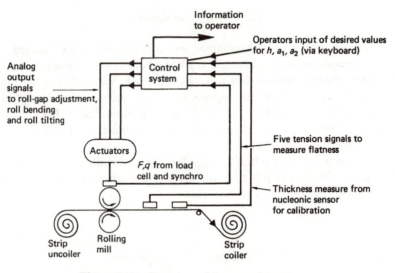

Figure 7.17 Overview of the complete scheme.

programs need to be developed on a larger machine equipped with the correct cross-assembler. Programs that have been developed can then be downloaded to the microprocessor configuration. Alternatively, a development system can be used in conjunction with a 68000 emulator. Again, once the programs have been developed and tested, they can be downloaded into the microcomputer. The degree of availability of powerful and expensive development tools must be considered as part of the design process. If such tools are not available, it will be necessary to include additional facilities for prototype development in the original design.

7.26 The complete scheme

Figure 7.17 gives an overview of the designed system. Eight input signals are scanned rapidly and the three output signals are manipulated to bring h, a_1, a_2 to their desired values as input from the operator via the keyboard. The actuators are updated every 0.1 s.

Exercises

7.1 Refer to the oven-control example where the process transfer function is given by

$$G(s) = \frac{1.63\,e^{-270s}}{(1 + 3480s)}.$$

Refer to Section 5.2.3 and choose a desired closed-loop transfer function $H(z)$ such that the poles of $H(z)$ are at $z = 0.55 \pm j0.22$.

Allow for a zero-order hold to precede the oven and, making reasonable simplifying approximations if necessary, design a digital controller $D(z)$ using the relation

$$D(z) = \frac{H(z)}{G(z)(1 - H(z))}$$

(a) for a sampling interval $T = 60$ s.
(b) for a sampling interval $T = 600$ s.

Check whether the controllers so obtained are open-loop stable.

Comment on the complexity, possibility of ill-conditioning and general practicability of the two controllers.

7.2 Refer to Section 7.7 and confirm the result that the inequality $T_s > 170$ μs must be satisfied if the accuracy specification is to be met.

7.3 (Read the comments on the inadequacy of Ziegler–Nichols models for digital design in Section 7.9.) Fit (by hill climbing, by graphical methods as in Reference L4, or by *ad hoc* methods) an approximate third-order transfer function to the response curve of Figure 7.4. Allow for the inclusion of a zero-order hold and design a controller using equation (7.7) to achieve dead-beat control. Choose a sampling interval of $T = 60$ s for the implementation. Simulate the closed-loop performance (a sampling interval of 5 s is recommended for the simulation) and comment critically on the control performance observed.

7.4 (Design exercise)   A d.c. motor drives a shaft whose angular position is to be controlled. The parameters of the system are:

Inertia of motor and load	1 kg m^2
Friction	negligible
Motor characteristic (torque per ampere)	5 Nm/A

(The motor is armature controlled and has a constant field.)

Motor e.m.f.	$e = c_1 \Phi \omega$ volts

($c_1 \Phi = 0.5$ and ω (rad/s) is the angular velocity of the motor shaft.)

(a) Derive the continuous transfer function of the system.
(b) The system, preceded by a zero-order hold, is to be controlled in closed loop to satisfy the requirements: $\zeta = 0.8$, $\omega_n = 10$ rad/s (ζ, ω_n have their usual meanings).

Produce two designs for digital controllers:
  (i) By continuous design techniques followed by discretization.
  (ii) By synthesis in the z plane.

Check, by analysis of the resultant closed-loop systems and by simulation, whether the designs satisfy the specification.

(c) Put forward a configuration for microcomputer implementation of each algorithm.

(d) Assuming an overall requirement for 0.5% accuracy, select the number of bits required for the interface devices and for the computation.

(e) Comment critically on the relative merits of the two design approaches, from the point of view of this particular design example.

(*Note*: Although, in practice, tachometer feedback might often be incorporated into this type of application, it is excluded here to increase the generality of the exercise.)

7.5 A temperature controller with output

$$q = \frac{Q}{2} + A(\theta_d - \theta)$$

is connected to a process described by the equation

$$\frac{d\theta}{dt} = \frac{1}{C}(k_1 q - k_2 \theta),$$

where Q, A, C, k_2, k_1 are constants, θ_d is desired temperature and θ is temperature.

Derive an expression for the error $\theta - \theta_d$ that occurs in the steady state. Explain the physical reason for the error and suggest how it may be minimized.

7.6 Figure 7.18 shows the open-loop unit step response and the open-loop unit ramp response of system A (shown in figure inset).

Assume that the arrangement is linear and that inertia and friction are the factors dominating dynamic behavior.

Devise and apply a computational technique to determine the transfer function of the arrangement from the responses.

7.7 Figure 7.19 shows the open-loop Bode magnitude plot for system A of Exercise 7.6. Use the Bode diagram to determine the system transfer function. Compare your result with the result of Exercise 7.6.

7.8 When an additional dynamic element is introduced into the arrangement of system A (Exercise 7.6), the open-loop unit step response of the new arrangement is as shown in Figure 7.20. Devise and apply a computational technique to determine the transfer function of the augmented arrangement from the response curve.

7.9 (Mini-product) Armed with the experiences of Exercises 7.6–7.8, obtain the transfer function of an available laboratory amplifer–servomotor combination.

Use the Bode diagram to choose a closed-loop bandwidth that can be attained within the constraints (probably current limits in the amplifier or motor). Decide also on a static accuracy requirement.

Select a suitable sampling frequency and a desired closed-loop transfer

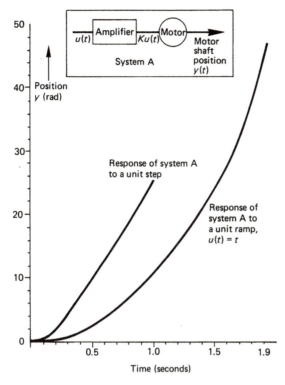

Figure 7.18

function $H(z)$. Remembering to allow for the inclusion of a sample and hold, calculate the transfer function $D(z)$ required for the controller.

Turn your attention to the implementation of the controller. Some questions to be considered are:

(a) How is the position of the motor shaft to be measured? Does your design implicitly require a tachometer to be available to measure shaft angular velocity?

(b) How fast and accurate does the A/D conversion need to be? Should a shaft encoder be preferred, avoiding the necessity for A/D conversion?

(c) How much time can be allowed for the algorithm to be computed, bearing in mind the destabilizing influence of delays in the control loop?

(d) What computing configuration with which realization will satisfy the speed requirement (c), and meet quantization noise requirements?

(e) How will the software be developed and tested?

(f) How will the user communicate with the system and be aware that it is performing as required?

Comment critically on the design and implementation procedures and the performance obtained.

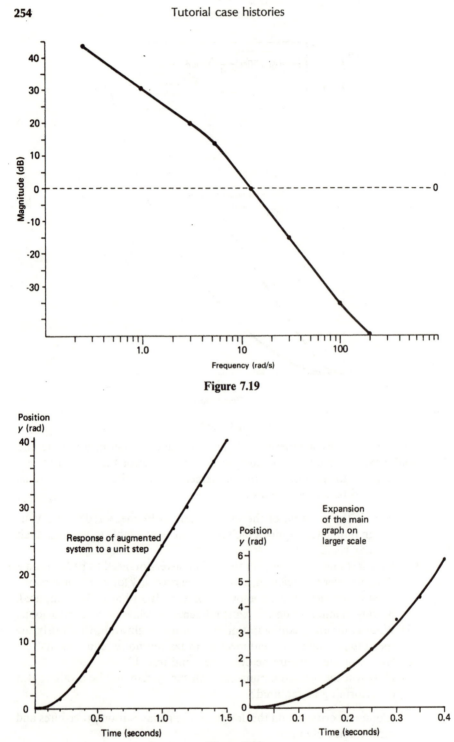

Figure 7.19

Figure 7.20

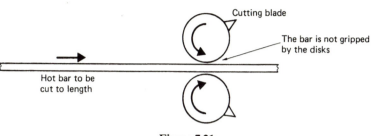

Figure 7.21

Comment briefly also on the economic viability of the design that you produced, indicating how the design might differ if (say) one hundred units needed to be produced.

7.10 Figure 7.21 shows a flying shear. It is used in the metals industries to cut fast-moving hot metal bars to lengths required by customers. The control problem is to accelerate the disks from rest so that when the blades meet the bar they are travelling at the correct linear speed and that they cut the bar at the required point. Sufficient torque has to be available to make the cut without incurring significant drop in disk speed during the cut. Finally, the disks have to be brought to rest in the correct starting position for the next cut.

Numerical data

Maximum speed of bars = 8 m/s.

Minimum length to be cut = 2 m.

Maximum thickness of bars = 0.4 m.

Length tolerance on cut bars = 0.02 m.

Diameter of cutting disks = 0.1 m.

Thickness of cutting disks = 0.05 m.

Additional inertia of drive motor and gear train = Allow 40% above inertia of disk.

Friction effects = negligible.

Resistance torque during the cutting phase = 5.6 N m.

During cutting, speed must not fall by more than 2% due to the resistance torque.

Accurate angular position measurements are available at the disks.

A tachometer on each disk shaft may be used to generate signals proportional to angular velocity.

Bar length measurement is available from a device that generates a pulse for each 0.001 m of bar length.

Bar speed cannot be assumed constant. In particular, accurate cutting cannot be obtained simply by operating the shear at regular time intervals calculated from the nominal bar speed.

Suggest a suitable control scheme and choose appropriate control parameters to meet the given performance and accuracy requirements. Sampling intervals and word lengths should be recommended. (Make and state any assumptions or approximations that are necessary during the design process.)

8

State-Variable Techniques

8.1 Introduction

This chapter establishes state-variable techniques for the representation and analysis of both continuous-time and discrete-time systems with an analogous development for the two cases. Canonical forms are introduced for the structural insight that they create and it is also indicated how canonical forms may be useful in control-system design. The chapter ends with a treatment of state estimation—a topic that is becoming increasingly important in practical applications.

8.2 The concept of state

Consider a mathematical model that consists of n linear time-invariant first-order ordinary differential equations:

$$\dot{x}_i = \sum_{j=1}^{n} a_{ij} x_j + \sum_{j=1}^{r} b_{ij} u_j. \tag{8.1}$$

Define the vectors

$$x = \begin{bmatrix} x_1 \\ \vdots \\ x_n \end{bmatrix}, \qquad u = \begin{bmatrix} u_1 \\ \vdots \\ u_r \end{bmatrix}.$$

If we are given $x(t_0)$ and $u(\tau)$ for all $\tau \geq t_0$, then we can determine $x(t)$ for any $t > t_0$. We do not require information $x(t)$ for $t < t_0$ since all necessary information is assumed to be contained in the vector $x(t_0)$. A vector having this property is called a *state vector*. It is often useful to consider the state vector as an element in an appropriate linear space, which is then defined as the *state space* for the system.

For a particular system Σ, the state space is not uniquely defined and many possible descriptions exist. However, the minimum possible dimension for the state space is fixed for any particular system Σ. (In the above example, no vector of dimension less than n can have the required properties needed by a state vector.)

8.3 Alternative system descriptions

A linear continuous multivariable process is shown in Figure 8.1. The r input variables $u_1, \ldots, u_r$ can be represented as a vector u in an r-dimensional linear space U. Similarly, the state vector x is an element in the state space X and the output vector y is an element in the output space Y. Table 8.1 summarizes this information.

The process will be referred to by the symbol Σ. Assume that the process Σ is time invariant; then it can be represented mathematically in three ways:

(a) By two linear mappings (we shall designate these ϕ and η) linking the three spaces U, X, Y. The process Σ can then be represented in *mapping form*

$$\Sigma = \{I, U, X, Y, \phi, \eta\}, \qquad (8.2)$$

where I is the space representing time.

(b) By a set of n ordinary differential equations of the form

$$\dot{x}_i = \sum_{j=1}^{n} a_{ij} x_j + \sum_{j=1}^{r} b_{ij} u_j, \qquad i = 1, \ldots, n \qquad (8.3)$$

Figure 8.1   A general model for a multi-variable system.

Table 8.1

	Vector			Space
u	r-dimensional	Input vector	U	Input space
x	n-dimensional	State vector	X	State space
y	m-dimensional	Output vector	Y	Output space

and a set of m algebraic equations of the form

$$y_i = \sum_{j=1}^{m} c_{ij}x_j, \qquad i = 1, \ldots, m. \tag{8.4}$$

The equations can be collected into *vector–matrix form*

$$\left. \begin{array}{l} \dot{x} = Ax + Bu, \\ y = Cx. \end{array} \right\} \tag{8.5}$$

The vector x is required to have the state property.

(c) By a set of $r \times m$ transfer functions, $g_{ij}(s)$, linking each of the inputs with each of the outputs. Define the *transfer matrix* $G(s)$ as the matrix whose elements are the $g_{ij}(s)$; then to represent Σ we can write

$$y(s) = G(s)u(s) \tag{8.6}$$

where

$$y(s) = \begin{bmatrix} y_1(s) \\ \vdots \\ y_m(s) \end{bmatrix}, \qquad u(s) = \begin{bmatrix} u_1(s) \\ \vdots \\ u_r(s) \end{bmatrix}$$

Equation (8.6) is the transfer matrix representation of the system.

8.4 The mapping representation of Σ

The mapping representation is valuable for developing insight into system fundamentals and the unification of the subject through simple broad theorems.

The description of the process Σ by the sextuplet $\{I, U, X, Y, \phi, \eta\}$ is completed by the following assumptions, or axioms.

1. $\phi: X \times U \to X, \qquad \phi(t_0, u(t), t_1, x(t_0)) = x(t_1)$.
2. $\eta: X \to Y, \qquad \eta(t_0, x(t_0)) = y(t_0)$.
   The mappings ϕ, η, operating as shown, are linear mappings. For a continuous-time process, I is replaced by the real line $\mathbb{R}^1$ representing the underlying continuous-time set. (For a discrete-time process, I is replaced by the set of integers, denoted Z.)
3. $\phi(t_0, u(t), t_2, x(t_0)) = \phi(t_1, u(t), t_2, (\phi(t_0, u(t), t_1, x(t_0))))$.
   This, the semigroup property, simply ensures that the solution at time t_2 is the same whether calculated in one large time step $t_2 - t_0$, or in two smaller time steps $t_1 - t_0$ and $t_2 - t_1$.
4. $\phi(t_0, u(t), t_0, x(t_0)) = x(t_0)$.
   This property is clearly necessary.
5. $\phi(t_0, u_1(t), t_1, x(t_0)) = \phi(t_0, u_2(t), t_1, x(t_0)), t_1 > t_0$.

provided that

$$u_1(t) = u_2(t) \text{ on the interval } [t_0, t_1].$$

This interesting axiom ensures that the system is causal (i.e. it cannot respond to a stimulus before the stimulus is received) and hence physically realizable. It also ensures that x really does have the property necessary in a state vector, as follows:

> Given a knowledge of the state vector $x(t_0)$ at time t_0 and a knowledge of the input on any time interval $[t_0, t]$, $t > t_0$, the state $x(t)$ can be determined.

The implication is that $x(t_0)$ contains sufficient information to make knowledge of events before time t_0 unnecessary, i.e. x is a state vector.

6. U is a space of piecewise continuous functions on I.

> One consequence of the axioms is that we can write

$$\phi(t_0, u(t), t_1, x(t_0)) = \phi(t_0, 0, t_1, x(t_0)) + \phi(t_0, u(t), t_1, 0).$$

The right-hand terms are called the zero-input response and the zero-state response respectively.

Let X be a linear space of finite dimension n; then Σ is called a *linear finite-dimensional system*.

Let X be a linear system of infinite dimension; then Σ is called a *linear infinite-dimensional system*.

To appreciate the simplicity of the axioms, they should be visualized in a geometric setting. For instance, axiom 3 can be visualized with the help of Figure 8.2 and axiom 5 with the help of Figure 8.3.

The study of systems through their mappings is most rewarding. The topic can be pursued further in Reference L3.

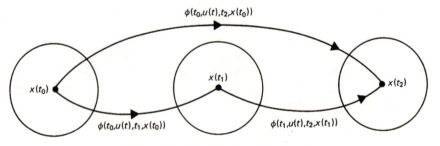

Figure 8.2 Visualization of axiom 3.

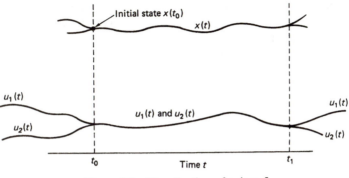

Figure 8.3   Visualization of axiom 5.

8.5 The modelling of continuous-time systems through state-space equations

8.5.1 A general nonlinear model

A general model for a multivariable system was shown in Figure 8.1. The system receives r input signals denoted $u_1, \ldots, u_r$ and gives out m output signals, denoted $y_1, \ldots, y_m$. It is convenient to define the vectors

$$u = \begin{bmatrix} u_1 \\ \vdots \\ u_r \end{bmatrix}, \qquad y = \begin{bmatrix} y_1 \\ \vdots \\ y_m \end{bmatrix}.$$

The *input vector u* carries information on the external influences (applied controls and disturbances) affecting the system. The *output vector y* represents the system measurements.

To complete the description, n internal states, $x_1, \ldots, x_n$ are assumed.

$$x = \begin{bmatrix} x_1 \\ \vdots \\ x_n \end{bmatrix}$$

is recognized as the *state vector* of the system. The state vector has been defined in Section 8.2.

The input, state and output vectors are considered to be members of the *input, state* and *output spaces* respectively, denoted U, X and Y.

With this background we are ready to represent the general multivariable system by the following set of equations:

$$\dot{x}_i = f_i(x_1, \ldots, x_n, u_1, \ldots, u_r, t), \qquad i = 1, \ldots, n \qquad (8.7)$$

$$y_i = g_i(x_1, \ldots, x_n, t), \qquad i = 1, \ldots, m,$$

in which the f_i and g_i are nonlinear functions and t represents time. We assume that $t \in I$, where $I = \mathbb{R}^1$. The inter-relation of the functions and the spaces can be expressed by the relations.

$$f_i: X \times U \times I \to X,$$
$$g_i: X \times I \to Y.$$

8.5.2 Linearization of the model

Multivariable control theory rests on a foundation of algebra and operator theory and the model (8.7) derived above must be linearized before proceeding further.

Suppose that the functions f_i, g_i defined in equation (8.7) are in fact linear; then necessarily (as a consequence of the representation theorem of real analysis) the following relations hold:

$$\dot{x}_i = a_{i1}(t)x_1 + \cdots + a_{in}(t)x_n + b_{i1}(t)u_1 + \cdots + b_{ir}(t)u_r, \qquad i = 1, \ldots, n, \quad (8.8)$$

$$y_i = c_{i1}(t)x_1 + \cdots + c_{in}(t)x_n, \qquad i = 1, \ldots, m.$$

Or in vector–matrix form

$$\dot{x} = \begin{bmatrix} a_{11}(t) & \cdots & a_{1n}(t) \\ \vdots & & \vdots \\ a_{n1}(t) & \cdots & a_{nn}(t) \end{bmatrix} x + \begin{bmatrix} b_{11}(t) & \cdots & b_{1r}(t) \\ \vdots & & \vdots \\ b_{n1}(t) & \cdots & b_{nr}(t) \end{bmatrix} u,$$

$$y = \begin{bmatrix} c_{11}(t) \\ \vdots \\ c_{m1}(t) \end{bmatrix} \cdots \begin{bmatrix} c_{1n}(t) \\ \vdots \\ c_{mn}(t) \end{bmatrix} x$$

(8.9)

or

$$\dot{x} = A(t)x + B(t)u, \qquad y = C(t)x.$$

Returning to the general case where the functions f, g are nonlinear then, provided that the functions are differentiable, local approximations to the matrices $A(t)$, $B(t)$, and $C(t)$ can be determined from the following relations (the operation is performed along nominal trajectories for x and u):

$$A(t) = \left(\frac{\partial f_i}{\partial x_j} \right)\Bigg|_{\substack{x, \\ u}} \qquad (8.10)$$

$$B(t) = \left(\frac{\partial f_i}{\partial u_j} \right)\Bigg|_{\substack{x, \\ u}} \qquad (8.11)$$

$$C(t) = \left(\frac{\partial g_i}{\partial x_j} \right)\Bigg|_{x}. \qquad (8.12)$$

8.6 *Calculation of time solutions using the transition matrix*

8.6.1 *The time-invariant case*

DEFINITION OF THE MATRIX EXPONENTIAL IN TERMS OF AN INFINITE SERIES
First, we define the *matrix exponential* e^{AT} as an infinite series:

$$e^{AT} = I + AT + \frac{1}{2!}(AT)^2 + \frac{1}{3!}(AT)^3 + \cdots,$$

the series being convergent for any finite T.

THE TRANSITION MATRIX
Given the equation

$$\dot{x} = Ax + Bu, \tag{8.13}$$

put $z = e^{-At}x$. Then (the reader should confirm the next step from first principles using the infinite series definition of e^{At})

$$\dot{z} = -A\,e^{-At}x + e^{-At}\dot{x}$$
$$= -A\,e^{-At}x + e^{-At}(Ax + Bu)$$
$$= e^{-At}Bu.$$

Integrate from t_0 to t

$$z(t) = z(t_0) + \int_{t_0}^{t} e^{-A\tau} Bu(\tau)\,d\tau,$$

$$e^{-At}x = e^{-At_0}x(t_0) + \int_{t_0}^{t} e^{-A\tau} Bu(\tau)\,d\tau,$$

$$x(t) = e^{A(t-t_0)}x(t_0) + \int_{t_0}^{t} e^{A(t-\tau)} Bu(\tau)\,d\tau. \tag{8.14}$$

The *transition matrix* $\Phi(t - t_0)$ is now defined such that

$$x(t) = \Phi(t - t_0)x(t_0) + \int_{t_0}^{t} \Phi(t - \tau)Bu(\tau)\,d\tau. \tag{8.15}$$

Numerical solution can be achieved by choosing a time step T, such that over every T, $u(t)$ can be considered constant. Then equation (8.15) can be integrated to yield

$$x(T) = \Phi(T)x(0) + A^{-1}(\Phi(T) - I)Bu(0),$$

and in general

$$x((k + 1)T) = \Phi(T)x(kT) + \Psi(T)u(kT) \tag{8.16}$$

where $\Psi(T)$ is defined by the relation

$$\Psi(T) = A^{-1}(\Phi(T) - I)B.$$

LAPLACE TRANSFORM APPROACH
Consider the equation

$$\dot{x} = Ax$$

in which A is an $n \times n$ matrix. Laplace transformation gives

$$sx(s) - x(0) = Ax(s),$$
$$(sI - A)x(s) = x(0),$$
$$x(t) = \mathcal{L}^{-1}\{(sI - A)^{-1}\}x(0),$$

from which, by uniqueness,

$$\Phi(t) = \mathcal{L}^{-1}\{(sI - A)^{-1}\}.$$

DIAGONALIZATION APPROACH
Let $\lambda_i, e_i, i = 1, \ldots, n$ be the (distinct) eigenvalues and eigenvectors respectively
of the system matrix A. Define the *modal matrix* E by the relation

$$E = (e_1, e_2, \ldots, e_n)$$

and define $v = E^{-1}x$. Then

$$E\dot{v} = AEv,$$

$$\dot{v} = E^{-1}AEv = E^{-1}(AE)v = E^{-1}\left(E\begin{bmatrix}\lambda_1 & & 0 \\ & \ddots & \\ 0 & & \lambda_n\end{bmatrix}\right)v = \Lambda v,$$

where Λ is a diagonal matrix

$$\begin{bmatrix}\lambda_1 & & 0 \\ & \ddots & \\ 0 & & \lambda_n\end{bmatrix}$$

of eigenvalues. Because of the diagonality of Λ, the transition matrix is
particularly simple and we can write

$$v(t) = \begin{bmatrix}e^{\lambda_1 t} & & 0 \\ \vdots & \ddots & \vdots \\ 0 & & e^{\lambda_n t}\end{bmatrix}v(0) \triangleq e^{\Lambda t}v(0).$$

Should we wish to return to the original system coordinates, we have to make the substitution $v = E^{-1}x$ to obtain

$$x(t) = E\,e^{At}E^{-1}x(0) = E\,e^{(E^{-1}AE)t}E^{-1}x(0),$$

from which

$$\Phi(t) = E\,e^{At}E^{-1}.$$

A variety of other methods is available for the calculation of the transition matrix but these three methods (infinite series, Laplace method, diagonalization method) will cover all our needs. Commenting briefly on the comparative merits of the three methods we note:

(a) The Laplace method is always applicable but its use entails the inverse Laplace transformation of n^2 terms which becomes unwieldy for large systems. The method is relatively difficult to mechanize because of the inverse transformation step.
(b) The diagonalization method only works unmodified for systems with distinct eigenvalues. It is relatively easy to mechanize.
(c) The series summation method is simple to mechanize although scaling and convergence problems may arise. For large t, very many terms have to be taken before convergence is obtained. The result from the series method is a matrix of numbers rather than of mathematical functions and this may limit the usefulness of the method for some applications.

Example Compute the transition matrix $\Phi(t)$ for the matrix

$$A = \begin{bmatrix} 0 & 1 \\ -4 & -5 \end{bmatrix}$$

by the three methods given above.

Laplace method

$$sI - A = \begin{bmatrix} s & -1 \\ 4 & s+5 \end{bmatrix}, \qquad (sI - A)^{-1} = \frac{1}{s(s+5)+4}\begin{bmatrix} s+5 & 1 \\ -4 & s \end{bmatrix},$$

$$\Phi(t) = \mathscr{L}^{-1}\left\{ \begin{matrix} \dfrac{s+5}{s^2+5s+4} & \dfrac{1}{s^2+5s+4} \\[3mm] \dfrac{-4}{s^2+5s+4} & \dfrac{s}{s^2+5s+4} \end{matrix} \right\}$$

$$= \frac{1}{3}\mathscr{L}^{-1}\begin{pmatrix} \left(\dfrac{4}{s+1} - \dfrac{1}{s+4}\right) & \left(\dfrac{1}{s+1} - \dfrac{1}{s+4}\right) \\[3mm] \left(\dfrac{-4}{s+1} + \dfrac{4}{s+4}\right) & \left(\dfrac{-1}{s+1} + \dfrac{4}{s+4}\right) \end{pmatrix}$$

$$= \frac{1}{3}\left(e^{-t}\begin{bmatrix} 4 & 1 \\ -4 & -1 \end{bmatrix} - e^{-4t}\begin{bmatrix} 1 & 1 \\ -4 & -4 \end{bmatrix} \right).$$

Diagonalization method Eigenvalues are found from the characteristic equation $\lambda^2 + 5\lambda + 4 = 0$ yielding $\lambda_1 = -4$, $\lambda_2 = -1$. Eigenvectors are

$$\begin{bmatrix} 1 \\ -4 \end{bmatrix}, \quad \begin{bmatrix} 1 \\ -1 \end{bmatrix}.$$

The modal matrix

$$E = \begin{bmatrix} 1 & 1 \\ -4 & -1 \end{bmatrix}, \qquad E^{-1} = \frac{1}{3}\begin{bmatrix} -1 & -1 \\ 4 & 1 \end{bmatrix}.$$

$$\Phi(t) = E\,e^{\Lambda t}E^{-1} = \frac{1}{3}\begin{bmatrix} 1 & 1 \\ -4 & -1 \end{bmatrix}\begin{bmatrix} e^{-4t} & 0 \\ 0 & e^{-t} \end{bmatrix}\begin{bmatrix} -1 & -1 \\ 4 & 1 \end{bmatrix}$$

$$= \frac{1}{3}\left(e^{-t}\begin{bmatrix} 4 & 1 \\ -4 & -1 \end{bmatrix} - e^{-4t}\begin{bmatrix} 1 & 1 \\ -4 & -4 \end{bmatrix}\right).$$

Series summation method
(a) With $t = 0.1$,

$$\Phi(0.1) = I + A(0.1) + \frac{A^2(0.1)^2}{2!} + \cdots.$$

Let S_R be the summation after R terms; then evaluating numerically we obtain the sequence in Table 8.2.
(b) With $T = 1$,

$$\Phi(1) = I + A + \frac{A^2}{2!} + \cdots,$$

we obtain the results shown in Table 8.3.

The disadvantage of poor convergence of this method is seen here.

Table 8.2

R	S_R	R	S_R
1	$\begin{bmatrix} 1 & 0 \\ 0 & 1 \end{bmatrix}$	5	$\begin{bmatrix} 0.983 & 0.0781 \\ -0.313 & 0.592 \end{bmatrix}$
2	$\begin{bmatrix} 1 & 0.1 \\ -0.4 & 0.5 \end{bmatrix}$	6	$\begin{bmatrix} 0.9830 & 0.7817 \\ -0.3127 & 0.5921 \end{bmatrix}$
3	$\begin{bmatrix} 0.98 & 0.075 \\ -0.3 & 0.605 \end{bmatrix}$	7	$\begin{bmatrix} 0.9830 & 0.7817 \\ -0.3127 & 0.5921 \end{bmatrix}$
4	$\begin{bmatrix} 0.983 & 0.0785 \\ -0.314 & 0.59 \end{bmatrix}$		

Table 8.3

R	S_R	R	S_R
1	$\begin{bmatrix} 1 & 0 \\ 0 & 1 \end{bmatrix}$	11	$\begin{bmatrix} 0.46 & 0.09 \\ -0.36 & 0.006 \end{bmatrix}$
2	$\begin{bmatrix} 1 & 1 \\ -4 & -4 \end{bmatrix}$	12	$\begin{bmatrix} 0.49 & 0.125 \\ -0.5 & -0.13 \end{bmatrix}$
3	$\begin{bmatrix} -1 & -1.5 \\ 6 & 6.5 \end{bmatrix}$	13	$\begin{bmatrix} 0.48 & 0.11 \\ -0.45 & -0.09 \end{bmatrix}$
4	$\begin{bmatrix} 2.3 & 2 \\ -8 & -7.6 \end{bmatrix}$	14	$\begin{bmatrix} 0.48 & 0.12 \\ -0.47 & -0.10 \end{bmatrix}$
5	$\begin{bmatrix} -1.17 & -1.5 \\ 6.16 & 6.5 \end{bmatrix}$	15	$\begin{bmatrix} 0.48 & 0.116 \\ -0.46 & -0.097 \end{bmatrix}$
6	$\begin{bmatrix} 1.6 & 1.3 \\ -5.2 & -4.8 \end{bmatrix}$	16	$\begin{bmatrix} 0.484 & 0.116 \\ -0.466 & -0.098 \end{bmatrix}$
7	$\begin{bmatrix} -0.22 & -0.6 \\ 2.38 & 2.78 \end{bmatrix}$	17	$\begin{bmatrix} 0.4844 & 0.1165 \\ -0.4660 & -0.0982 \end{bmatrix}$
8	$\begin{bmatrix} 0.85 & 0.49 \\ -1.95 & -1.58 \end{bmatrix}$	18	$\begin{bmatrix} 0.4844 & 0.1165 \\ -0.4661 & -0.09821 \end{bmatrix}$
9	$\begin{bmatrix} 0.31 & -0.05 \\ 0.21 & 0.58 \end{bmatrix}$	19	$\begin{bmatrix} 0.4844 & 0.1165 \\ -0.4661 & -0.09821 \end{bmatrix}$
10	$\begin{bmatrix} 0.55 & 0.18 \\ -0.74 & -0.37 \end{bmatrix}$		

8.6.2 The time-varying case

Consider the equation

$$\dot{x}(t) = A(t)x(t), \qquad x(t) = x_0 \tag{8.17}$$

where $x \in \mathbb{R}^n$ and A is a square matrix of absolutely integrable functions defined on (t_0, t_1); that is,

$$a_{ij}(t) \in L_1(t_0, t_1) \qquad \text{for } i, j \leqslant n.$$

Notice that the elements of $A(t)$ are not required to be continuous for the system to be well posed.

In fact if $\|A(t)\| < g(t)$, $\forall t$, while $\int_{t_0}^{t_1} g(t)\,dt < \infty$, then equation (8.17) has a unique solution over (t_0, t_1) that is continuous in t.

The solution of equation (8.17) can still be written $x(t) = \Phi(t, t_0)x_0$, where Φ is the transition matrix associated with $A(t)$.

However, the transition matrix is more difficult to determine than for the time-invariant case, and before discussing this aspect we need more theoretical background concerning the transition matrix.

First we define the *fundamental matrix*.

Given an equation

$$\dot{x}(t) = A(t)x(t), \qquad (8.18)$$

consider the associated equation

$$\dot{X} = A(t)X \qquad (8.19)$$

where X is an $n \times n$ matrix.

Given $X(t_0) = X_0$, with det $X_0 \neq 0$, equation (8.19) can be solved to yield a solution $X(t)$, and $X(t)$ is called the *fundamental matrix* associated with equation (8.19). It satisfies the condition

$$\det X(t) \neq 0, \qquad \forall t.$$

Now let $X(t_0) = I$; then the solution $X(t)$ of (8.19) is called the *transition matrix* of $A(t)$. (This is the general definition of the transition matrix Φ as opposed to that given in equation (8.15), which applies only to the time-invariant case. Note carefully that in the time-invariant case, Φ is a function of an interval, whereas here, Φ is a function of initial and final times.) The transition matrix has the important characterizing property

$$\frac{\partial}{\partial t}(\Phi(t, t_0)) = A(t)\Phi(t, t_0), \qquad \forall t \geqslant t_0, \qquad (8.20)$$

$$\Phi(t_0, t_0) = I.$$

The transition matrix transforms the initial condition $x(t_0)$ into the state $x(t)$ and gives the complete solution of the autonomous equation (8.17). To show this, differentiate the solution

$$x(t) = \Phi(t, t_0)x(t_0),$$

yielding

$$\dot{x}(t) = \dot{\Phi}(t, t_0)x(t_0),$$

using (8.20)

$$\dot{x}(t) = A(t)\Phi(t, t_0)x(t_0) = A(t)x(t),$$

which is the same as equation (8.17).

As we have seen

$$\left.\begin{array}{r} \dot{\Phi}(t, t_0) = A(t)\Phi(t, t_0), \\ \Phi(t_0, t) = I \end{array}\right\} \qquad (8.21)$$

has the solution $\Phi(t, t_0)$ which is the transition matrix we require.

The solution of equation (8.21) is given by the *Peano–Baker* series.

$$\Phi(t, t_0) = I + \int_{t_0}^{t} A(\tau_1) \, d\tau_1 + \int_{t_0}^{t} A(\tau_1) \int_{t_0}^{\tau_1} A(\tau_2) \, d\tau_2 \, d\tau_1 + \cdots. \quad (8.22)$$

Theorem　If A is a matrix of constant coefficients, then

$$\Phi(t, t_0) = I + A(t - t_0) + \frac{A^2(t - t_0)^2}{2!} + \cdots. \quad (8.23)$$

Proof　Since A is constant we can write

$$\Phi(t, t_0) = I + A \int_{t_0}^{t} d\tau_1 + A^2 \int_{t_0}^{t} \int_{\tau_0}^{\tau_1} d\tau_2 \, d\tau_1 + \cdots \quad (8.24)$$

but it is important to note that in general

$$\Phi(t, t_0) \neq \exp\left(\int_{t_o}^{t} A(\tau) \, d\tau \right). \quad (8.25)$$

However, the equality does hold if

$$A(t) \int_{t_0}^{t} A(\tau) \, d\tau = \int_{t_0}^{t} A(\tau) \, d\tau \, A(t), \quad (8.26)$$

i.e. if the matrices $(A(t), \int_{t_0}^{t} A(\tau) \, d\tau)$ commute.

These matrices commute in case the elements of $A(t)$ are time-invariant or if $A(t)$ is a diagonal matrix or in case $A(t)$ can be decomposed into a constant matrix M and a scalar $\alpha(t)$ so that $A(t) = \alpha(t)M$.

It is only in the case where the above matrices commute that $A(t)$ can be brought out of the integral during the derivation to yield

$$\frac{d}{dt} \left(\exp\left(\int_{t_0}^{t} A(\tau) \, d\tau \right) \right) = A(t) \exp\left(\int_{t_0}^{t} A(\tau) \, d\tau \right). \quad (8.27)$$

In the general case, the transition matrix is not the exponential of the integral of A.

Therefore, for linear time-varying problems, $\Phi(t, t_0)$ has to be determined by numerical evaluation of the expression (8.22). For simple cases, however, equation (8.22) can be expanded and solved analytically.

Not only is this much more time consuming than the evaluation of Φ for time-invariant systems, but also the same matrix cannot be used repetitively to advance the solution as in the time-invariant case.

If the system is time-varying in a known deterministic manner, it may be possible to make a transformation such that the system appears invariant with respect to a new substituted variable. Cases where such transformation can be made are understandably rare.

Section 8.6.3 describes the periodically varying case, where such a transformation can be made with advantage.

8.6.3 The periodically time-varying case

In the modelling of rotating devices, such as radar antennae, periodically time-varying parameters are encountered.

Consider a model of the form

$$\dot{x}(t) = A(t)x(t), \qquad A(t + T) = A(t), \tag{8.28}$$

where T is the period of variation.

The transition matrix $\Phi(t, t_0)$ is periodic in T. Define a matrix R by the relation

$$e^{RT} = \Phi(T, 0). \tag{8.29}$$

Such an R exists if $\Phi(T, 0)$ is nonsingular, since Φ can then be diagonalized to give

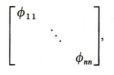

so that R could be chosen as

$$\begin{bmatrix} \ln \phi_{11} & & \\ & \ddots & \\ & & \ln \phi_{nn} \end{bmatrix}.$$

Define an operator $P(t)$ by

$$P^{-1}(t) = \Phi(t, 0)e^{-Rt},$$

$$\Phi(t, t_0) = \Phi(t, 0)\Phi(0, t_0)$$

$$= P^{-1}(t)e^{Rt}e^{-Rt_0}P(t_0)$$

$$= P^{-1}(t)e^{R(t - t_0)}P(t_0).$$

From equation (8.28),

$$x(t) = \Phi(t, t_0)x(t_0)$$

$$= P^{-1}(t)e^{R(t - t_0)}P(t_0)x(t_0).$$

Put $z(t) = P(t)x(t)$; then

$$P^{-1}(t)z(t) = P^{-1}(t)e^{R(t - t_0)}P(t_0)P^{-1}(t_0)z(t_0)$$

$$z(t) = e^{R(t - t_0)}z(t_0).$$

This equation has a time-invariant transition matrix and is the solution of the transformed version of equation (8.28),

$$z(t) = e^{R(t-t_0)} z(t_0). \tag{8.30}$$

This procedure is known as the *Floquet–Lyapunov transformation*.

8.7 Properties of the transition matrix

Identity property

$$\Phi(t_0, t_0) x(t_0) = x(t_0).$$

Inverse property If

$$\Phi(t, t_0) x(t_0) = x(t),$$

then

$$\Phi(t_0, t) x(t) = x(t_0)$$

or

$$\Phi(t_0, t)(\Phi(t, t_0) x(t_0)) = x(t_0).$$

$\Phi(t_0, t)$ is defined as the *inverse* of $\Phi(t, t_0)$, denoted $\Phi^{-1}(t, t_0)$. (For the time-invariant case, $\Phi^{-1}(t) = (e^{At})^{-1} = e^{-At} = \Phi(-t)$.)

Semigroup property

$$\Phi(t, t_1)\Phi(t_1, t_0) x(t_0) = \Phi(t, t_1)(\Phi(t_1, t_0) x(t_0))$$
$$= (\Phi(t, t_1)\Phi(t_1, t_0)) x(t_0)$$
$$= x(t).$$

Differentiation

$$\frac{d}{dt}(\Phi(t, t_0)) = A(t)\Phi(t, t_0).$$

8.8 Relation between the transfer-matrix description and the vector–matrix description

Assume that the process can be described by the equations

$$\left.\begin{array}{l} \dot{x} = Ax + Bu \\ y = Cx \end{array}\right\} \tag{8.31}$$

and by the transfer matrix $G(s)$.

Assuming zero initial conditions in equation (8.31) and Laplace transforming, we obtain

$$y(s) = C(sI - A)^{-1}Bu(s), \qquad (8.32)$$

from which, by uniqueness of solution, it follows that

$$G(s) = C(sI - A)^{-1}B. \qquad (8.33)$$

8.9 Equivalent systems

Two system representations

$$\dot{x} = Ax + Bu \qquad \dot{z} = Hz + Ju$$
$$y = Cx \qquad y = Kz$$

are said to be *equivalent* if

$$H = QAQ^{-1}, \qquad J = QB, \qquad K = CQ^{-1},$$

where Q is any $n \times n$ nonsingular matrix with real elements. (We see at once that a diagonalized representation, as used for instance in Section 8.6.1, is equivalent to the original system with the model matrix E playing the role of Q.)

The impulse responses of the systems are given by

$$\mathscr{L}^{-1}\{C(sI - A)^{-1}B\} \qquad \text{and} \qquad \mathscr{L}^{-1}\{K(sI - H)^{-1}J\}$$

or

$$C\,e^{At}B \qquad \text{and} \qquad K\,e^{Ht}J.$$

However,

$$K\,e^{Ht}J = CQ^{-1}Q\,e^{At}Q^{-1}QB = C\,e^{At}B, \qquad (8.34)$$

i.e. the impulse responses (and hence the transfer functions) of equivalent systems are identical. Thus, to one transfer function correspond many alternative state representations. A positive result is that $G(s)$ fixes the zero-state response but not the zero-input response.

The set $\{A, B, C\}$ is called a *realization* of $G(s)$ if

$$C(sI - A)^{-1}B = G(s).$$

The realization is *minimal* if there exists no other realization $\{A', B', C'\}$ for which

$$\dim A' < \dim A.$$

8.10 System realization

Section 8.8 showed that, to a particular system realization $\{A, B, C\}$ there corresponds a transfer matrix

$$G(s) = C(sI - A)^{-1}B.$$

We now consider the converse problem. Given a particular $G(s)$, does there always exist a corresponding realization $\{A, B, C\}$? If such a realization exists, is it unique? In general, how are realizations $\{A, B, C\}$ to be constructed from a knowledge of $G(s)$?

8.10.1 Existence

A realization of $G(s)$ exists provided that $G(s)$ satisfies the following conditions:

(a) for each term of $G(s)$, the degree of the numerator does not exceed that of the denominator;
(b) each term of $G(s)$ is a rational function of s.

8.10.2 Uniqueness

It is clear from the discussion on equivalent systems, Section 8.9, that the state-space realization corresponding to a particular $G(s)$ cannot be unique.

This leads us to ask the weaker question. Are all the realizations corresponding to a particular $G(s)$ equivalent to each other? More formally, does $G(s)$ define a unique equivalence class of realizations?

It is easy to show by counter-example (Exercise 8.9) that the answer to this last question is also no. However, it is true that $G(s)$ defines a unique equivalence class of minimal realizations.

8.11 Stability

A system is defined to be *asymptotically stable* if for any initial vector $x(t_0)$ and assuming zero input

$$\|x(t)\| \to 0 \qquad \text{as } t \to \infty \tag{8.35}$$

where $\| \ \|$ is a measure of the magnitude of the vector. One suitable measure is given by

$$\|x\| = (\sum x_i^2)^{1/2}. \tag{8.36}$$

A system is defined to be *stable* if given an initial vector $x(t_0)$ satisfying $\|x(t_0)\| < \delta$, there exists an $\varepsilon > 0$ such that $\|x(t_0)\| < \delta$ implies that

$$\|x(t)\| < \varepsilon \qquad \text{for all } t \geqslant t_0. \tag{8.37}$$

8.11.1 Stability tests for continuous-time systems

Let $\Phi(t)$ be the transition matrix of a system Σ whose state description is given by $\{A, B, C\}$. The system Σ is asymptotically stable if

$$\|\Phi(t)\| \to 0 \qquad \text{as } t \to 0.$$

The system Σ is stable if

$$\|\Phi(t)\| \text{ remains bounded for all } t \geqslant 0.$$

The system Σ is asymptotically stable if each eigenvalue λ_i of the matrix A satisfies

$$\mathscr{R}(\lambda_i) < 0, \qquad i = 1, \dots, n.$$

The system Σ is stable if each eigenvalue satisfies

$$\mathscr{R}(\lambda_i) \leqslant 0, \qquad i = 1, \dots, n.$$

8.12 Reachability, controllability, observability and reconstructibility for continuous-time systems

Reachability   A state x_1 is defined to be *reachable* if some control $u(t)$ applied over a finite interval (t_0, t_1), $t_1 > t_0$, transfers the system from an initial state $x(t_0) = 0$ to the state $x(t_1) = x_1$. A system is defined to be *reachable* if every state is reachable.

Controllability   A state x_1 is defined to be *controllable* if some control $u(t)$ applied over a finite interval (t_0, t_1), $t_1 > t_0$, transfers the system from an initial state $x(t_0) = x_1$ to the state $x(t_1) = 0$. A system is defined to be *controllable* if every state is controllable.

Observability   A state $x(t_0)$ of a system is *observable* if it can be determined from a knowledge of the values of the system outputs $y(t)$ over a finite interval (t_0, t_1), $t_1 > t_0$. A system is *observable* if every state is observable.

Reconstructibility   A state $x(t_0)$ is *reconstructible* if it can be determined from a knowledge of the values of the system output $y(t)$ over a finite interval (t_1, t_0), $t_1 < t_0$. For continuous-time systems, reachability is equivalent to controllability and observability is equivalent to reconstructibility. The definitions will be seen to differ when discrete-time systems are under consideration.

8.12.1 Controllability and observability tests for continuous-time, time-invariant systems by the matrices Q_c, Q_o

Define the matrices

$$Q_c = (B, AB, \ldots, A^{n-1}B),$$

$$Q_o = \begin{bmatrix} C \\ CA \\ \vdots \\ CA^{n-1} \end{bmatrix}.$$

Then the n-dimensional system with state-space description $\{A, B, C\}$ is completely controllable (completely observable) if and only if rank Q_c (rank Q_o) $= n$.

The proof of this theorem is omitted. However, the analogous theorem for discrete-time systems is proved.

8.12.2 Controllability and observability tests for continuous-time, time-invariant systems by the diagonalization method

Given a system described by the equations

$$\begin{aligned} \dot{x} &= Ax + Bu, \\ y &= Cx. \end{aligned} \tag{8.38}$$

Then provided that A has distinct eigenvalues, the system can be diagonalized by putting $x = Ez$ in equation (8.38), where E is the modal matrix (see Section 8.6.1), to yield

$$\left.\begin{aligned} \dot{z} &= E^{-1}Aez + E^{-1}Bu \\ y &= CEz. \end{aligned}\right\} \tag{8.39}$$

Since the matrix $E^{-1}AE$ is diagonal, the controllability and observability of the transformed system (equation (8.39)) can be tested as follows (proof by inspection):

The system is uncontrollable if any row of the matrix $E^{-1}B$ consists solely of zero entries.

The system is unobservable if any column of the CE matrix consists solely of zero entries.

Controllability and observability are invariant properties under the transformation $x = Ez$. To see this define matrices

$$Q_{c_1} = (B, AB, \ldots, A^{n-1}B),$$

$$Q_{c_2} = (E^{-1}B, (E^{-1}AE)E^{-1}B, \ldots, (E^{-1}AE)^{n-1}E^{-1}B).$$

Since Rank $E^{-1} = $ Rank $E = n$, then Rank $Q_{c_1} = $ Rank Q_{c_2} and the invariance is proved.

Thus the tests based on diagonalization can be used to check whether the original system is uncontrollable or unobservable.

8.13 The unforced state equation in discrete time

The general (i.e. time-varying) unforced discrete-time equation has the form

$$x(k + 1) = A_k x(k), \qquad x(0) \text{ given.} \tag{8.40}$$

8.13.1 Existence and uniqueness of solution

The equation has one and only one solution for each integer k if and only if the matrices A_k are nonsingular.

Proof

$$x(k) = A_{k-1}A_{k-2} \cdots A_0 x(0), \qquad k \geqslant 0. \tag{8.41}$$

If any of the matrices A_i, $0 \leqslant i \leqslant k - 1$ is singular, then by arguments based on the rank of mappings or otherwise, the solution is not unique—hence it is necessary that all matrices A_i, $0 \leqslant i \leqslant k - 1$ are nonsingular.

It is also true that

$$x(k) = A_k^{-1}A_{k-1}^{-1} \cdots A_{-1}^{-1}x_0, \qquad k \leqslant 0. \tag{8.42}$$

And from (8.41) and (8.42) it follows that nonsingularity of the matrices A_k is sufficient to ensure that a unique solution exists.

8.13.2 The transition matrix for the discrete-time equation

Define

$$\Phi(k, k_0) = A_{k-1}A_{k-2} \cdots A_{k_0}, \qquad k > k_0 \tag{8.43}$$

and

$$\Phi(k_0, k_0) = I. \tag{8.44}$$

The transition matrix can be used to generate solutions of equation (8.40) according to the relation

$$x(k) = \Phi(k, k_0)x(k_0), \qquad k \geqslant k_0. \tag{8.45}$$

The transition matrix has the properties

$$\Phi(k_0, k_0) = I \qquad \text{(stated above and clearly necessary)}$$

$$\Phi(k_2, k_1)\Phi(k_1, k_0) = \Phi(k_2, k_0), \qquad k_0 \leqslant k_1 \leqslant k_2,$$

while if all the matrices A_k are nonsingular for all k, then the expression holds for all k_0, k_1, k_2 regardless of their ordering.

For constant nonsingular matrices A

$$\Phi(k - k_0) = A^{k-k_0},$$

$$\Phi(k_1 + k_2) = \Phi(k_1)\Phi(k_2), \qquad \text{for all } k_1, k_2,$$

$$\Phi(k) = \Phi^{-1}(-k),$$

and in this case, $\Phi(k)$ can be calculated by inverse $\mathscr{Z}$ transformation as

$$\Phi(k) = \mathscr{Z}^{-1}\{zI - A^{-1}z\}. \tag{8.46}$$

8.14 Solution of the discrete-time state equation with forcing

Given the equation

$$x(k + 1) = Ax(k) + Bu(k), \tag{8.47}$$

we can write successively

$$x(1) = Ax(0) + Bu(0),$$

$$x(2) = A(Ax(0) + Bu(0)) + Bu(1),$$

until we can produce the desired result

$$x(k) = A^k x(0) + \sum_{j=0}^{k-1} A^j Bu(k - j - 1). \tag{8.48}$$

In some ways this equation is rather easier than the equivalent continuous-time equation. For instance, if we know A and B and are given a sequence $\{x(k)\}$, it is a simple matter to calculate a necessary control sequence $\{u(k)\}$ to steer the system through the given state sequence.

8.15 Obtaining the $\mathscr{L}$ transform equivalent of the state equation

Given the single-input single-output equations

$$\left.\begin{array}{l} x(k+1) = Ax(k) + Bu(k) \\ y(k) = Cx(k). \end{array}\right\} \qquad (8.49)$$

$\mathscr{L}$ transformation gives

$$zx(z) - zx(0) = Ax(z) + Bu(z)$$

$$x(z) = (zI - A)^{-1}zx(0) + (zI - A)^{-1}Bu(z)$$

Compare with equation (8.48) and it can be seen that

$$\mathscr{L}\{A^k\} = (zI - A)^{-1}z. \qquad (8.50)$$

But also from the definition of $\mathscr{L}$ transformation,

$$\mathscr{L}\{A^k\} = \sum_{k=0}^{\infty} A^k z^{-k} = I + Az^{-1} + A^2 z^{-2} + \cdots. \qquad (8.51)$$

From equation (8.49), we obtain

$$y(z) = Cx(z). \qquad (8.52)$$

If we form the transfer function $y(z)/u(z)$, say $G(z)$, then in terms of the above equations, we have

$$G(z) = C(zI - A)^{-1}B, \qquad (8.53)$$

and if we use the series expansion of equation (8.53), then we obtain

$$G(z) = \sum_{k=1}^{\infty} Ca^{k-1}Bz^{-k}. \qquad (8.54)$$

8.16 Stability tests

Let a discrete-time system Σ be described by the equation

$$x(k+1) = Ax(k). \qquad (8.55)$$

Theorem   The system Σ is asymptotically stable if each of the eigenvalues λ_i of A satisfies

$$|\lambda_i| < 1.$$

Proof   When A has distinct eigenvalues we can write equation (8.55) in terms of a new variable $q = E^{-1}x$, where E is the modal matrix, as

$$q(k) = (E^{-1}AE)^k q(0) = \begin{bmatrix} \lambda_1^k & & 0 \\ & \ddots & \\ 0 & & \lambda_n^k \end{bmatrix} q(0) \qquad (8.56)$$

and the statement is clearly true.

(The theorem applies also to the case where A has repeated eigenvalues. A modification is required to the proof, using the result that $\|A^k\| \to 0$ as $k \to \infty$, if each eigenvalue λ_i of A satisfies $|\lambda_i| < 1$.)

8.17 Reachability, controllability, observability and reconstructibility for discrete-time systems

The properties of reachability, controllability, observability and reconstructibility are defined for discrete systems exactly as for continuous-time systems (Section 8.12). However, there are some important differences in the tests for these conditions between the discrete and continuous cases.

For instance, in a time-invariant continuous system, controllability implies reachability—a continuous controllable system can be transferred from any arbitrary state to any other arbitrary state in some interval of time. For a discrete-time system to have the reachability property the condition is also required that the system matrix A is nonsingular.

If the discrete-time equation has been obtained from the continuous-time system equation, $\dot{x} = Fx$, then since $A = e^{FT}$, A always satisfies $\det A \neq 0$. However, discrete-time equations that represent inherently discrete phenomena do not necessarily satisfy this condition.

Consider the system

$$\left. \begin{array}{l} x(k+1) = Ax(k) + Bu(k), \\ y(k) = Cx(k). \end{array} \right\} \qquad (8.57)$$

It has the solution

$$x(k) = A^k x(0) + A^{k-1}Bu(0) + \cdots + ABu(k-2) + Bu(k-1)$$
$$= A^k x(0) + (B, AB, \ldots, A^{k-1}B)(u(k-1), \ldots, u(0))^T$$
$$y(k) = Cx(k). \qquad (8.58)$$

Test for reachability   For reachability, set $x(0) = 0$; then any state in the n-dimensional state space can be reached in n steps provided that the vectors $(B, AB, \ldots, A^{n-1}B)$ span the state space. This leads to the definition:

The system of equations is *reachable* if rank $(B, AB, \ldots, A^{n-1}B) = n$.

It also follows that the sequence

$$u(n-1) = (B, AB, \ldots, A^{n-1}B)^{-1}(x(n) - A^n x(0)),$$
$$\vdots$$
$$u(0) \qquad\qquad (8.59)$$

will transfer the single-input system from state $x(0)$ to state $x(n)$.

Test for controllability   Referring to equation (8.58), we set $x(k) = 0$ and note that for controllability the equation that must be satisfied is

$$A^n x(0) = -(B, \ldots, A^{n-1}B)(u(n-1), \ldots, u(0))^T,$$
$$x(0) = -(A^{-n}B, A^{-n+1}B, \ldots, A^{-1}B)(u(n-1), \ldots, u(0))^T. \qquad (8.60)$$

The condition for controllability then follows as:

The system of equations is controllable if

$$\text{rank } (A^{-n}B, \ldots, A^{-1}B) = n. \qquad (8.61)$$

(*Note*: If the matrix A is nonsingular, the conditions for controllability and reachability are identical.)

Tests for observability and reconstructibility   The following conditions apply:

The system Σ is observable if the matrix

$$\begin{bmatrix} C \\ CA \\ \vdots \\ CA^{n-1} \end{bmatrix}$$

has rank n.

The system is reconstructible if the matrix

$$\begin{bmatrix} CA^{-n} \\ CA^{-n+1} \\ \vdots \\ CA^{-1} \end{bmatrix}$$

has rank n.

Again, it can be seen that the two conditions are equivalent if the matrix A is nonsingular.

8.18 Canonical state-space representations

8.18.1 Introduction

Given a system Σ, we know that there are many possible state-space representations $\{A, B, C\}$ corresponding with Σ. In a general sense, all state-space representations can be thought of as representable by Figure 8.4.

Particular representations may arise in the first instance because of the physical background to the original problem. However, it is often advantageous to transform an arbitrary representation into one of the canonical representations that are to be described.

The advantages of the canonical representations are that they allow the system structure to be easily understood. (This is in contrast to the use of alternative representations in Chapter 5, where the motivation was to find realizations that, considered as algorithms, might be superior in terms of ease of programming, insensitivity to parameter errors or insensitivity to quantization noise.)

Let a single-input–single-output system Σ have some arbitrary but fixed state-space representation which we designate by the suffix A.

$$\left.\begin{aligned} x_A(k+1) &= A_A x(k) + B_A u(k), \\ y(k) &= C_A x(k). \end{aligned}\right\} \tag{8.62}$$

Assume also that the system Σ can be represented by the difference equation

$$(z^n + a_{n-1}z^{n-1} + \cdots + a_0)y(z) = (b_{n-1}z^{n-1} + \cdots + b_0)u(z), \tag{8.63}$$

which can be written in the concise form

$$E(z, a)y(z) = F(z, b)u(z). \tag{8.64}$$

First we consider the simpler equation

$$E(z, a)q(z) = u(z). \tag{8.65}$$

E^{-1} can be realized in two alternative ways. They are illustrated in Figures 8.5 and 8.6 for a third-order version of equation (8.63). From Figure 8.6, it follows

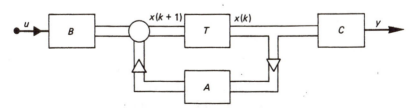

Figure 8.4   Discrete-time state-variable representation.

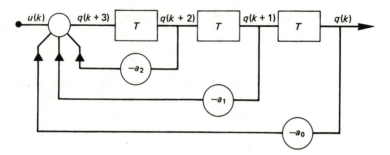

Figure 8.5   Realization of equation (8.65) for a third-order system.

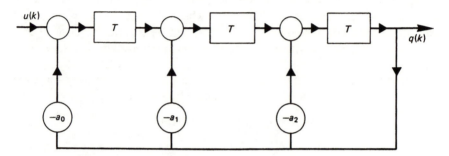

Figure 8.6   Alternative realization of equation (8.65) for a third-order system.

that

$$u(z) = (z^3 + a_2 z^2 + a_1 z + a_0)q(z).$$

Returning now to the general equation (equation (8.64)), it can be written either as

$$y(z) = F(z, b)(E(z, a)^{-1}u(z)) \tag{8.66}$$

or as

$$y(z) = E(z, a)^{-1}(F(z, b)u(z)). \tag{8.67}$$

The alternative ways of realizing E^{-1} in conjunction with equations (8.66) and (8.67) yield four different canonical realizations.

It should be understood that many alternative realizations are possible and that the names given here are by no means completely standard.

After the four canonical realizations have been given for single-input–single-output systems, the generalization to multi-input–multi-output systems is considered. Finally, the Jordan canonical form is presented.

8.18.2 The reachability canonical form

From Figure 8.7 we can write

$$x_R(k+1) = \begin{bmatrix} 0 & 0 & -a_0 \\ 1 & 0 & -a_1 \\ 0 & 1 & -a_2 \end{bmatrix} x_R(k) + \begin{bmatrix} 1 \\ 0 \\ 0 \end{bmatrix} u(k)$$

$$= A_R x_R(k) + B_R u(k), \tag{8.68}$$

$$y(k) = C_R x_R(k) = (C_{R_1}, C_{R_2}, C_{R_3}) x_R(k). \tag{8.69}$$

To fix values for C_R we note from equation (8.63) that

$$y(z) = (b_2 z^2 + b_1 z + b_0) q(z).$$

Then by careful use of the figure it can be seen that

$$C_R = (b_2, \; b_1 - a_2 b_2, \; b_0 - a_1 b_2 + a_2^2 b_2 - a_2 b_1).$$

If any difficulty is experienced with the derivation of C_R it is suggested that Exercise 8.27 should first be completed, to allow labelling of the variables within Figure 8.6. y is immediately seen to be a linear combination of the intermediate variables and C_R is chosen accordingly.

The reachability matrix for an arbitrary system $\{A, B, C\}$ is given by

$$R = (B, \; AB, \ldots, \; A^{n-1}B)$$

and here by

$$R_R = (B_R, \; A_R B_R, \ldots, \; A_R^{n-1} B_R) = I. \tag{8.70}$$

The substitution $x = R x_R$ transforms the arbitrary system representation

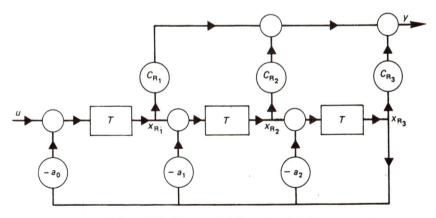

Figure 8.7 The reachability canonical form.

$\{A, B, C\}$ into the reachability form, that is

$$A_R = R^{-1}AR, \qquad B_R = \begin{bmatrix} 1 \\ 0 \\ \vdots \\ 0 \end{bmatrix}, \qquad C_R = CR.$$

The coefficients of the matrix A_R can be computed conveniently from the characteristic equation of the matrix A.

Example   Conversion of an arbitrary system to reachability form   A system is described by the equations

$$x(k + 1) = \begin{bmatrix} 1 & 1 & 0 \\ -1 & -2 & -1 \\ 0 & 1 & -1 \end{bmatrix} x(k) + \begin{bmatrix} 0 \\ 1 \\ 1 \end{bmatrix} u(k) = Ax(k) + Bu(k),$$

$$y(k) = (1 \quad 0 \quad 1)x(k) = Cx(k).$$

To convert to the reachability form, A_R can be calculated from the relation, $A_R = R^{-1}AR$, where R is the reachability matrix $R = (B, AB, A^2B)$ of the given system. However, it is easier to calculate the characteristic equation and then to use equation (8.68):

$$(zI - A) = \begin{bmatrix} z & -1 & 0 \\ 1 & z + 2 & 1 \\ 0 & -1 & z + 1 \end{bmatrix}$$

leading to the characteristic equation

$$z^3 + 3z^2 + 4z + 1 = 0.$$

Thus

$$A_R = \begin{bmatrix} 0 & 0 & -1 \\ 1 & 0 & -4 \\ 0 & 1 & -3 \end{bmatrix}, \qquad B_R = \begin{bmatrix} 1 \\ 0 \\ 0 \end{bmatrix},$$

$$C_R = CR = (B, AB, A^2B)$$

$$= (1 \quad 0 \quad 1) \begin{bmatrix} 0 & 1 & -3 \\ 1 & -3 & 5 \\ 1 & 0 & -3 \end{bmatrix} = (1 \quad 1 \quad -6).$$

(In Exercise 8.22, the reader is asked to confirm that the two representations correspond with the same transfer function.)

8.18.3 The controllability canonical form (phase-variable form) (Figure 8.8)

$$x_c(k + 1) = \begin{bmatrix} 0 & 1 & 0 \\ 0 & 0 & 1 \\ -a_0 & -a_1 & -a_2 \end{bmatrix} x_c(k) = \begin{bmatrix} 0 \\ 0 \\ 1 \end{bmatrix} u(k) \qquad (8.71)$$

$$= A_c x_c(k) + B_c u(k),$$

$$y(k) = (b_0, b_1, b_2) x_c(k). \qquad (8.72)$$

The controllability canonical form can be written down by inspection from a given difference equation.

We can transform an arbitrary representation $\{A, B\}$ into the controllability representation $\{A_c, B_c\}$ by inspection of the characteristic equation. From A, B, A_c, B_c we can then construct a transformation L between the two representations that we may use to obtain C_c, or to return a control design undertaken in terms of A_c, B_c to the real-world coordinates of the A, B system.

To proceed we form

$$Q_c = (B, AB, \ldots, A^{n-1}B),$$

$$Q'_c = (B_c, A_c B_c, \ldots, A_c^{n-1} B_c)$$

and seek a transformation L such that $x = Lx_c$. We note that

$$Ax(k) + Bu(k) = x(k + 1) = Lx_c(k + 1)$$

$$= LA_c x_c(k) + LB_c u(k) = ALx_c(k) + Bu(k),$$

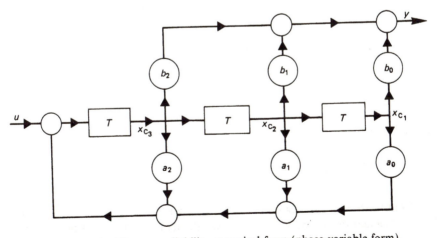

Figure 8.8   The controllability canonical form (phase-variable form).

from which

$$A = LA_cL^{-1}, \qquad B = LB_c,$$

therefore,

$$Q_c = (LB_c, LA_cL^{-1}LB_c, \ldots) = LQ'_c.$$

For a controllable single-input–single-output system, Q_c, Q'_c are square and invertible. Therefore, we can construct L or L^{-1} from the expressions

$$L = Q_c(Q'_c)^{-1}, \qquad L^{-1} = Q'_cQ_c^{-1}.$$

Example   Transform the system

$$x(k+1) = \begin{bmatrix} 1 & 1 \\ 3 & -2 \end{bmatrix} x(k) + \begin{bmatrix} 1 \\ 1 \end{bmatrix} u(k) = Ax(k) + Bu(k)$$

to controllability canonical form and determine the transformation L, linking the original representation with the canonical form.

A_c can be written down from inspection of the characteristic equation

$$(z-1)(z+2) - 3 = z^2 + z - 5 = 0.$$

We obtain immediately

$$x_c(k+1) = \begin{bmatrix} 0 & 1 \\ 5 & -1 \end{bmatrix} x_c(k) + \begin{bmatrix} 0 \\ 1 \end{bmatrix} u(k)$$

$$= A_cx_c(k) + B_cu(k),$$

$$Q_c = (B, AB) = \begin{bmatrix} 1 & 2 \\ 1 & 1 \end{bmatrix},$$

$$Q'_c = (B_c, A_cB_c) = \begin{bmatrix} 0 & 1 \\ 1 & -1 \end{bmatrix},$$

$$L = Q_c(Q'_c)^{-1} = \begin{bmatrix} 1 & 2 \\ 1 & 1 \end{bmatrix}\begin{bmatrix} 1 & 1 \\ 1 & 0 \end{bmatrix} = \begin{bmatrix} 3 & 1 \\ 2 & 1 \end{bmatrix},$$

$$L^{-1} = \begin{bmatrix} 1 & -1 \\ -2 & 3 \end{bmatrix}.$$

We may check this by verifying that $A = LA_cL^{-1}$:

$$LA_cL^{-1} = \begin{bmatrix} 3 & 1 \\ 2 & 1 \end{bmatrix}\begin{bmatrix} 0 & 1 \\ 5 & -1 \end{bmatrix}\begin{bmatrix} 1 & -1 \\ -2 & 3 \end{bmatrix} = \begin{bmatrix} 1 & 1 \\ 3 & -2 \end{bmatrix} = A.$$

Continuation of the example to show an approach to control system design
Having reached this point, we cannot resist the temptation to continue to show how useful the controllability canonical form is in the design of feedback

controllers, although we are being premature, since the topic is not treated until Section 8.21.

Suppose then that we have to use feedback by setting $u = Dx$ to ensure that the resulting closed-loop system has poles at $z = 0$, and $z = 1$. We need the closed-loop characteristic equation (recall that it is the same for all the different representations of the same system)

$$z(z - 1) = 0.$$

Let D_c be the necessary feedback matrix in the coordinates of the controllability form; then we can write down by inspection of the required closed-loop equation

$$A_c + B_c D_c = \begin{bmatrix} 0 & 1 \\ 0 & 1 \end{bmatrix}.$$

Since A_c, B_c are known, the elements d_1, d_2 of D_c must be chosen to satisfy

$$\begin{bmatrix} 0 & 1 \\ 5 & -1 \end{bmatrix} + \begin{bmatrix} 0 & 0 \\ d_1 & d_2 \end{bmatrix} = \begin{bmatrix} 0 & 1 \\ 0 & 1 \end{bmatrix},$$

from which

$$D_c = (-5 \quad 2).$$

It now remains to determine D using the relation

$$D = D_c L^{-1} = (-5 \quad 2) \begin{bmatrix} 1 & -1 \\ -2 & 3 \end{bmatrix} = (-9 \quad 11).$$

Check

$$(A + BD) = \begin{bmatrix} -8 & 12 \\ -6 & 9 \end{bmatrix}$$

which has the characteristic equation

$$z^2 - z = 0$$

as required.

Notice carefully the line of reasoning that we have used:

1. We are given a process description in the form of state-space equations.
2. We are required to devise a state-variable feedback strategy to place the closed-loop poles at given locations.
3. From (2) we can construct a characteristic equation having the required closed-loop poles as its roots. (We are aware that the characteristic equation of a system is invariant under a similarity transformation.)
4. We transform the process equations from 1 into a convenient canonical form.
5. By inspection, we choose the coefficients of a feedback matrix D_c such that the required characteristic equation will be obtained for the closed-loop system.

State-variable techniques

6. D_c needs to be returned to the real-world original coordinates. We therefore determine a matrix $\dot{L}$ that can achieve the necessary transformation of D_c into the required feedback matrix D.
7. We check that the matrix D, when used to provide feedback, does in fact achieve the desired effect.

8.18.4 The observability canonical form

See Figure 8.9 for a diagram depicting this form.

$$x_O(k+1) = \begin{bmatrix} 0 & 1 & 0 \\ 0 & 0 & 1 \\ -a_0 & -a_1 & -a_2 \end{bmatrix} x_O(k) + \begin{bmatrix} b_{01} \\ b_{02} \\ b_{03} \end{bmatrix} u(k) \qquad (8.73)$$

$$= A_O x_O(k) + B_O u(k)$$

$$y(k) = C_O \ x_O(k) = (1, 0, 0) x_O(k). \qquad (8.74)$$

The matrices A_O, B_O, C_O are related to the matrices $\{A_R, B_R, C_R\}$ in the reachability representation by

$$A_O = A_R^T, \qquad B_O = C_R^T, \qquad C_O = B_R^T.$$

Define

$$D = \begin{bmatrix} C \\ CA \\ \vdots \\ CA^{n-1} \end{bmatrix};$$

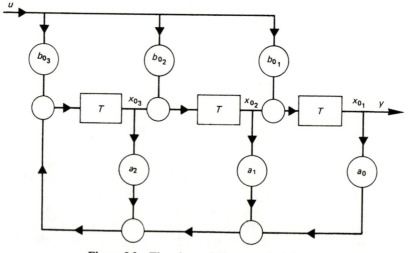

Figure 8.9 The observability canonical form.

then given an arbitrary representation $\{A, B, C\}$, the observability form can be obtained from the expressions

$$A_O = D^{-1}AD, \qquad B_O = D^{-1}B$$

(C_O is given by equation (8.74)).

8.18.5 The reconstructibility canonical form

Refer to Figure 8.10, from which the representation below follows:

$$x_p(k+1) = \begin{bmatrix} 0 & 0 & -a_0 \\ 1 & 0 & -a_1 \\ 0 & 1 & -a_2 \end{bmatrix} x_p(k) + \begin{bmatrix} b_0 \\ b_1 \\ b_2 \end{bmatrix} u(k) \qquad (8.75)$$

$$= A_p x_p(k) + B_p u(k),$$

$$y(k) = (0 \quad 0 \quad 1) x_p(k) = C_p x_p(k). \qquad (8.76)$$

The representation can be written down by inspection from a transfer function. Like the observability canonical form it has the simplest possible output matrix. Other properties follow analogously to those of the three canonical forms described previously.

The four canonical forms described above have the considerable advantage that their patterns are simple and their coefficients always real.

8.18.6 State equations for multi-input–multi-output processes

A process with n inputs and n outputs can be represented in discrete time by a set of n^2 difference equations linking the n input variables $u_1, \ldots, u_n$ with the

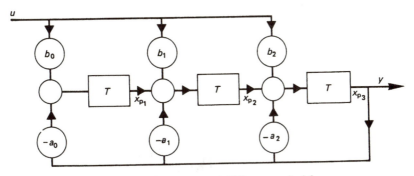

Figure 8.10 The reconstructibility canonical form.

n output variables $y_1, \ldots, y_n$. Define

$$
u = \begin{bmatrix} u_1 \\ \vdots \\ u_n \end{bmatrix}, \qquad y = \begin{bmatrix} y_1 \\ \vdots \\ y_n \end{bmatrix};
$$

then n^2 difference equations can be written with the aid of matrices A_i, B_i ($i = 1, \ldots, n$) and the operator z in the form:

$$
(Iz^n + A_{n-1}z^{n-1} + \cdots + A_0)y(z) = (B_{n-1}z^{n-1} + \cdots + B_0)u(z). \quad (8.77)
$$

Analogous to the derivations for single-input–single-output systems, state equations can be written. For instance, analogous to the derivation in Section 8.18.3, we obtain

$$
\begin{bmatrix} x_1(k+1) \\ \vdots \\ x_n(k+1) \end{bmatrix} = \begin{bmatrix} 0 & I & \cdots & 0 \\ \vdots & \vdots & & \vdots \\ -A_0 & -A_1 & \cdots & -A_{n-1} \end{bmatrix} \begin{bmatrix} x_1(k) \\ \vdots \\ x_n(k) \end{bmatrix} + \begin{bmatrix} 0 \\ \vdots \\ I \end{bmatrix} \begin{bmatrix} u_1(k) \\ \vdots \\ u_n(k) \end{bmatrix}, \quad (8.78)
$$

$$
y(k) = (B_0, \ldots, B_{n-1}) \begin{bmatrix} x_1(k) \\ \vdots \\ x_n(k) \end{bmatrix} \quad (8.79)
$$

as a state representation in which (note) each of the x_i is itself an n vector.

Example The example will resolve any problems of interpreting the notation of this section. A process is described by the equations

$$
y_1(k+2) + y_1(k) + y_2(k+1) = u_1(k) + u_2(k+1) + 2u_2(k),
$$
$$
2y_1(k) + y_2(k+2) = u_1(k+2) + u_2(k).
$$

$\mathscr{Z}$ transforming produces the equations

$$
z^2 y_1(z) + z y_2(z) + y_1(z) = z u_2(z) + u_1(z) + 2u_2(z),
$$
$$
z^2 y_2(z) + 2y_1(z) = z u_1(z) + u_2(z).
$$

In matrix form (as equation (8.78))

$$
\left(\begin{bmatrix} 1 & 0 \\ 0 & 1 \end{bmatrix} z^2 + \begin{bmatrix} 0 & 1 \\ 0 & 0 \end{bmatrix} z + \begin{bmatrix} 1 & 0 \\ 2 & 0 \end{bmatrix} \right) y(z) = \left(\begin{bmatrix} 0 & 1 \\ 1 & 0 \end{bmatrix} z + \begin{bmatrix} 1 & 2 \\ 0 & 1 \end{bmatrix} \right) u(z),
$$

leading to the state-space equations.

$$\begin{bmatrix} x_1(k+1) \\ x_2(k+1) \end{bmatrix} = \left(\begin{bmatrix} 0 & 0 \\ 0 & 0 \\ 1 & 0 \\ 2 & 0 \end{bmatrix} \begin{bmatrix} 1 & 0 \\ 0 & 1 \\ 0 & 1 \\ 0 & 0 \end{bmatrix} \right) \begin{bmatrix} x_1(k) \\ x_2(k) \end{bmatrix} + \begin{bmatrix} 0 & 0 \\ 0 & 0 \\ 1 & 0 \\ 0 & 1 \end{bmatrix} \begin{bmatrix} u_1(k) \\ u_2(k) \end{bmatrix}$$

$$\begin{bmatrix} y_1(k) \\ y_2(k) \end{bmatrix} = \begin{bmatrix} 1 & 2 & 0 & 1 \\ 0 & 1 & 1 & 0 \end{bmatrix} \begin{bmatrix} x_1(k) \\ x_2(k) \end{bmatrix}.$$

8.18.7 The Jordan canonical form

Suppose that a transfer function $G(z)$ can be decomposed into partial fractions to yield

$$G(z) = \alpha_0 + \sum_{i=1}^{n} \frac{\alpha_i}{z - z_i} \tag{8.80}$$

(α_0 is zero if the degree of the denominator exceeds that of the numerator).
Define a state vector x such that

$$y(z) = \alpha_0 u(z) + \sum_{i=1}^{n} \alpha_i x_i(z), \tag{8.81}$$

$$x_i(z) = \frac{1}{z - z_i} u(z), \tag{8.82}$$

which in the time domain becomes

$$x_i(k+1) = z_i x_i(k) + u(k), \qquad i = 1, \ldots, n$$

or in vector–matrix form

$$x(k+1) = \begin{bmatrix} z_1 & 0 & \cdots & 0 \\ 0 & z_2 & \cdots & 0 \\ \vdots & \vdots & & \vdots \\ 0 & 0 & \cdots & z_n \end{bmatrix} x(k) + \begin{bmatrix} 1 \\ 1 \\ \vdots \\ 1 \end{bmatrix} u(k), \tag{8.83}$$

$$y(k) = (\alpha_1, \alpha_2, \ldots, \alpha_n) x(k) + \alpha_0 u(k). \tag{8.84}$$

The z_i are the roots of the characteristic equation of $G(z)$. In case some of the roots are repeated, the form of the partial fractions is modified in a manner that is presumed well known to the reader.

For instance if the real root z_1 has multiplicity p, then $G(z)$ has the expansion

$$G(z) = \alpha_0 + \frac{\alpha_1}{(z - z_1)^p} + \frac{\alpha_2}{(z - z_1)^{p-1}} + \cdots + \frac{\alpha_p}{(z - z_1)} + \frac{\alpha_{p+1}}{(z - z_2)} + \cdots + \frac{\alpha_n}{(z - z_m)}.$$

$$(8.85)$$

Define

$$x_1(z) = \frac{1}{z - z_1} x_2(z) + \frac{1}{(z - z_1)^p} u(z)$$

$$x_2(z) = \frac{1}{z - z_1} x_3(z) + \frac{1}{(z - z_1)^{p-1}} u(z)$$

$$\vdots$$

$$x_p(z) = \frac{1}{z - z_1} u(z)$$

$$x_{p+1}(z) = \frac{1}{z - z_2} u(z)$$

$$\vdots$$

$$x_n(z) = \frac{1}{z - z_m} u(z).$$

In the time domain the following vector–matrix equation corresponds to equation (8.85).

a $p \times p$ block

$$x(k+1) = \begin{pmatrix} z_1 & 1 & 0 & 0 & \cdots & 0 & & & \\ 0 & z_1 & 1 & 0 & \cdots & 0 & & 0 & \\ 0 & 0 & z_1 & 1 & \cdots & 0 & & & \\ \vdots & \vdots & \vdots & \vdots & & \vdots & & & \\ 0 & 0 & 0 & 0 & \cdots & z_1 & & & \\ & & & & & & z_2 & & \\ & & 0 & & & & & \ddots & \\ & & & & & & & & z_m \end{pmatrix} x(k) + \begin{pmatrix} 0 \\ \vdots \\ 0 \\ 1 \\ 1 \\ \vdots \\ 1 \end{pmatrix} \begin{matrix} p-1 \\ \text{zeros} \end{matrix} \; u(k) \quad (8.86)$$

$$y(k) = (\alpha_1, \ldots, \alpha_n) x(k) + \alpha_0 u(k). \qquad (8.87)$$

These diagonal or block-diagonal representations are called the *Jordan forms*. Jordan forms decouple the modes of a system and allow the dynamic behavior of a system to be appreciated by inspection.

8.19 The state-variable approach to control-system design

Suppose that a single-input–single-output process can be described by the discrete-time state equations (referred to below as the model)

$$x(k + 1) = Ax(k) + Bu(k), \brace y(k) = Cx(k).$$ (8.88)

The *control problem* can be defined in a very general way as: Determine a control sequence, $u(0)$, $u(1)$,... such that the output sequence $y(0)$, $y(1)$,... behaves in some desired manner. If y has to be kept constant, this defines a *regulator problem*. If y has to follow a preset trajectory, this defines a *tracking problem*. If the control sequence has to be chosen such that y behaves in some desired manner while, at the same time, some profit function $J(u, y)$ is maximized, this defines an *optimal control problem*.

All the above problems are *open-loop control problems* in that the control sequence is determined from a prior knowledge of the model with no account taken of the actual system response.

In practice, no model is complete, nor can its parameters be specified exactly. Realistic control can only be achieved by *feedback control* in which the control sequence $u(0)$, $u(1)$,... is calculated from a knowledge of the measured output sequence $y(0)$, $y(1)$,... .

One group of powerful feedback control algorithms requires that the control sequence be calculated from a knowledge of the state vector $x(0)$, $x(1)$,... . Such a strategy is called a *state-variable feedback strategy*. It is central to state-variable control theory that if the model is linear and the cost function J is quadratic, then the optimal control sequence can be generated as a linear function of the state vector $x(k)$.

When the state vector x is not accessible or is corrupted by noise, but is required for feedback, then it can be estimated by a *state observer* (noise-free case) or a *state estimator* (noisy case). When both the model parameters and the state vector are to be estimated, a combined *state and parameter estimator* can be used.

8.20 Design of non-interacting controllers

In general, a change in any one input to a multivariable process causes changes in all the process outputs. The magnitude and dynamic characteristic of each of the output changes is of course governed by the coefficients in the process equations. In practical applications, the interaction often makes the task of a process operator difficult; if he needs to change the magnitude of one process output while leaving the remainder unchanged, it will not be obvious how such a requirement is to be achieved.

A non-interacting control system overcomes the difficulty outlined above by compensating the interactions in the process with equal and opposite cancelling interactions in the controller so that, overall, a state of non-interaction is achieved.

Achieving an ideal state of non-interaction requires exact dynamic matching of process interactions and controller compensations. Below, we have referred to such a strategy as dynamic decoupling. A less ambitious, though still useful, strategy is to neglect dynamics and achieve non-interaction only in the steady state. Such a strategy is also outlined below.

In this section we shall need to use the *Nyquist contour*. This is the path of a point that moves up the imaginary axis of the complex plane, that sweeps round to include all finite singularities in the right half plane and that returns to the origin of the complex plane along the negative imaginary axis.

DESIGN OF A CONTROLLER FOR DYNAMIC DECOUPLING
Refer to Figure 8.11 and assume that the two processes with transfer functions g_{11}, g_{22} respectively are to be controlled as shown.

Assume also that, because of interaction between the processes (indicated by dotted lines), the expected responses are not obtained.

The control problem is then to design a scheme that will give the performance as it would be obtained if interaction were absent. We proceed as follows.

Denote by $G(s)$ the transfer matrix

$$G(s) = \begin{bmatrix} g_{11}(s) & g_{12}(s) \\ g_{21}(s) & g_{22}(s) \end{bmatrix} \tag{8.89}$$

the off-diagonal terms represent the interaction effects—they must be modelled to allow us to compensate their effects.

Denote by $D(s)$ the controller that is to be designed

$$D(s) = \begin{bmatrix} d_{11}(s) & d_{12}(s) \\ d_{21}(s) & d_{22}(s) \end{bmatrix}. \tag{8.90}$$

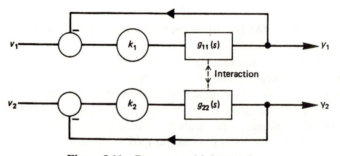

Figure 8.11 Processes with interaction.

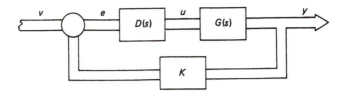

Figure 8.12 Feedback control of a multivariable process.

The configuration is shown in Figure 8.12. It represents the two interacting systems of Figure 8.11, together with the controller which is to achieve compensation. K is a diagonal feedback matrix, and y is related to v by the equation

$$y(s) = (I + GDK)^{-1}GDv(s). \tag{8.91}$$

It is clear that if

$$(I + GDK)^{-1}GD$$

is a diagonal matrix, then the interaction will be removed and each v_i will affect only the corresponding y_i as required.

Referring again to Figure 8.12, the closed-loop transfer matrix of the system can be represented as

$$H(s) = (1 + Q(s)K)^{-1}Q(s), \qquad \text{where } Q(s) = G(s)D(s). \tag{8.92}$$

If $H(s)^{-1}$ exists, then it is given by

$$H(s)^{-1} = Q(s)^{-1}(1 + Q(s)K) = Q(s)^{-1} + K$$

$$= D(s)^{-1}G(s)^{-1} + K. \tag{8.93}$$

The matrix K is diagonal and $H(s)$ is also diagonal provided that

$$D(s)^{-1}G(s)^{-1} = F(s),$$

where $F(s)$ is any diagonal matrix.

This leads to a requirement that the controller should satisfy the equation

$$D(s) = G(s)^{-1}F(s)^{-1}. \tag{8.94}$$

DESIGN USING THE PRINCIPLE OF DIAGONAL DOMINANCE

The design of a controller to achieve complete diagonalization of $H(s)$ as just outlined turns out to be very difficult to achieve in practice. However, many of the benefits of complete diagonalization will still be obtained provided that $H(s)$ is *diagonally dominant* (to be defined shortly). In particular, if $H(s)$ is diagonally dominant, then stability of individual loops guarantees stability of the overall system—this is an important property since, once the condition is satisfied, individual loops can be treated by single-input–single-output techniques. The condition is stated formally in the following theorem.

Stability theorem   Let the matrix $I + Q(s)$ corresponding with a closed-loop multivariable system Σ be diagonally dominant (either row or column dominant) for every s on the Nyquist contour. Let each of the diagonal transfer functions $q_{ii}(s)$ satisfy single-loop conditions for asymptotic stability; then the multivariable closed-loop system is asymptotically stable.

Diagonal dominance—definition   An $n \times n$ matrix A with complex entries a_{ij} is said to be diagonally (row) dominant if for $i = 1, \ldots, n$ the condition holds that

$$|a_{ii}| > \sum_{\substack{j=1 \\ j \neq i}}^{n} |a_{ij}|.$$

Similarly, A is said to be diagonally (column) dominant if for $i = 1, \ldots, n$,

$$|a_{ii}| > \sum_{\substack{j=1 \\ j \neq i}}^{n} |a_{ji}|.$$

Checking for diagonal dominance   Diagonal dominance may be checked for graphically by superimposing *Gershgorin circles* on the Nyquist plots of the diagonal transfer functions at a sufficient number of selected frequencies along the Nyquist contour. A Gershgorin circle centered at a point s_1 on the complex plane has radius

$$r_i(s_1) = \sum_{\substack{j=1 \\ j \neq i}}^{n} |q_{ij}(s_1)| \qquad \text{for the row-dominance check,}$$

$$r_i(s_1) = \sum_{\substack{j=1 \\ j \neq i}}^{n} |q_{ji}(s_1)| \qquad \text{for the column-dominance check.}$$

Assume that Gershgorin circles have been plotted along each of the frequency response graphs of the diagonal transfer functions of the matrix $I + Q(s)$. Then $I + Q(s)$ is diagonally dominant if none of the $(-1, 0)$ points is in or on a Gershgorin circle.

Provided that the matrix $Q(s)$ satisfies $\lim_{s \to \infty} Q(s) = 0$, then Gershgorin circles need to be plotted only for values of s along the imaginary axis rather than on the whole of the Nyquist contour.

Having established the basic idea, it is best for convenience of working to continue the development in terms of inverse Nyquist plots.

Recall that the closed-loop transfer matrix was given by

$$H(s) = (I + Q(s)K)^{-1}Q(s)$$

and

$$H(s)^{-1} = Q(s)^{-1} + K,$$

and with unity feedback

$$H(s)^{-1} = Q(s)^{-1} + I. \tag{8.95}$$

Restatement of the stability theorem   Let $Q(s)^{-1} + I$ and $Q(s)^{-1}$ be diagonally dominant on the Nyquist contour; then system stability depends only on the stability of the individual loops. The diagonal dominance condition is satisfied if no Gershgorin circle includes the origin or the $(-1, 0)$ point of the complex plane.

Achieving diagonal dominance   To achieve diagonal dominance of H, it is usual practice to start by achieving zero-frequency diagonalization, i.e. by choosing an initial design for the controller D that ensures that $H(s)_{s=0}$ is diagonal. For this initial choice of D, the degree of interaction at nonzero frequencies is then displayed graphically in the form of an inverse Nyquist array, using a widely available computer package (original version due to H. H. Rosenbrock). On the inverse Nyquist diagrams, Gershgorin circles indicate the upper bounds of interaction between loops. By a process of informed, interactive trial and error the interactions are reduced (by modification of D) until diagonal dominance is achieved. Reference O2 gives further information.

Design to achieve steady-state decoupling   We have seen in the previous section how the controller D may be chosen to diagonalize H at zero frequency. Physically, such a controller compensates the interactions by injecting equal and opposite signals, but without regard to the matching of the transient effects.

For many applications, this type of control is sufficient, since often the degree of process interaction is not too great and often the transients are not badly mismatched.

With such a compensating controller, a process operator is able to change the desired value v_i of any process output, knowing that in the steady state, only the corresponding output y_i changes.

Decoupling by state feedback   Decoupling may also be achieved by state-variable feedback. We illustrate for discrete-time systems.

The multivariable discrete-time process described by the equations

$$\left.\begin{array}{l} x(k + 1) = Ax(k) + Bu(k), \\ y(k) = Cx(k) \end{array}\right\} \tag{8.96}$$

and with state feedback given by

$$u(k) = Dx(k) \tag{8.97}$$

is non-interacting provided that

$$D = -(CB)^{-1}(CA - \Lambda C), \tag{8.98}$$

where

$$\Lambda = \begin{bmatrix} \lambda_1 & \cdots & 0 \\ \vdots & \ddots & \vdots \\ 0 & \cdots & \lambda_n \end{bmatrix}$$

is a diagonal matrix containing the eigenvalues of the matrix A on its leading diagonal.

Proof Set D according to equation (8.98) in the expression

$$y(k + 1) = C(A + BD)x(k)$$

to yield

$$\begin{aligned} y(k + 1) &= C(A - B(CB)^{-1}(CA - \Lambda C))x(k) \\ &= C(A - BB^{-1}C^{-1}(CA - \Lambda C))x(k) \\ &= C(A - (A - C^{-1}\Lambda C))x(k) \\ &= C\Lambda x(k) = \Lambda Cx(k) = \Lambda y(k). \end{aligned}$$

The system is seen to be decoupled as required.

8.21 Design to achieve a specified closed-loop characteristic equation

Control techniques, some of which will be outlined below, allow, within the limits of physical realizability, any process whatsoever to be given any desired closed-loop dynamic performance.

The techniques basically amount to pole-shifting by appropriately weighted state-variable feedback.

CONTROL DESIGN BASED ON STATE-VARIABLE FEEDBACK
Assume that a discretized single-input–single-output continuous process is described by the equations

$$\left. \begin{aligned} x(k + 1) &= \Phi(T)x(k) + \Psi(T)u(k), \\ y(k) &= Cx(k). \end{aligned} \right\} \tag{8.99}$$

State-variable feedback consists in replacing all or part of the input vector u by a linear combination of states, Dx, where D is a diagonal matrix to be chosen. The equation becomes, assuming all of the vector u is replaced,

$$x(k + 1) = \Phi(T)x(k) - \Psi(T)Dx(k). \tag{8.100}$$

The equation can be $\mathscr{L}$ transformed (neglecting initial condition effects) to yield

$$(zI - \Phi - \Psi D)x(z) = 0. \tag{8.101}$$

Suppose now that the closed-loop poles of the system are required to be at given locations $\alpha_1, \alpha_2, \ldots, \alpha_n$ in the z plane.

Control design then consists in choosing D so that

$$zI - \Phi - \Psi D = (z - \alpha_1)(z - \alpha_2)\ldots(z - \alpha_n). \tag{8.102}$$

We first note that, for an nth-order process, a suitable matrix D always exists to allocate the poles to n arbitrary locations in the z plane, provided that the process is controllable.

The calculation is particularly easy in those cases where the process equations, (8.99), are in one of the canonical forms (Section 8.18), since then the characteristic equation can be written down by inspection. The following section makes this clear.

Let the system to be controlled be transformed into the controllability canonical form (Section 8.18.3):

$$x(k+1) = \begin{bmatrix} 0 & 1 & \cdots & 0 \\ \vdots & \vdots & & \vdots \\ 0 & 0 & \cdots & 1 \\ -a_0 & -a_1 & & -a_{n-1} \end{bmatrix} x(k) + \begin{bmatrix} 0 \\ \vdots \\ 0 \\ 1 \end{bmatrix} u(k). \tag{8.103}$$

State feedback is introduced by setting

$$u(k) = Dx(k) = (d_0, d_1, \ldots, d_{n-1})x(k). \tag{8.104}$$

The state equation then becomes

$$x(k+1) = (A + BD)x(k)$$

$$= \begin{bmatrix} 0 & 1 & 0 & \cdots & 0 \\ \vdots & & & & \vdots \\ 0 & & & \cdots & 1 \\ (-a_0 + d_0) & & \cdots & (-a_{n-1} + d_{n-1}) \end{bmatrix} x(k). \tag{8.105}$$

The characteristic equation is

$$\det(zI - (A + BD)) = 0, \tag{8.106}$$

$$(a_0 - d_0) + \cdots + (a_{n-1} - d_{n-1})z^{n-1} + z^n = 0.$$

The state feedback matrix D can be chosen to modify the characteristic equation as required. In particular the closed-loop poles can be located where required.

Example The process

$$x(k+1) = \begin{bmatrix} 0 & 1 \\ -0.2 & 1 \end{bmatrix} x(k) + \begin{bmatrix} 0 \\ 1 \end{bmatrix} u(k)$$

is to have a double pole at $z = 0$ when under closed-loop control. By comparing with (8.103), we find $a_0 = 0.2$, $a_1 = 1$. We require $z^2 = 0$ as the closed-loop characteristic equation; that is, we require $d_0 = 0.2$, $d_1 = -1$, to make $D = (0.2, -1)$. Check

$$A + BD = \begin{bmatrix} 0 & 1 \\ 0 & 0 \end{bmatrix},$$

and the object is accomplished. (See also the example in Section 8.18.3.)

Note that if the state feedback matrix D is chosen so that in equation (8.106)

$$a_i - d_i = 0, \qquad i = 0, \ldots, n-1,$$

then the characteristic equation of the closed-loop system becomes

$$z^n = 0,$$

and this would lead to finite settling time (dead-beat) control as was shown in Section 5.2.4.

MODAL CONTROL BY STATE FEEDBACK

Let the system to be controlled have distinct eigenvalues and be in the Jordan form (Section 8.18.7):

$$x(k+1) = \begin{bmatrix} \lambda_1 & \cdots & 0 \\ \vdots & \ddots & \vdots \\ 0 & \cdots & \lambda_n \end{bmatrix} x(k) + Bu(k)$$

$$= \Lambda x(k) + Bu(k). \tag{8.107}$$

Let state feedback be applied by setting

$$u(k) = Dx(k). \tag{8.108}$$

The state equation becomes

$$x(k+1) = (\Lambda + BD)x(k). \tag{8.109}$$

The closed-loop eigenvalues can be moved by choice of the feedback matrix D. If the matrix BD is diagonal, then the closed-loop eigenvalues or modes can be moved independently. This depends on the form of B. The technique loses much of its simplicity when generalized to multi-input–multi-output processes.

Notice finally that, in common with all design methods involving a transformation of form, the implementation must be considered with care. Either the control design must be transformed to real-world coordinates or measured state variables must first be transformed to Jordan form and the calculated control vector must be inverse transformed before application to the actual system.

8.22 A brief introduction to optimal control

The most mathematically sophisticated approach to the design of a control system for some process Σ requires that we first form a scalar-valued cost function $J(x, u, \dots)$ that realistically quantifies all the process factors of importance. (For instance, in a process that produces product from raw materials, J might map variables such as process yield, deviation of product from specification, process throughput rate and energy efficiency into a single number J which, when minimized, ensures optimal overall profitability of the process.)

Given the equations of the process Σ and the cost function J, it is the role of optimal control theory to determine a control policy $u(k)$, $k = 0, \dots, n$ (n may be finite or infinite) which will achieve given control objectives and simultaneously minimize the cost function J.

In general, the control aim may require the state to follow a given trajectory, to rendezvous with a given trajectory or to reach some specified state in some specified time or in minimum time.

The cost function in general contains elements representing the effort, energy or peak amplitudes of the control signals and it may also incorporate some or all of the control aims such as the cost associated with divergence of the state from a desired trajectory.

Many practical difficulties arise with the approach just outlined, including the following:

(a) The approach requires an accurate model and a well-specified cost function— neither is readily available for a real process.
(b) The general problem cannot be solved analytically, so that an implementation would require massive continuous real-time computation.
(c) Given linear process equations and severely restricting the allowable functions J, does allow the required control policy to be calculated. The policy so obtained produces $u(k)$ as a function of $n - k$ and is very difficult to implement realistically.

However, provided that the process equations are linear and the cost function J is restricted to be simply a weighted sum of squares of its arguments, then the problem has a straightforward analytic solution and the optimal policy so obtained (and this is most important) may be generated by a linear time-varying matrix, operating on the state vector x of the process. In other words the optimal policy may be generated by state feedback. In the case where the control horizon is infinite (as in a continuously operating as opposed to a batch process) the required state feedback matrix is constant. The problem, restricted as we have just described, is often called the (infinite horizon) *quadratic regulator problem.*

TECHNIQUES AVAILABLE FOR CALCULATION OF OPTIMAL CONTROL STRATEGIES

The classical tool for solution of optimization problems is the calculus of variations. An extensive and distinguished literature exists and it can be argued that the topic is worthy of study on intellectual grounds. However, the calculus of variations cannot deal with the discontinuous functions that arise frequently in control applications.

Bellman's dynamic programming approach leads to numerical techniques which, supported by sufficient computing power, offer the best possibility for solution of complex optimization problems.

Pontryagin's maximum principle solves problems without discontinuities in a very similar series of steps to those obtained when using the calculus of variations. However, the method is more general than the calculus of variations in that it works also with discontinuous control functions. We have chosen the maximum principle as the preferred method to illustrate the calculation of optimal control.

A STATEMENT OF PONTRYAGIN'S MAXIMUM PRINCIPLE

The background to the maximum principle may be found in Reference P3 or alternatively in many specialist texts on optimization.

Let Σ be a process described by the equations

$$\dot{x}_i = f_i(x_1, \ldots, x_n, u_1, \ldots, u_r, t) = f_i(x, u, t)) \tag{8.110}$$

and let J be a cost function $J: X \to \mathbb{R}^1$, where X is the state space and $\mathbb{R}^1$ is the real line and where J can be expressed in the form

$$J = \sum_{i=1}^{n} \alpha_i x_i(T). \tag{8.111}$$

n auxiliary *adjoint variables* $p_1, \ldots, p_n$ are introduced and a system Hamiltonian H is defined by the equation

$$H = \pm \sum_{i=1}^{n} p_i f_i(x, u, t) \tag{8.112}$$

(with the plus sign being taken for maximization). The adjoint variables are required to satisfy the condition

$$\dot{p}_i = -\frac{\partial H}{\partial x_i}, \qquad i = 1, \ldots, n. \tag{8.113}$$

If any target values have been designated for a particular state $x_i(T)$, then $p_i(T)$ is free. However, if any particular ith state has a free condition at time

T, then the corresponding adjoint variable $p_i(T)$ must satisfy

$$p_i(T) = -\alpha_i. \qquad (8.114)$$

The maximum principle then states that J will be minimized (maximized) by a choice of control input $u(t)$ that maximizes (minimizes) the Hamiltonian function at every instant t in the interval $(0, T)$.

For problems where the control function is continuous, the Hamiltonian is maximized by setting $\partial H/\partial u_i = 0$ for $i = 1, \ldots, n$. In general, though, the maximization of H cannot be achieved by the methods of ordinary calculus because of the presence of constraints on the control and the consequent discontinuities.

APPLICATION OF THE PONTRYAGIN MAXIMUM PRINCIPLE TO CONTROL PROBLEMS

To apply the principle to a control problem, the Hamiltonian is augmented to become

$$H = \frac{\partial J}{\partial t}(x_1, \ldots, x_n, u_1, \ldots, u_r, t)$$

$$+ \sum_{i=1}^{n} p_i f_i(x_1, \ldots, x_n, u_1, \ldots, u_r, t) \qquad (8.115)$$

and a dummy state variable x_{n+1} often needs to be introduced.

The solution then follows using the approach outlined. The following example illustrates this.

Example A process is described by the equations

$$\begin{bmatrix} \dot{x}_1 \\ \dot{x}_2 \end{bmatrix} = \begin{bmatrix} -1 & 0 \\ 1 & 0 \end{bmatrix}\begin{bmatrix} x_1 \\ x_2 \end{bmatrix} + \begin{bmatrix} 1 \\ 0 \end{bmatrix}u,$$

$$J = \int_0^T \left(x^T \begin{bmatrix} 0 & 0 \\ 0 & 1 \end{bmatrix} x + au^2 \right) dt,$$

$$H = -(x_2^2 + au^2) + p_1(u - x_1) + p_2 x_1,$$

$$\frac{\partial H}{\partial u} = -2au + p_1,$$

leading to $u = p_1/2a$.

The equations for the adjoint variables are obtained (see equation (8.113)) as

$$\dot{p}_1 = p_1 - p_2,$$

$$\dot{p}_2 = 2x_2.$$

J has to be expressed in the form

$$J = \sum_{i=1}^{n+1} \alpha_i x_i(T).$$

This is achieved by introducing the dummy variable x_3 defined so that

$$\dot{x}_3(T) = \frac{\partial J}{\partial t},$$

so that $x_3(T) = J$. Therefore $\alpha_1 = \alpha_2 = 0$, $\alpha_3 = 1$, so that final conditions for the adjoint variables are

$$p_1(T) = 0, \qquad p_2(T) = 0.$$

The value found for u can be substituted into the process equations to yield

$$\left.\begin{aligned} \dot{x}_1 &= -x_1 + p_1/2a, \\ \dot{x}_2 &= x_1. \end{aligned}\right\}$$

In this very simple example, the state equations and the adjoint equations can be successively substituted to yield

$$\frac{d^4 x_2}{dt^4} - \frac{d^2 x_2}{dt^2} + \frac{x_3}{a} = 0,$$

which has the solution

$$x_2 = e^{-c_1 t}(c_2 e^{j\theta t} + c_3 e^{-j\theta t}) + e^{c_1 t}(c_4 e^{j\theta t} + c_5 e^{-j\theta t})$$

(in which θ depends on a).

From the original equations

$$x_1 = \dot{x}_2 = e^{-c_1 t}((-c_1 + j\theta)c_2 e^{j\theta t} + (-c_1 - j\theta)c_3 e^{-j\theta t})$$
$$+ e^{c_1 t}((c_1 + j\theta)c_4 e^{j\theta t} + (c_1 - j\theta)c_5 e^{-j\theta t}),$$

while the control u can be found from the original equations as

$$u(t) = \dot{x}_1(t) + x_1(t).$$

The equations for x_1 and x_2 can be simplified to the forms

$$x_2(t) = c_6 e^{-c_1 t} \sin(\theta t + \phi),$$
$$x_1(t) = c_6 e^{-c_1 t}(\theta \cos(\theta t + \phi) - c_1 \sin(\theta t + \phi)).$$

Let $t = 0$; then

$$x_2(0) = c_6 \sin \phi,$$
$$x_1(0) = c_6 \theta \cos \phi - c_1 c_6 \sin \phi,$$
$$\dot{x}_1(0) = c_6(-\theta^2 \sin \phi - c_1 \theta \cos \phi) - c_1 c_6(\theta \cos \phi - c_1 \sin \phi).$$

Let us now assume that T is infinite, i.e. the control horizon is infinite. Then the control policy becomes time invariant, represented by

$$u(0) = \dot{x}_1(0) + x_1(0).$$

Finally, we know, since the problem is linear with a quadratic cost function, that the control may be generated by a feedback strategy. Seeking this, we proceed as follows: Substituting the given expression for $\dot{x}_1(0)$ yields

$$u(0) = -\theta^2 c_6 \sin\phi - 2\theta c_1 c_6 \cos\phi + c_1^2 c_6 \sin\phi + x_1(0)$$

$$= x_1(0)(1 - 2c_1) + c_6(\theta^2 + c_1^2)x_2(0)$$

$$= k_1 x_1(0) + k_2 x_2(0),$$

and this is the required optimal feedback strategy.

SOLUTION OF THE QUADRATIC REGULATOR PROBLEM BY MEANS OF THE
MATRIX RICCATI EQUATION

The matrix Riccati equation offers a systematic method for the solution of the quadratic regulator problem.

We assume, as will usually be the case, that the process to be controlled operates in continuous time and that an optimal digital controller is to be designed.

Since optimality is the aim, we shall not begin by transforming the process equations into discrete approximations. To do so would mean that the controller so obtained would not be optimal with respect to the original equations.

The process Σ is therefore represented by the continuous-time equations

$$\dot{x}(t) = Ax(t) + Bu(t)$$

and the cost function is defined in the form

$$J = \int_0^T (x^T P x + u^T Q u)\,\mathrm{d}t.$$

Define

$$x^T R x = \min_u \int_0^T (x^T P x + u^T Q u)\,\mathrm{d}t.$$

It can then be shown, using the Pontryagin method, that the problem is equivalent to the minimization problem

$$\min_u \{x^T P x + u^T Q u + (Ax + Bu)^T R x + x^T R(Ax + Bu)\} + x^T \dot{R} x = 0.$$

Differentiating and simplifying yields the equation

$$Qu + B^T R x = 0$$

or

$$u = -Q^{-1}B^T Rx. \tag{8.116}$$

Substitution and simplification produces the equation

$$\dot{R} = RBQ^{-1}B^T R - B - C^T R - RC, \tag{8.117}$$

which is called a matrix Riccati equation.

The equation can be solved for R numerically under the given condition $R(T) = 0$.

The required feedback control can then be implemented using equation (8.116). Notice carefully that R varies with time so that in general, the feedback matrix $-Q^{-1}B^T R$ is time-varying.

In case $T = \infty$, i.e. the control horizon is infinite, then R is constant and a constant feedback matrix results.

8.23 State estimation

8.23.1 The reasons for using state estimation

Most of the sophisticated and ambitious feedback controllers that can be designed using state-space techniques require a knowledge of the magnitude of the state vector of the system that is to be controlled.

Often the state vector is not available (inaccessible state) or only partially available, in which case an estimator has to be used to reconstruct the state.

Even when the state vector is available for measurement, it is always more or less corrupted by noise and there is still a need for estimation of the 'true' state vector of the system.

Of very great practical importance is the estimation of important physical variables within a process, from more easily available peripheral variables. Examples are the estimation of temperature inside large blocks of metal, based on other available measurements, and the estimation of biomass activity in fermentation processes, based on gas analysis and other measurements.

8.23.2 State observer

Under the ideal conditions that we have an exact model of an observable process Σ and that the signals are noise free, the states of Σ can be determined exactly using an observer.

The observer is simply a model of the process. It is assumed that the initial state of the process is unknown. The only problem is therefore to set up the observer such that the initial error in specifying the states tends to zero; that is, such that

$$\hat{x}(k) - x(k) \to 0 \qquad \text{as } k \to \infty$$

where the superscript indicates the state of the observer.

This convergence is assured by driving the state of the observer until the output $\hat{y}(k)$ agrees with the process output $y(k)$.

Let the process be described by the equations

$$\left. \begin{array}{l} x(k+1) = Ax(k) + Bu(k), \\ y(k) = Cx(k). \end{array} \right\} \tag{8.118}$$

Then the observer has the equation

$$\hat{x}(k+1) = A\hat{x}(k) + Bu(k) + R(y(k) - C\hat{x}(k)). \tag{8.119}$$

The last term performs the correction to ensure that $\hat{y}(k) = y(k)$ for k sufficiently large.

Define the observer error as $\tilde{x}(k) = x(k) - \hat{x}(k)$. From the process and model equations

$$\tilde{x}(k+1) = A(x(k) - \hat{x}(k)) - RC(x(k) - \hat{x}(k))$$
$$= (A - RC)\tilde{x}(k), \tag{8.120}$$

clearly this last equation must be asymptotically stable if the observer is to be effective. Additionally R can be chosen by any of the normal control design techniques to meet particular performance objectives in the observer.

In applications where some of the state variables can be measured directly, the remaining (non-measurable) state variables may be determined using a so-called *reduced order observer* that is obtained through obvious modifications to the above derivation.

8.23.3 The principle of state estimation

Consider a process Σ for which input and output measurements are available and where the state is to be estimated.

A model of the process is available in the form

$$\left. \begin{array}{l} x(k+1) = Ax(k) + Bu(k), \\ y(k) = Cx(k). \end{array} \right\} \tag{8.121}$$

(*Notation*: We denote the best estimate of the state at time k based on measurements up to time j by $\hat{x}(k|j)$.)

Assume that, somehow, we have obtained the estimate $\hat{x}(j|j)$. Using equation (8.121) we can make a one-step ahead prediction as

$$\hat{x}(j+1|j) = A\hat{x}(j|j) + Bu(j). \tag{8.122}$$

We can also predict the system output

$$\hat{y}(j + 1|j) = C\hat{x}(j + 1|j) \tag{8.123}$$

and when time $j + 1$ is reached we can set

$$e(j + 1) = y(j + 1) - \hat{y}(j + 1|j). \tag{8.124}$$

e is a measure of the goodness of the prediction. If e is nonzero we can use it to improve the estimate of the state vector.

$$\hat{x}(j + 1|j + 1) = \hat{x}(j + 1|j) + \alpha e(j + 1), \tag{8.125}$$

where α is a coefficient to be chosen.

Equation (8.125) might be regarded as the basic archetypal estimator algorithm. It remains to choose the gain. The choice of the gain α is difficult.

In general α needs to be time-varying with its value dependent on the statistical properties of modelling errors and measurement noise. Should these statistical properties be known, then techniques due to R. E. Kalman can be used to specify α and the estimator becomes known as the Kalman filter—its operation is outlined below.

We note before leaving this section that the estimator as described is a dynamic element. It is convenient to study its dynamic behavior in terms of estimation error. Define

$$\tilde{x} = x - \hat{x}; \tag{8.126}$$

then $\tilde{x}$ is the error made by the estimator.

From the equations of the model and the estimator

$$\begin{aligned}
\tilde{x}(k + 1) &= x(k + 1) - \hat{x}(k + 1|k + 1) \\
&= Ax(k) + Bu(k) - (A\hat{x}(k|k) + Bu(k) \\
&\quad + \alpha(C(Ax(k) + Bu(k)) - C(A\hat{x}(k|k) + Bu(k)))) \\
&= (I + \alpha C)A\tilde{x}(k). \tag{8.127}
\end{aligned}$$

The stability and rate of convergence of the estimator depend on the choice of α. Clearly it will be the aim that $\tilde{x}(k) \to 0$ as $k \to \infty$ with rapid convergence.

8.23.4 The Kalman filter

Let the process be modelled by the equations

$$\left.\begin{aligned}
x(j + 1) &= Ax(j) + Bu(j) + Ew(j), \\
y(j) &= v(j).
\end{aligned}\right\} \tag{8.128}$$

Equation (8.128) represents a linear process with time-invariant parameters disturbed by noise inputs w and v acting on the process and on the output respectively. The covariance matrices of these noise signals are defined as Q and R respectively.

The problem is:

Given the noise-corrupted data, y, determine the best estimate $\hat{x}$, in a least-squares sense, of the process internal state vector x.

HEURISTIC DERIVATION OF THE KALMAN GAIN

The noise vectors v and w are assumed to have idealized properties of independence and complete randomness ('whiteness'). Their covariance matrices are the (diagonal) matrices Q and R respectively.

A one-step-ahead prediction is made as follows:

$$\left.\begin{array}{l} x(j|j-1) = A\hat{x}(j-1|j-1) + Bu(j-1), \\ \hat{y}(j) = C\hat{x}(j|j-1). \end{array}\right\} \tag{8.129}$$

The random vectors do not appear in the prediction since their expected values are by definition zero.

When a measurement $y(j)$ is made, the prediction error $\tilde{y}(j)$ can be calculated by the expression

$$\tilde{y}(j) = y(j) - \hat{y}(j). \tag{8.130}$$

We now argue that the optimal estimate $\hat{x}(j|j)$ will be of the form

$$\hat{x}(j|j) = \hat{x}(j|j-1) + K\tilde{y}(j), \tag{8.131}$$

where K is the Kalman gain matrix. From above,

$$\hat{x}(j|j) = \hat{x}(j|j-1) + K\{y(j) - C\hat{x}(j|j-1)\} \tag{8.132}$$

and by substitution

$$\hat{x}(j|j) = A\hat{x}(j-1|j-1) + Bu(j-1) + K\{y(j) - C\hat{x}(j|j-1). \tag{8.133}$$

Equation (8.133) is the state estimation algorithm. The Kalman gain matrix K needs to be calculated before it can be used.

The state estimation error is defined

$$\tilde{x}(j) = x(j) - \hat{x}(j|j) \tag{8.134}$$

which can be expressed

$$\tilde{x}(j) = (I - KC)(A\tilde{x}(j-1) + Ew(j-1)) - Kv(j), \tag{8.135}$$

where I is the identity matrix.

A covariance matrix $P(j)$ is defined as

$$P(j) = \mathscr{E}\{\tilde{x}(j)\tilde{x}(j)^T\}, \tag{8.136}$$

where $\mathscr{E}$ indicates expected value. P is a matrix that indicates the 'goodness' of the state estimation.

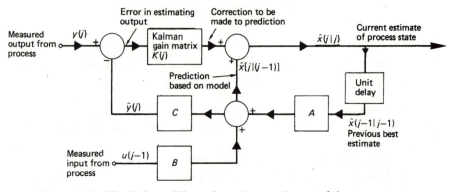

Figure 8.13   The Kalman Filter—it produces estimates of the process state.

Equation (8.135) is substituted into equation (8.133) and it yields a quadratic in the matrix K which can be solved after some manipulation to yield, together with earlier equations, the final algorithm:

$$P(j) = (I - K(j)C)M(j), \qquad (8.137)$$

where

$$M(j) = AP(j-1)A^T + EQE^T, \qquad (8.138)$$

$$K(j) = M(j)C^T(CM(j)C^T + R)^{-1}, \qquad (8.139)$$

$$\hat{x}(j|j) = (I - K(j)C)\{A\hat{x}(j-1|j-1) + Bu(j-1)\} + K(j)y(j). \qquad (8.140)$$

The equation for $\hat{x}$ is the on-line estimator. It makes use of the three equations above it, but they require no process data input and so can be pre-computed if necessary.

For the example treated here (linear, time-invariant process, noise assumptions satisfied) the matrices P and K tend to a steady state as j increases. Figure 8.13 shows a block diagram of the filter algorithm.

When, as is often the case, the noise characteristics fail to satisfy the whiteness assumption, the situation may still be handled by postulating a fictitious white noise source feeding into a noise model. This noise model is then considered to be part of the process and the enhancement allows the original assumptions to be satisfied.

To obtain a feeling for the essential structure of the Kalman algorithm we treat below the restricted case of a single-input–single-output process and illustrate the operation with a simple numerical example.

THE KALMAN ALGORITHM FOR A SINGLE-INPUT–SINGLE-OUPUT PROCESS
(This development follows reference T1).

Consider the process

$$\left.\begin{array}{l} x(k + 1) = ax(k) + bu(k) + ew(k), \\ y(k) = cx(k) + v(k), \end{array}\right\} \qquad (8.141)$$

where the symbols have the same meanings, translated to a single-input–single-output context, as in equation (8.128).

The problem is:

Given the sequence of measurements $\{y(0), \ldots, y(k)\}$ (which we denote by S_k) and a knowledge of the constants a, b, e, c, and of the variances q, r of the noise signals w and v respectively, obtain a best estimate for the state $x(k)$, in a least squares sense.

Define the estimation error by $\tilde{x}(t)$, i.e.

$$\tilde{x}(k) = x(k) - \hat{x}(k|k),$$

$x(k)$, because of its perturbation by noise inputs, is a random variable. We define the mean and variance of $x(k)$ by $\mu(k)$, $\sigma^2(k)$; then

$$\mathscr{E}\{x(k)\} = \mu(k)$$

$$\mathscr{E}\{(x(k) - \mu(k))^2\} = \sigma^2(k). \qquad (8.142)$$

The derivation proceeds by minimization of $\tilde{x}(k)^2$,

$$\tilde{x}(k)^2 = x(k)^2 - 2x(k)\hat{x}(k|k) + \hat{x}(k|k)^2.$$

Expected values are substituted for $x(k)$, $x(k)^2$ to yield

$$\mathscr{E}\{\tilde{x}(k)^2|S_k\} = \mathscr{E}\{x(k)^2|S_k\} - 2\hat{x}(k|k)\mathscr{E}\{x(k)|S_k\} + \hat{x}(k|k)^2. \quad (8.143)$$

Now

$$\mu(k) = \mathscr{E}\{x(k)|S_k\} \qquad (8.144)$$

and

$$\begin{aligned}
\sigma(k)^2 &= \mathscr{E}\{(x(k) - \mu(k))^2|S_k\} \\
&= \mathscr{E}\{x(k)^2|S_k\} - 2\mu(k)E\{x(k)|S_k\} + \mu(k)^2 \\
&= \mathscr{E}\{x(k)^2|S_k\} - 2\mu(k)^2 + \mu(k)^2, \qquad (8.145)
\end{aligned}$$

from which

$$\mathscr{E}\{x(k)^2|S_k\} = \sigma(k)^2 + \mu(k)^2. \qquad (8.146)$$

Substituting this last result into equation (8.143) yields

$$\begin{aligned}
\mathscr{E}\{\tilde{x}(k)^2|S_k\} &= \hat{x}(k|k)^2 - 2\mu(k)\hat{x}(k|k) + \sigma(k)^2 + \mu(k)^2 \\
&= (\hat{x}(k|k) - \mu(k))^2 + \sigma(k)^2. \qquad (8.147)
\end{aligned}$$

The expected value of the estimation error $\tilde{x}(k)$ may be minimized by setting

$$\hat{x}(k|k) = \mu(k).$$

The estimation algorithm can then be shown to be

$$\hat{x}(k|k) = \hat{x}(k|k-1) + K(k)(y(k) - c\hat{x}(k|k-1))$$

$$= a\hat{x}(k-1|k-1) + bu(k-1)$$

$$+ K(k)(y(k) - c(a\hat{x}(k-1|k-1) + bu(k-1))). \quad (8.148)$$

Notice that the prediction does not contain the term in $ew(k)$ since the expected value of $w(k)$ is zero.

The algorithm is serviced by the following equations:

$$K(k) = \frac{m(k)c}{c^2 m(k) + r} \quad (8.149)$$

(The Kalman gain).

$$m(k) = a^2 p(k-1) + e^2 q \quad (8.150)$$

(The variance of the one step ahead prediction).

$$p(k) = (1 - K(k)c)m(k) \quad (8.151)$$

(The variance of the random variable $x(k)$).

The last two equations may be combined into the single equation

$$m(k) = \frac{a^2 r(m(k-1))}{c^2 m(k-1) + r} + e^2 q. \quad (8.152)$$

The algorithm needs to be started by setting in a value for $\hat{x}(0|-1)$ and for $m(0)$ (physically $m(0) = \mathscr{E}\{x(0)^2\}$).

Notice that the equations (8.149)–(8.151) contain no measurement information and that therefore they can be calculated *a priori* if required. In fact K, the Kalman gain, reaches a steady state fairly rapidly and for many applications the filter is most conveniently implemented using only the steady-state value.

Numerical example   In equation (8.148) and its sequel, let $a = 0.5$, $b = 0$, $c = 1$, $e = 1$, $q = 3$, $r = 1$.

$$x(k+1) = 0.5x(k) + w(k),$$

$$y(k) = x(k) + v(k).$$

For these values we have the equations

$$K(k) = \frac{m(k)}{1 + m(k)},$$

$$m(k) = \frac{0.25m(k-1)}{1 + m(k-1)} + 3.$$

Table 8.4 lists the first four values of k, $m(k)$, $K(k)$.

Table 8.4

k	m(k)	K(k)
0	3	0.75
1	3.1875	0.761
2	3.1903	0.761
3	3.1903	0.761

In the steady state the algorithm is

$$\hat{x}(k|k) = 0.25\hat{x}(k-1|k-1) + 0.761(y(k) - 0.25\hat{x}(k-1|k-1))$$

$$\hat{x}(k|k) = 0.05975\hat{x}(k-1|k-1) + 0.761y(k).$$

Taking $\mathscr{L}$ transforms,

$$z\hat{x}(z) = 0.05975\hat{x}(z) + zy(z),$$

$$\frac{\hat{x}(z)}{y(z)} = \frac{z}{z - 0.05975}.$$

Compare this with the transfer function $x(z)/w(z)$ of the process

$$\frac{x(z)}{w(z)} = \frac{z}{z - 0.5}.$$

We can see that the filter is some eight times faster than the process—a desirable characteristic.

8.23.5 Variants of the Kalman filter

Before the Kalman filter can be put into operation, the statistical properties (designated Q and R) of the process noise and measurement noise need to be specified numerically. In practical cases, there is rarely sufficient accurate information on these sources of noise, and herein lies one of the major difficulties that impede the application of the filter.

When the statistical properties have been accurately specified, the estimation errors made during the operation of the filter have Gaussian characteristics.

Using information from the sequence of estimation errors, it is possible to modify the values of Q and R during filter operation. With this added facility the estimator is called an *adaptive Kalman filter*.

The Kalman filter needs to contain a mathematical model of the process for which it is the estimator. In actual cases, a completely time-invariant model is unlikely to agree with the real process over a long time period. Thus, some or all of the model parameters may need periodic updating. Such a facility can be incorporated into the Kalman filter. Any parameter that needs to be estimated can be designated as an (inaccessible) dummy state variable and estimated along with the state vector.

Such an algorithm would normally be called a *combined state and parameter estimator*. See however the following paragraph on nonlinearity.

Many important processes are inherently nonlinear and nonlinearity will nearly always be introduced when (see above) process parameters are designated as dummy state variables. The Kalman filter can still be applied provided that the process equations are linearized at every time step. A different linear model has therefore to be used at each time step during the operation of the filter. The amount of computation is very considerably increased. The estimator incorporating this facility is called the *extended Kalman filter*.

8.24 The separation theorem

Consider the process described by the equations

$$x(k + 1) = A(k)x(k) + B(k)(u(k) + w(k)), \atop y(k) = C(k)x(k) + p(k), \Bigg\} \tag{8.153}$$

where $w(k)$ and $p(k)$ are independent Gaussian random noise sequences.

Because of the presence of the noise sequences, no exact measurements are available—this is a stochastic control problem. Optimal control for such a problem consists in minimizing the expected value of some index of performance.

Provided that the process equations are linear and that the optimal control problem is posed in terms of a quadratic performance index, it can be shown that state estimation and optimal control do not interact. (This result is known as the *separation theorem*.) This means that the state estimates from a stochastic estimator can be fed directly into an optimal control law, which is identical to a deterministic optimal control algorithm.

Thus, the control of the system described by equation (8.153) can be achieved, for instance, by producing an estimate $\hat{x}$ of the state vector x using the Kalman filter. The feedback loop can be completed by operating on $\hat{x}$ with an algorithm based on matrix Riccati techniques.

The overall control achieved will be identical to that obtained from taking an integrated approach to the combined optimal estimation–optimal control problem.

THE SEPARATION THEOREM—ILLUSTRATION

Let a process Σ have input, state and output variables u, x, y and be described by the equation

$$\dot{x}(t) = Ax(t) + Bu(t). \tag{8.154}$$

Discretization yields

$$x(k + 1) = \Phi(T)x(k) + \Psi(T)u(k). \tag{8.155}$$

A state-variable feedback strategy is to be implemented by setting

$$u(k) = -Dx(k), \tag{8.156}$$

where D is a known feedback matrix.

Assume next that, in the absence of measurements of the state vector x, it is proposed to use an on-line estimate $\hat{x}$ to achieve closed-loop control. The system equation now has the form

$$x(k + 1) = \Phi(T)x(k) - \Psi(T)D\hat{x}(k). \tag{8.157}$$

Turning now to the estimator, it has an equation of the form

$$\tilde{x}(k + 1) = (I + \alpha C)A\tilde{x}(k) \tag{8.158}$$

where $\tilde{x}(k)$ indicates as before the error between state $x(k)$ and estimate $\hat{x}(k)$,

$$\tilde{x}(k) = x(k) - \hat{x}(k). \tag{8.159}$$

The dynamic equations of the process with its control system and estimation system are then given by

$$x(k + 1) = \Phi(T)x(k) - \Psi(T)D(x(k) - \tilde{x}(k)), \tag{8.160}$$

$$\tilde{x}(k + 1) = (I + \alpha C)A\tilde{x}(k), \tag{8.161}$$

or

$$\begin{bmatrix} x(k + 1) \\ \tilde{x}(k + 1) \end{bmatrix} = \begin{bmatrix} \Phi(T) - \Psi(T)D & \Psi(T)D \\ 0 & (I + \alpha C)A \end{bmatrix} \begin{bmatrix} x(k) \\ \tilde{x}(k) \end{bmatrix}. \tag{8.162}$$

And the characteristic equation that governs the overall system is

$$\det(\lambda I - (\Phi(T)D))\det(\lambda I - (I + \alpha C)A) = 0. \tag{8.163}$$

That is, the poles of the composite system are just the poles of the two separate systems and there is no interaction. The estimation can be designed to produce an optimal estimate $\hat{x}$, the optimal control system can be designed as though $\hat{x}$ were the true state vector and the overall composite system will itself be optimal, although it has been designed in two separate exercises (see Figure 8.14).

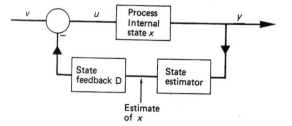

Figure 8.14 Illustration of the separation principle.

Worked example 8.1

A system has the equation

$$\ddot{y} + 8\dot{y} + 7y = 0, \qquad \left.\begin{aligned} y(0) &= 100 \\ \dot{y}(0) &= 0. \end{aligned}\right\} \tag{W1}$$

Setting $x_1 = y$, $x_2 = \dot{y}$, express the system in state variable form

$$\dot{x} = Ax. \tag{W2}$$

Determine the eigenvalues of the system. Is the system stable? Give reasons. Is the system oscillatory? Give reasons.

Determine the transition matrix for the system.

With the help of the transition matrix or otherwise, determine the time T^* at which x_2 takes on its greatest absolute value.

Using a time step of 1 second and again using the transition matrix, express equation (W2) in the form

$$x(k) = Px(k - 1) \tag{W3}$$

What is the relation between the eigenvalues of the matrix P and those of the matrix A? Check your assertion numerically.

Use a time step of T^* to recalculate P for equation (W3) and hence determine the maximum absolute velocity.

Return to the original equation (W1). Solve it analytically and use the result to sketch the transient behavior of y and $\dot{y}$.

Worked solution 8.1

$x_1 = y$, therefore $\dot{x}_1 = \dot{y} = x_2$ and $\ddot{y} = \dot{x}_2$. Equation (W1) can be written as two equations

$$\dot{x}_1 = x_2$$
$$\dot{x}_2 = -7x_1 - 8x_2$$

where

$$\begin{bmatrix} \dot{x}_1 \\ \dot{x}_2 \end{bmatrix} = \begin{bmatrix} 0 & 1 \\ -7 & -8 \end{bmatrix} \begin{bmatrix} x_1 \\ x_2 \end{bmatrix}$$

or

$$\dot{x} = Ax.$$

This is the system in state-variable form.

To find the eigenvalues, we need to solve the characteristic equation

$$\det(\lambda I - A) = 0.$$

Now

$$\lambda I - A = \begin{bmatrix} \lambda & 0 \\ 0 & \lambda \end{bmatrix} - \begin{bmatrix} 0 & 1 \\ -7 & -8 \end{bmatrix}$$

$$= \begin{bmatrix} \lambda & -1 \\ 7 & \lambda + 8 \end{bmatrix}$$

with determinant $\lambda^2 + 8\lambda + 7$ and the characteristic equation is

$$\lambda^2 + 8\lambda + 7 = 0.$$

Hence the eigenvalues are the solutions of this equation, i.e.

$$\lambda_1 = -1, \qquad \lambda_2 = -7$$

Since both eigenvalues have negative real parts, the system is asymptotically stable.

Since no eigenvalue has an imaginary component, the system is not oscillatory. Each eigenvector e_i must satisfy $(\lambda_i I - A)e_i = 0$ by definition. So we must have, putting $\lambda_i = \lambda_1 = -1$,

$$\begin{bmatrix} -1 & -1 \\ 7 & 7 \end{bmatrix}\begin{bmatrix} e_{11} \\ e_{21} \end{bmatrix} = 0$$

or

$$e_{11} - e_{12} = 0.$$

Put $e_{11} = 1$ then $e_{21} = -1$. Similarly, putting $\lambda_i = \lambda_2 = -7$, we obtain

$$\begin{bmatrix} -7 & -1 \\ 7 & 1 \end{bmatrix}\begin{bmatrix} e_{12} \\ e_{22} \end{bmatrix} = 0$$

or

$$-7e_{12} - e_{22} = 0$$

putting $e_{12} = 1$ yields $e_{22} = -7$.

The modal matrix

$$E = \begin{bmatrix} e_{11} & e_{12} \\ e_{21} & e_{22} \end{bmatrix} = \begin{bmatrix} 1 & 1 \\ -1 & -7 \end{bmatrix}.$$

The inverse of E is found to be

$$E^{-1} = \frac{1}{6}\begin{bmatrix} 7 & 1 \\ -1 & -1 \end{bmatrix}.$$

Now

$$\Phi(t) = E\begin{bmatrix} e^{\lambda_1 t} & 0 \\ 0 & e^{\lambda_2 t} \end{bmatrix}E^{-1}$$

(see Section 8.6.1). Hence

$$\Phi(t) = \begin{bmatrix} 1 & 1 \\ -1 & -7 \end{bmatrix} \begin{bmatrix} e^{-t} & 0. \\ 0 & e^{-7t} \end{bmatrix} \begin{bmatrix} 7 & 1 \\ -1 & -1 \end{bmatrix} \frac{1}{6}$$

is the required transition matrix.

Considering only x_2, we have

$$x_2(t) = \tfrac{1}{6}(\phi_{21}(t)x_1(0) + \phi_{22}x_2(0)).$$

But $x_2(0) = \dot{y}(0)$ is zero in this question. Hence

$$x_2(t) = \tfrac{1}{6}(-7\,e^{-t} + 7\,e^{-7t})x_1(0)$$

and we can find maxima or minima by differentiating, i.e. setting

$$\dot{x}_2(t) = \frac{100}{6}(7\,e^{-t} - 49\,e^{-7t})$$

to zero. This leads to the condition for maximum or minimum as

$$7\,e^{-t} = 49\,e^{-7t}$$

$$e^{-t} = 7\,e^{-7t},$$

$$-t = \ln 7 - 7t,$$

$$6t = \ln 7,$$

$$t = 0.3243 = T^*$$

as required.

Putting $t = 1$ into the transition matrix yields

$$\Phi(1) = \begin{bmatrix} 0.42904 & 0.06116 \\ -0.428 & -0.06025 \end{bmatrix}$$

and this is the matrix P required by the question, since

$$x(k) = \Phi(1)x(k-1) = Px(k-1).$$

Suppose that A were put in diagonal form A', it would have the values

$$A' = \begin{bmatrix} \lambda_1 & 0 \\ 0 & \lambda_2 \end{bmatrix}$$

and P would have the form

$$P' = \begin{bmatrix} e^{\lambda_1 t} & 0 \\ 0 & e^{\lambda_2 t} \end{bmatrix} = \begin{bmatrix} e^{\lambda_1} & 0 \\ 0 & e^{\lambda_2} \end{bmatrix}$$

since $t = 1$ in this particular case.

Now eigenvalues are invariant under diagonalization, hence we can make the statement that

$$\lambda_i(P) = e^{\lambda_i}(A') = e^{\lambda_i}(A)$$

in this case.

Check

The eigenvalues of P are 0.36788 and 9.1188×10^{-4} and these same values are found by calculating $e^{\lambda_1 t}$, $e^{\lambda_2 t}$ for $\lambda_1 = -1$, $\lambda_2 = -7$, $t = 1$.

To recalculate P for a time step T^*, we use the relation

$$P(T^*) = \Phi(T^*) = \frac{1}{6}\begin{bmatrix} 4.958 & 0.61973 \\ -4.338 & 7.96 \times 10^{-5} \end{bmatrix}$$

$$= \begin{bmatrix} 0.8263 & 0.1033 \\ -0.7230 & 1.327 \times 10^{-5} \end{bmatrix}.$$

The absolute maximum velocity is then found as

$$x_2(T^*) = \begin{bmatrix} -0.723 & 1.327 \times 10^{-5} \end{bmatrix}\begin{bmatrix} 100 \\ 0 \end{bmatrix}$$

$$= -72.3.$$

The analytical solution of equation (W1) is found to be

$$y = -16.67\,e^{-7t} + 116.67\,e^{-t}$$
$$\dot{y} = (7 \times 16.67)\,e^{-7t} - 116.67\,e^{-t}.$$

The transient behavior is sketched below in Figure 8.15.

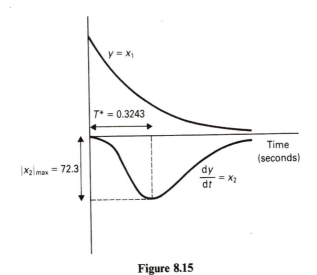

Figure 8.15

Worked example 8.2

Figure 8.16 shows a sample-and-hold device sampling every 0.2 seconds and feeding into a process of transfer function

$$G(s) = \frac{4}{s(s+2)}$$

(a) Obtain a state-space representation for $G(s)$ in the form

$$\left.\begin{array}{l} \dot{x} = Ax + Bu \\ y = Cx \end{array}\right\} \quad \text{equation (W4)}$$

(b) Determine an expression for the transition matrix $\Phi(t)$ for equation (W4)
(c) Noting that $u(t)$ is merely $v(t)$ held piecewise constant over sampling intervals, obtain a numerical representation for the whole system in the form

$$\left.\begin{array}{l} x((k+1)T) = Px(kT) + Qu(kT) \\ y(kT) \quad\quad = Rx(kT) \end{array}\right\} \quad \text{equation (W5)}$$

(d) Determine the input sequence $u(0)$, $u(1)$ that will transfer the system of equation (W4) from the state $\binom{0}{0}$ to the state $\binom{1}{0}$ in just two sampling periods.

Worked solution 8.2

(a) $(s^2 + 2s)y(s) = 4u(s)$. Let $x_1 = y$, $x_2 = \dot{y}$ then $\dot{x}_1 = x_2$, $\dot{x}_2 = -2x_1 + 4u$.
Hence

$$\begin{bmatrix} \dot{x}_1 \\ \dot{x}_2 \end{bmatrix} = \begin{bmatrix} 0 & 1 \\ 0 & -2 \end{bmatrix} \begin{bmatrix} x_1 \\ x_2 \end{bmatrix} + \begin{bmatrix} 0 \\ 1 \end{bmatrix} u$$

(b)
$$y = [4 \quad 0] \begin{bmatrix} x_1 \\ x_2 \end{bmatrix}$$

$$\Phi = \mathscr{L}^{-1}\{(sI - A)^{-1}\},$$

$$sI - A = \begin{bmatrix} s & -1 \\ 0 & s+2 \end{bmatrix},$$

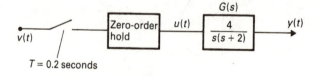

Figure 8.16

$$(sI - A)^{-1} = \begin{bmatrix} \dfrac{1}{s} & \dfrac{1}{(s+2)s} \\ 0 & \dfrac{1}{s+2} \end{bmatrix},$$

$$\Phi(t) = \begin{bmatrix} 1 & \frac{1}{2}(1 - e^{-2t}) \\ 0 & e^{-2t} \end{bmatrix}.$$

(c) In the case of piecewise constant u, $P = \Phi(T)$ with $T = 0.2$ and

$$Q = A^{-1}(\Phi(T) - I)B$$

or alternatively

$$Q = \int_0^T e^{At} B \, dt$$

$$R = [4 \quad 0]$$

(R being equal to matrix C in the continuous case)

$$P(T) = \begin{bmatrix} 1 & \frac{1}{2}(1 - e^{-0.4}) \\ 0 & e^{-0.4} \end{bmatrix}$$

$$Q(T) = \int_0^{0.2} \begin{bmatrix} \frac{1}{2}(1 - e^{-2t}) \\ e^{-2t} \end{bmatrix} dt$$

$$= \begin{bmatrix} \frac{1}{2}t + \frac{1}{4}e^{-2t} \\ -\frac{1}{2}e^{-2t} \end{bmatrix}\Bigg|_0^{0.2}.$$

Thus, we have

$$x((k+1)T) = \begin{bmatrix} 1 & 0.1648 \\ 0 & 0.6703 \end{bmatrix} x(kT) + \begin{bmatrix} 0.0176 \\ 0.1648 \end{bmatrix} u(kT)$$

$$y(kT) = [4 \quad 0]x(kT).$$

(d) We have

$$x(1) = P(T)x(0) + Q(T)u(0)$$
$$x(2) = P(T)x(1) + Q(T)u(1)$$
$$= P(T)(P(T)x(0) + Q(T)u(0)) + Q(T)u(1)$$
$$= P(T)^2 + P(T)Q(T)u(0) + Q(T)u(1)$$

In which $x(0)$ is zero, hence

$$x(2) = P(T)Q(T)u(0) + Q(T)u(1)$$

$$\triangleq \begin{bmatrix} 1 \\ 0 \end{bmatrix}.$$

We require the following relation to hold

$$\begin{bmatrix} 1 \\ 0 \end{bmatrix} = \begin{bmatrix} 0.0447 & 0.0176 \\ 0.1105 & 0.1648 \end{bmatrix} \begin{bmatrix} u(0) \\ u(1) \end{bmatrix}$$

or

$$\begin{bmatrix} u(0) \\ u(1) \end{bmatrix} = \begin{bmatrix} 0.0447 & 0.0176 \\ 0.1105 & 0.1648 \end{bmatrix}^{-1} \begin{bmatrix} 1 \\ 0 \end{bmatrix}$$

$$= \begin{bmatrix} 30.33 \\ -20.33 \end{bmatrix}.$$

The sequence

$$u(0) = 30.33,$$
$$u(1) = -20.33$$

will accomplish the required state transference.

Worked example 8.3

An n-state single-input system G has the equation

$$x((k+1)T) = Px(kT) + Qu(kT) \qquad \text{(W6)}$$

(a) Derive an expression for the sequence $\{u(0), \dots, u(nT)\}$ that will drive the system of equation (W1) from a known initial state $x(0)$ to a known final state $x(nT)$.
(b) State or derive any conditions on the expression derived in (a) that could prevent it having a unique solution.
(c) A feedback controller D, operating on the error $v - y$, where

$$y = (1 \quad 0) \begin{bmatrix} x_1 \\ x_2 \end{bmatrix} = Rx(k)$$

is to be designed to drive the system from state $x(0)$ to state $x(nT)$ (see Figure 8.17). Suggest a formula by which the transfer function $D(z)$ of the controller might be determined.

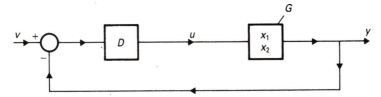

Figure 8.17

(d) Now assume $v = 1$,

$$x(0) = \begin{bmatrix} 0 \\ 0 \end{bmatrix},$$

the desired state

$$x(2) = \begin{bmatrix} 1 \\ 0 \end{bmatrix},$$

and that P, Q are given by

$$P = \begin{bmatrix} 0.8 & 0.8 \\ 0 & 0.8 \end{bmatrix},$$

$$Q = \begin{bmatrix} 0.1 \\ 0.1 \end{bmatrix},$$

use the approach developed in (c) to design a controller D to achieve the objective.

(e) Calculate the closed-loop transfer function of the system as designed and plot the poles and zeros in the complex plane. Comment on the location of the poles and zeros in relation to the poles of the system G.

Worked solution 8.3

(a) By repeated substitution, we can show that

$$x(nT) = P^n x(0) + P^{n-1}Qu(0) + \cdots + PQu((n-1)T) + Qu(nT)$$

or

$$[P^{n-1}Q, P^{n-2}Q, \ldots, PQ, Q] \begin{bmatrix} u(0) \\ u(T) \\ \vdots \\ u(nT) \end{bmatrix} = x(nT) - P^n x(0)$$

or

$$\begin{bmatrix} u(0) \\ \vdots \\ u(nT) \end{bmatrix} = \Delta^{-1}(x(nT) - P^n x(0))$$

where

$$\Delta = [P^{n-1}Q, P^{n-2}Q, \ldots, PQ, Q]$$

(b) Δ can be recognised as the system's controllability matrix. Clearly if Δ^{-1} does not exist, the sequence for u will not be calculable.

Hence, the expression is only guaranteed to have a unique solution if the system of equation (W6) is controllable.

(c) Assuming that the system is controllable, and given $x(0)$, $x(nT)$ we can calculate a sequence $\{u(kT), k = 0, \ldots, n\}$ to drive the state from $x(0)$ to any given state $x(nT)$. With a knowledge of the u sequence and using equation (W6), we can next determine the intermediate terms in the sequence $\{x(kT), k = 0, \ldots, n\}$ we can then go on to determine the sequence $\{y(kT), k = 0, \ldots, n\}$, using the relation

$$u = (1 \quad 0)\begin{bmatrix} x_1 \\ x_2 \end{bmatrix}$$

Finally, we know that the sequence of inputs $\{e(kT), k = 0, \ldots, n\}$ to the controller must be given by the relation $e = v - y$.

So that we may determine the $\mathscr{Z}$ transform $D(z)$ of the required controller using

$$D(z) = \frac{u(z)}{e(z)}$$

for a particular case.

(d) Using the approach derived above for the particular case given, we have

$$x(0) = \begin{bmatrix} 0 \\ 0 \end{bmatrix}, \qquad x(1) = ?, \qquad x(2) = \begin{bmatrix} 1 \\ 0 \end{bmatrix},$$

using the approach developed in (a) we obtain

$$\begin{bmatrix} 1 \\ 0 \end{bmatrix} = \begin{bmatrix} 0.8 & 0.8 \\ 0 & 0.8 \end{bmatrix}\begin{bmatrix} 0.1 \\ 0.1 \end{bmatrix}u(0) + \begin{bmatrix} 0.1 \\ 0.1 \end{bmatrix}u(1),$$

$$\begin{bmatrix} 1 \\ 0 \end{bmatrix} = \begin{bmatrix} 0.16 & 0.1 \\ 0.08 & 0.1 \end{bmatrix}\begin{bmatrix} u(0) \\ u(1) \end{bmatrix}$$

from which

$$u(0) = 12.5,$$

$$u(1) = -10.$$

Check

$$x(1) = \begin{bmatrix} 1.25 \\ 1.25 \end{bmatrix}, \qquad x(2) = \begin{bmatrix} 2 & -1 \\ 1 & -1 \end{bmatrix} = \begin{bmatrix} 1 \\ 0 \end{bmatrix}.$$

So

$$y(0) = 0, \qquad y(1) = 1.25, \qquad y(2) = 1.$$

Then

$$e(0) = 1 - y(0) = 1,$$
$$e(1) = 1 - y(1) = -0.25,$$
$$e(2) = 1 - y(2) = 0.$$

The transfer function of the required controller is given by

$$D(z) = \frac{u(0) + u(1)z^{-1}}{e(0) + e(1)z^{-1}}$$

$$= \frac{12.5 - 10z^{-1}}{1 - 0.25z^{-1}} = \frac{12.5z - 10}{z - 0.25}.$$

The transfer function $G(z)$ of the process (P, Q, R) is given by

$$G(z) = R(zI - P)^{-1}Q \qquad \text{(see Section 8.13.2)}$$

$$= (1 \quad 0) \begin{bmatrix} z - 0.8 & -0.8 \\ 0 & z - 0.8 \end{bmatrix}^{-1} \begin{bmatrix} 0.1 \\ 0.1 \end{bmatrix}$$

$$= \frac{1}{(z - 0.8)^2}(1 \quad 0) \begin{bmatrix} z - 0.8 & 0.8 \\ 0 & z - 0.8 \end{bmatrix} \begin{bmatrix} 0.1 \\ 0.1 \end{bmatrix}$$

$$= \frac{1}{(z - 0.8)^2}(z - 0.8 \quad 0.8) \begin{bmatrix} 0.1 \\ 0.1 \end{bmatrix}$$

$$= \frac{0.1z}{(z - 0.8)^2}.$$

The closed-loop transfer function of the system is

$$\frac{G(z)D(z)}{1 + G(z)D(z)} = \frac{\dfrac{0.1z}{(z - 0.8)^2}\dfrac{(12.5z - 10)}{(z - 0.25)}}{1 + G(z)D(z)}$$

$$= \frac{z(1.25z - 1)}{(z - 0.8)^2(z - 0.25) + z(1.25z - 1)}$$

State-variable techniques

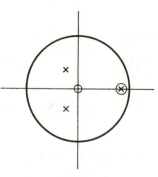

Figure 8.18

$$= \frac{z(1.25z - 1)}{(z^2 - 1.6z + 0.64)(z - 0.25) + 1.25z^2 - z}$$

$$= \frac{z(1.25z - 1)}{z^3 - 1.6z^2 + 0.64z - 0.25z^2 + 0.4z - 0.16 + 1.25z^2 - z}$$

$$= \frac{z(1.25z - 1)}{z^3 - 0.6z^2 + 0.04z - 0.16}$$

$$= \frac{1.25z(z - 0.8)}{(z - 0.8)(z^2 + 0.2z + 0.2)} \qquad \text{(by trial and error calculations)}$$

$$= \frac{z(1.25z - 1)}{(z - 0.8)(z + 0.1 \pm j\sqrt{0.19})}$$

This is the required closed-loop transfer function. The pole–zero plot for the transfer function is shown in Figure 8.18. Compared with the double pole at $z = 0.8$ of the system G, the closed loop system has a single pole cancelled by a zero at that point. The closed loop system obtains its rapid response from the effect of the complex pole pair at

$$z = \pm - 0.1 \pm j\sqrt{0.19}$$

Exercises

8.1 By differentiation of equation (8.14), obtain the equation $\dot{x} = Ax + Bu$ and hence confirm the result given in the text.

8.2 State whether each of the assertions below is true or false. Support your statements by proof or counter examples. In each case Φ is the transition matrix corresponding to a square constant A matrix.

(a) $\Phi(-t) = \Phi(t)^{-1}$.

(b) $\dfrac{d}{dt} = \Phi(t) = A\Phi(t)$.

(c) If $A = \begin{bmatrix} 1 & 2 \\ 3 & 1 \end{bmatrix}$, then $\Phi(t) = \begin{bmatrix} e^t & e^{2t} \\ e^{3t} & e^t \end{bmatrix}$.

(d) If $A = \begin{bmatrix} 5 & 0 \\ 0 & -3 \end{bmatrix}$, then $\Phi(t) = \begin{bmatrix} e^{5t} & 0 \\ 0 & e^{-3t} \end{bmatrix}$.

(e) $\Phi(t_2 - t_1)\Phi(t_1 - t_0) = \Phi(t_2 - t_0)$ for $t_0 \leqslant t_1 \leqslant t_2$.

(f) $\Phi(t_2 - t_1)\Phi(t_1 - t_0) = \Phi(t_2 - t_0)$ whatever the values of t_0, t_1, t_2.

(g) Let $C_1 C_2$ be two square matrices of the same dimension; then $e^{C_1 t} e^{C_2 t} = e^{(C_1 + C_2)t}$.

8.3 Let A be a 2×2 constant skew-symmetric matrix (i.e. it satisfies $A = -A^T$, where A^T signifies the transpose of A). Show that the solution of the equation

$$\begin{bmatrix} \dot{x}_1(t) \\ \dot{x}_2(t) \end{bmatrix} = A \begin{bmatrix} x_1(t) \\ x_2(t) \end{bmatrix}, \qquad \begin{bmatrix} x_1(0) \\ x_2(0) \end{bmatrix} \text{given},$$

satisfies

$$\sqrt{[x_1(t)^2 + x_2^2(t)]} = \text{constant}$$

for all values of t.

What is the physical significance of this result?

8.4 Compute the transition matrix

$$\Phi(t)|_{t=0.1 \text{ and } t=1}$$

for the matrix

$$A = \begin{bmatrix} 0 & 1 \\ -4 & -5 \end{bmatrix}$$

by the three methods of Section 8.6.1.

8.5 A second-order system governed by the equation

$$\frac{d^2 y}{dt^2} + 3\frac{dy}{dt} + 2y = u(t)$$

has applied to it the input signal

$$u(t) = 1 \qquad \text{when } 0 \leqslant t < 1,$$

$$u(t) = 0 \qquad \text{when } t \geqslant 1.$$

At $t = 0$, the system was at rest at $y = 0$. By using the transition matrix, determine y after 2 seconds. Check your result by any other method.

8.6 Derive the solution of the general time-varying differential equation

$$\dot{x}(t) = A(t)x(t) + B(t)u(t). \tag{1}$$

Show that if, in equation (1), $A(t)$ satisfies the condition

$$A(t + T) = A(t),$$

for some constant T and for all time t, then the system can be transformed into an equivalent time-invariant system.

Show that when A and B are constant matrices, equation (1) has the solution

$$x(t) = e^{At}x(0) + \int_0^t e^{A(t-\tau)} Bu(\tau)\, d\tau.$$

8.7 The system described by equation (1) is to be controlled by a linear controller of the form given in equations (2), (3). Derive conditions on the matrices C, D for non-interacting control to be achieved such that each element v_i of the state vector is affected by one and only one element v_i in the v vector

$$\dot{x} = Ax + Bu. \tag{1}$$

$$\dot{u} = Cu + De. \tag{2}$$

$$e = v - x. \tag{3}$$

Where x, u, e, v are n-dimensional vectors.

8.8 Two matrices A, A' are said to be similar if there exists a nonsingular matrix Q such that $A = QA'Q^{-1}$. Show that similar matrices share the same eigenvalues.

8.9 All equivalent state-space representations of a particular system have the same transfer function, so it is clear that a state-space representation corresponding to a particular $G(s)$ cannot be unique. We therefore ask: Are all the state-space representations that correspond to a particular $G(s)$ equivalent? Investigate by determining $G(s)$ for the two systems:

(a) $\dot{x} = \begin{bmatrix} 0 & 1 \\ -6 & -5 \end{bmatrix} x + \begin{bmatrix} 0 \\ 1 \end{bmatrix} u,$

$y = (2, 1)x.$

(b) $\dot{x} = -3x + u,$

$y = x.$

8.10 The output y of the system described by equation (1) is measured at two different times t_1 and t_2 during the decay of a transient. Under what conditions on a, b would it be possible to calculate the initial state from these two measurements?

$$\dot{x}(t) = \begin{bmatrix} -2 & 0 \\ 0 & -3 \end{bmatrix} x(t), \tag{1}$$

$$y(t) = [a \quad b] x(t).$$

Indicate how such a calculation might proceed.

The system described by equation (2) is known to be uncontrollable. Calculate the values of a and b.

$$\dot{x}(t) = \begin{bmatrix} 0 & 6 \\ 1 & -6 \end{bmatrix} x(t) + \begin{bmatrix} b & 1 \\ 1 & a \end{bmatrix} u(t). \tag{2}$$

8.11 A process can be described by the vector–matrix equation

$$\dot{x} = Ax + Bu, \qquad y = Cx.$$

Define the transfer matrix of this process and express it in terms of the matrices A, B, C.

In the single-input–single-output case, how are the eigenvalues of the matrix A related to the poles of the transfer function?

8.12 A process is described by the equation

$$\dot{x} = \begin{bmatrix} 0 & 1 \\ -3 & -4 \end{bmatrix} x + \begin{bmatrix} 1 & 1 \\ 0 & 1 \end{bmatrix} u.$$

Determine a state feedback matrix that will move the closed-loop poles to the positions -1, -10.

8.13 Show that the system $\dot{x} = Ax + Bu$ can be driven from an initial state x_0 to a final state x_1 by the input signal

$$u(t) = (e^{-At}B)^T \left(\int_0^{t_1} (e^{-At}B)(e^{-At}B)^T \, dt \right)^{-1} (e^{-At_1}x_1 - x_0) \qquad (0 \leqslant t \leqslant t_1).$$

Hence or otherwise determine a control that will drive the system

$$\dot{x}(t) = \begin{bmatrix} 0 & 1 \\ 0 & 0 \end{bmatrix} x(t) + \begin{bmatrix} 0 \\ \frac{1}{2} \end{bmatrix} u(t),$$

$$x(0) = \begin{bmatrix} 0 \\ 0 \end{bmatrix},$$

to a final condition

$$x(1) = \begin{bmatrix} 1 \\ 1 \end{bmatrix}.$$

8.14 Derive conditions on the matrices A, B in the equation

$$\dot{x} = Ax + Bu, \qquad y = Cx$$

to ensure controllability and observability.

  A system is characterized by

$$\dot{x} = \begin{bmatrix} 1 & 1 \\ 1 & -1 \end{bmatrix} x + \begin{bmatrix} -1 \\ \alpha \end{bmatrix} u.$$

Determine the value(s) of α for which it is uncontrollable.

  Determine whether, by the use of state-variable feedback, an uncontrollable second-order system might be made controllable.

8.15 Figure 8.19 shows three interconnected water tanks. Each tank has unit cross-section area. For each of the three tanks

$$\left.\begin{array}{ll} x_i & \text{is the water level,} \\ u_i & \text{is the incoming flow rate,} \\ q_i & \text{is the outgoing flow rate.} \end{array}\right\} \quad i = 1, 2, 3.$$

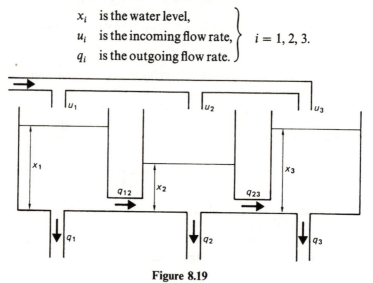

Figure 8.19

Each of the outflows is proportional to water level, i.e., $q_i = k_i x_i$. The cross flows q_{12}, q_{23} are proportional to level differences

$$q_{12} = k_{12}(x_1 - x_2), \qquad q_{23} = k_{23}(x_2 - x_3).$$

Describe the system by a vector–matrix equation of the form

$$\dot{x} = Ax + Bu.$$

  Assuming that all the constants are strictly positive, investigate the controllability of the system:

(a) By input u_1 acting alone.
(b) By input u_2 acting alone.

Comment on the physical significance of your result.

As an extension to the above problem, the state x is required to follow an arbitrary given trajectory $x_d(t)$, $t \geqslant t_0$. Is controllability of the system sufficient to ensure that such trajectory following can be achieved? If it is not sufficient, derive any additional condition that has to be satisfied.

8.16 A system with transfer function

$$G_1(s) = \frac{1}{(1 + sT_1)}$$

is put in series with a compensator with transfer function

$$G_2(s) = \frac{(1 + sT_1)}{(1 + sT_2)}.$$

(The constants T_1, T_2 are different and are both positive.)

Will the resultant system $G(s) = G_1(s)G_2(s)$ be (a) controllable (b) observable?

What conclusions can you draw from your results?

8.17 (a) A system is described by the equation

$$\begin{bmatrix} \dot{x}_1 \\ \dot{x}_2 \end{bmatrix} = \begin{bmatrix} -2 & 0 \\ -3 & -1 \end{bmatrix} \begin{bmatrix} x_1 \\ x_2 \end{bmatrix}.$$

What must be the initial state at $t = 0$ for the system to be in the state $x_1 = 1$, $x_2 = 0$ at a given time t_1?

(b) For which values of p is the system

$$\dot{x}_1 = 3x_1 + p(x_2 + u),$$
$$\dot{x}_2 = (p - 1)x_1 + 2x_2 + u$$

controllable?

When p is chosen such that the system is uncontrollable, to which values x is it still possible to drive the system from the origin?

8.18 A system with transfer function

$$G(s) = \frac{1}{s^2 + 3s + 2}$$

is preceded by a sample and zero-order hold device. Determine the discrete state equations of the (open-loop) system when the sampling interval is 5 seconds.

Express the state of the system at the kth sampling instant in terms of the initial conditions at time zero and the control applied over the period $(0, k)$.

8.19 Given that

$$x(z) = \sum_0^\infty x(k)z^{-k},$$

show that $\mathscr{Z}\{x(k + 1)\} = zx(z) - zx_0$.

A multivariable continuous system with constant coefficients is described by the vector–matrix equation

$$\dot{x} = Ax + Bu.$$

It receives input signals that are piecewise constant on sampling intervals of length T seconds.

Transform the system equation into an equation in z, including initial condition effects. Use your result to show that

$$\Phi(kT) = \mathscr{Z}^{-1}\{(zI - \Phi(T))^{-1}z\},$$

where $\mathscr{Z}^{-1}$ indicates inverse $\mathscr{Z}$ transformation.

8.20 Show that the transfer function corresponding with a single-input–single-output discrete-time system

$$x(k + 1) = Ax(k) + Bu(k),$$
$$y(k) = Cx(k),$$

is

$$G(z) = C(zI - A)^{-1}B.$$

8.21 A system is described by the difference equations

$$y_1(k - 2) + 2y_1(k - 1) + 2y_1(k) - 0.5y_2(k - 2) + y_2(k - 1) + y_2(k)$$
$$= -u_1(k - 1) + 0.5u_1(k - 2) + u_2(k - 1),$$

$$0.1y_1(k - 2) - 3y_1(k - 1) + y_1(k) + y_2(k)$$
$$= -5u_1(k - 1) + 2u_1(k - 2) - u_2(k - 1) - 0.1u_2(k - 2).$$

Express the system equations in state-variable form.

8.22 Show that the two system representations $\{A, B, C\}$, $\{A_R, B_R, C_R\}$ in the example in Section 8.18.2 correspond with the same transfer function $G(z)$.

Confirm also that the reachability matrix R_R of the system is equal to the identity matrix.

8.23 A system described by the equation

$$x(k) = \begin{bmatrix} 1 & p \\ 0 & 1 \end{bmatrix} x(k - 1) + \begin{bmatrix} 0 \\ 1 \end{bmatrix} u(k - 1)$$

has the initial condition

$$x(0) = \begin{bmatrix} 0 \\ 0 \end{bmatrix}.$$

Determine a sequence of control signals that will drive the state to

$$x(k) = \begin{bmatrix} 1 \\ 4 \end{bmatrix}$$

at $k = 2$.

8.24 Show that the nth-order single-input–single-output system

$$x(k + 1) = Ax(k) + Bu(k)$$

may be brought from an arbitrary initial condition to the origin in at most n steps provided that the system is reachable.

Let

$$A = \begin{bmatrix} -0.5 & 0.1 \\ 0 & -0.5 \end{bmatrix}, \qquad B = \begin{bmatrix} 1 \\ 1 \end{bmatrix}, \qquad x(0) = \begin{bmatrix} 4 \\ 4 \end{bmatrix}.$$

Determine the control sequence to bring the state to the origin in two steps.

8.25 Show that if the matrix $(B, AB, \ldots, A^{n-1}B)$ has rank n, then the state-variable system $\{A, B\}$ is controllable. Let $\{A, B\}$, $\{A^1, B^1\}$ be two representations of the same system Σ. Let Q, Q^1 be the respective controllability matrices. Construct a mapping D that relates the two representations.

A system is described by the equation

$$\dot{x} = \begin{bmatrix} -2 & -1 & 0 \\ 0 & -1 & 1 \\ 1 & -1 & -1 \end{bmatrix} x + \begin{bmatrix} 1 \\ 0 \\ -1 \end{bmatrix} u.$$

Transform the system to phase-variable form. The system is put into closed loop by setting $u = Kx$. It is required to have the closed-loop poles at -1, $-1 \pm j$. Determine the matrix K to achieve this.

8.26 Express the transfer function

$$G(z) = \frac{z^2 - 1.1z + 0.3}{z^3 - 1.1z^2 + 0.26z - 0.016}$$

in each of the four canonical forms of Sections 8.18.2 to 8.18.5 and in the Jordan form.

8.27 Refer to equation (8.65), $(E(z, a)q(z))$ and Figure 8.6 and confirm that the relation

$$u(z) = (z^3 + a_2 z^2 + q_1 z + a_0)q(z)$$

holds for this configuration. Mark on the figure the values at the entry and exit of each of the delay blocks—the annotated figure will be useful in Section 8.18.2.

8.28 Derive an observability criterion for the single-input–single-output discrete-time linear system $\{A, B, C\}$.

For the system

$$x(k + 1) = \begin{bmatrix} 0 & -0.06 \\ 1 & 0.5 \end{bmatrix} x(k) + \begin{bmatrix} -0.1 \\ 1 \end{bmatrix} u(k),$$

$$y(k) = \begin{bmatrix} 0 & 1 \end{bmatrix} x(k),$$

design an observer to reconstruct the state that does not appear in the output. Check by simulation that the observer works and in particular that it rapidly converges to the correct value from an incorrect initial estimate, even during a process transient.

8.29 A single-input–single-output process described by the equations

$$\dot{x} = Ax + Bu,$$

$$y = Cx,$$

is subjected for a long time to the input signal shown in Figure 8.20.

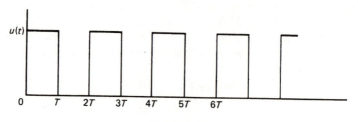

Figure 8.20

Let y^*, y_* be the upper and lower limits respectively reached in the steady state by the response $y(t)$. Using the transition matrix, derive expressions for y^* and y_*. Calculate y^* and y_* for the process with

$$A = \begin{bmatrix} 0 & 1 \\ -2 & -3 \end{bmatrix}, \qquad B = \begin{bmatrix} 0 \\ 1 \end{bmatrix}, \qquad C = \begin{bmatrix} 1 & 0 \end{bmatrix}, \qquad T = 0.1.$$

8.30 A process has the discrete-time description

$$x(k+1) = \begin{bmatrix} 0.1 & 0.2 & 0.3 \\ 1 & 0 & 0 \\ 0 & 1 & 0 \end{bmatrix} x(k) + \begin{bmatrix} 1 \\ 0 \\ 0 \end{bmatrix} u(k)$$

determine a state-variable feedback control to locate the closed-loop poles at $z = 0.5$, $0.2 \pm j0.5$.

8.31 A single-input–single-output process is described by the equations

$$\dot{x} = ax + bu + w,$$

$$y = cx + dv,$$

in which

$$a = b = \mathscr{E}\{v\} = \mathscr{E}\{w\} = 0,$$

$$c = d = 1, \qquad \mathscr{E}\{v^2\} = l, \qquad \mathscr{E}\{w^2\} = m$$

($\mathscr{E}$ indicates expected value).

Design a steady-state Kalman filter that will work in discrete time to estimate the process state. Give the $\mathscr{Z}$ transform of the estimator equation.

8.32 Refer to the numerical example of Section 8.23.4 and modify the values q and r as follows:

(a) $q = 10, r = 1$.
(b) $q = 1, r = 1$.
(c) $q = 1, r = 3$.
(d) $q = 1, r = 10$.

For each case, calculate the steady-state Kalman gain K and sketch the form of relation for this example between the steady-state Kalman gain and the ratio q/r.

9

Control of Large-Scale Systems

9.1 Introduction

The ultimate aim of industrial control must surely be to achieve efficient overall control of an entire plant or of a group of plants.

Developments in hardware have been such as to encourage the design of interlinked control systems. In particular, microprocessor-based controllers have the data-processing power and the communication facilities to allow ambitious linked control schemes to be realized.

A theoretical framework is needed to support the analysis of large-scale problems and the design of interlinked control systems.

The establishment of this framework is the subject of a considerable on-going research effort. We shall content ourselves with outlining the approaches available at the more practicable end of the spectrum.

A typical industrial plant is made up of a number of processes linked by material, energy and information flows. Each process has its own local control requirements. The plant has an overall set of control aims that are to some extent dependent on factors (such as raw material prices) external to the plant. Figure 9.1 illustrates the configuration as described.

How is a control scheme to be designed for a plant such as that in Figure 9.1, so that the overall aims can be achieved?

A rather unrealistic solution is to feed all the information to a central point and to optimize the control such that every detailed action within every process contributes ideally to the achievement of the given overall aims. The approach is unrealistic since it requires an exact and complete mathematical description of every aspect of the plant and a (rarely available) statement of an overall performance index.

Real plants have of course to be controlled under conditions of partial uncertainty and imperfect measurement.

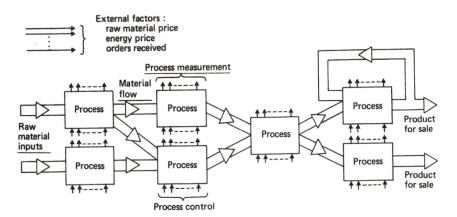

Figure 9.1   A typical industrial plant.

The way forward is to decompose the problem into sub-problems. Using this approach, sub-goals must be defined in such a way that the overall aim is achieved so far as possible.

Any individual control loop can be designed using the methods put forward in the earlier chapters and the set of controls for a particular process within the plane can be designed using methods put forward in Chapter 8.

The chief difficulty perhaps lies in deciding local aims (involving physical variables) which, when achieved, will ensure the achievement of the overall aims (defined in much more general terms).

9.2 Initial approaches to system decomposition

In attempting to understand a large and complex system it is natural to represent the overall system in terms of interconnected subsystems. A diagram showing the subsystems and their interconnection will allow a broad appreciation of system operation, since the details are hidden inside the subsystems.

In a particular case, suitable subsystems may already physically exist. In other cases they may be defined artificially—even a homogeneous system could, if required, be divided into subsystems. Such subdivision can be performed according to different alternative criteria. For instance, geographical location, function within the system, function in a control systems sense or relative time scale could be possible criteria for subdivision.

We illustrate the use of the four criteria by a simple example. A factory takes in scrap glass, melts it in furnaces, makes bottles and inspects and despatches them. It has three parallel process routes *A, B, C* each performing the same sequence outlined.

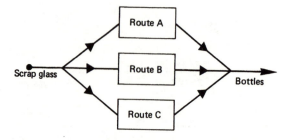

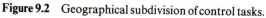

Figure 9.2   Geographical subdivision of control tasks.

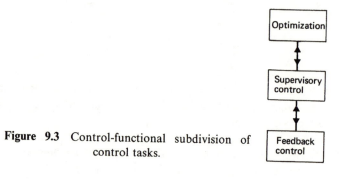

Figure 9.3   Control-functional subdivision of control tasks.

Figure 9.4   Process-functional subdivision of control tasks.

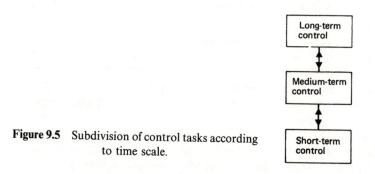

Figure 9.5   Subdivision of control tasks according to time scale.

If geographical location is used as a subdivision criterion, three subsystems result corresponding to the three process routes (Figure 9.2). A subdivision according to control systems function might result in the configuration shown in Figure 9.3. A process-functional subdivision (Figure 9.4) would group all similar physical activities together. Figure 9.5 illustrates a subdivision according to the time horizons of the decision-making involved. Often such a subdivision tends to produce a very similar configuration to that of Figure 9.3.

9.3 Hierarchical control

Some of the most natural and useful structures resulting from subdivision are *hierarchical structures*. The most familiar example of a hierarchical structure is probably a business organization. At the lowest level of the hierarchy, several individuals perform tasks within a functional unit. Some number of functional units form, let us say, a division and the divisions taken together form the organization. In such an arrangement, the elements at each level, except the highest, receive targets and instructions and report their achievement back to the level above.

Much of the early and by now classical literature on hierarchical systems was concerned with decomposition and control in a rather idealized setting. By contrast, recent work in this field takes into account the inevitable imperfection of models, the effects of noise, the availability of feedback and the constraints imposed by real processes and the time limitations of real-time computation. The two main approaches are *multilayer control*, where subdivision is according to complexity of control task, the *multilevel control*, involving the concept of local tasks coordinated by supervisory controllers.

9.4 Multilayer control

In a typical multilayer control system (Figure 9.6), the lowest level is that of *regulation*. The task of control at this level is to ensure that, so far as possible, each output variable y_i is maintained equal to the corresponding desired value, v_i.

The second level is that of *optimization*. The task of control here is to determine the best values $v_i(t)$ to ensure that some overall measure, say J, of system performance is maximized. The optimization routines make use of a model M containing a vector, $\hat{a}$, of system parameters.

The third level is that of *parameter adaptation*. Based on comparison of model predictions with results actually obtained, the parameter vector $\hat{a}$ is modified. The aim of such adaptation is to maintain the validity of the model, despite changes in the system.

It is seen that, at the lowest level in the hierarchy, simple frequent decisions have to be made. As the level in the hierarchy increases, so the decisions become more complex, but in general the frequency at which decisions need to be made decreases.

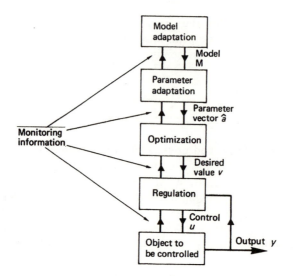

Figure 9.6 A typical multilayer control system.

As the diagram indicates, information to be used directly (commands) is passed downwards in the hierarchy, while information to be used indirectly (monitoring information) is passed upwards.

The fourth level is that of *model adaptation*. Based on long-term comparisons of model performance against results obtained, the model structure may be updated and a new model M made available to the parameter adaptation layer.

The advantage of the multilayer configuration, compared with the alternative of a monolithic integrated controller, is that it simplifies the analysis, computation and information-flow requirements at the cost of some loss of performance.

Another type of multilayer control was illustrated in Figure 9.5. Here, short-term control is based on detailed but short-term predictions. Long-term control is based on long-term predictions using a broadly based model that omits the details. The philosophy of this approach is that:

(a) Frequent short-term control requires many details to be taken into account but predictions into the far distant future are not required.
(b) Long-term control requires long-term predictions. In making such predictions, details do not need to be taken into account since:
  (i) Current details will have little effect on the distant future.
  (ii) Details in the distant future would be of little interest even if they could be predicted.

We can imagine that the upper layer in the system produces a long-term plan that the lower layer is then required to implement.

One interesting class of problems is where the system must be steered from an initial state $x(t_0)$ at time t_0 to a prescribed state x_d at a final time t_f.

In those cases where the time interval $t_f - t_0$ is very large compared with system time constants, the following observations are useful:

(a) Control of the initial period soon after t_0 requires detailed prediction of the complete state vector from known initial conditions.
(b) An intermediate steady-state period where the state is influenced very little by either $x(t_0)$ or x_d is often encountered.
(c) A final period can be expected where the state trajectory is influenced by x_d.

9.5 Multilayer optimization

The *overall problem* is: maximize

$$J = \int_{t_0}^{t_f} q^1(x^1(t), u^1(t), w^1)) \, dt \tag{9.1}$$

subject to

$$\dot{x}^1(t) = f^1(x^1(t), u^1(t), w^1(t)). \tag{9.2}$$

$x^1(t_0)$ is given and $x^1(t_f)$ is free or is specified according to the type of problem.

DECOMPOSITION OF THE PROBLEM
   Top layer   Maximize

$$J_3 = \int_{t_0}^{t_f} q^3(x^3(t), u^3(t), w^3(t)) \, dt \tag{9.3}$$

subject to

$$\dot{x}^3(t) = f^3(x^3(t), u^3(t), w^3(t)). \tag{9.4}$$

$x^3(t_0)$ is given and $x^3(t_f)$ is free or specified according to the type of problem.

The variables in the top layer formulation are simplified aggregations of the variables in the overall formulation.

   Intermediate layer   Maximize

$$J_2 = \int_{t_0}^{t_2} q^2(x^2(t), u^2(t), w^2(t)) \, dt \tag{9.5}$$

subject to

$$\dot{x}^2(t) = f^2(x^2(t), u^2(t), w^2(t)). \tag{9.6}$$

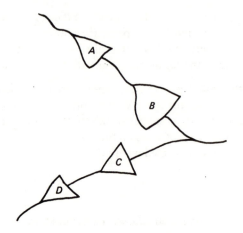

Figure 9.7 The four-reservoir problem:
location of reservoirs.

$x^2(t_0)$ is given but $x^2(t_2)$ has to be determined from $x^3(t_2)$ using some function γ that is strongly linked with the model simplification procedure that was used in moving from the intermediate to the highest layer. Let $\gamma_{23}(\gamma_{12})$ be a model simplification function linking x^2 with x^3 (x^1 with x^2); then $x^2(t_2)$ has to be chosen so that

$$x^3(t_2) = \gamma_{23}x^2(t_2).$$

Lowest layer   Analogously to the intermediate layer: Maximize

$$J_1 = \int_{t_0}^{t_1} q^1(x^1(t), u^1(t), w^1(t))\, \mathrm{d}t \tag{9.7}$$

subject to

$$\dot{x}^1(t) = f^1(x^1(t), u^1(t), w^1(t)); \tag{9.8}$$

$x^1(t_0)$ is given and

$$x^1(t_1) = \gamma_{12}x^2(t_1). \tag{9.9}$$

This short horizon optimization determines the real control actions that need to be implemented.

   Example†   A water authority has four reservoirs, A, B, C, D (Figure 9.7) from which it supplies customers. The authority has the possibility to make bulk purchases or bulk sales (Figure 9.8).

† Based with permission on material in Reference F2.

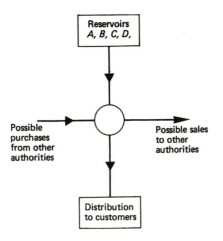

Figure 9.8 The four-reservoir problem:
interaction with external factors.

Define x^3 as the total volume of water in the four reservoirs. The authority aims to have a total water stock of x_d^3 at time t_f, i.e. it requires that $x^3(t_f) = x_d^3$. Using an aggregated model an optimal trajectory is generated such that the required condition is satisfied. Now define x^2 so that

$$x_1^2 = \text{volume of water stored in reservoirs } A \text{ and } B,$$

$$x_2^2 = \text{volume of water stored in reservoirs } C \text{ and } D.$$

The vector $x^2(t_2)$ has to be determined from the known value $x^3(t_2)$ such that

$$x_1^2(t_2) + x_2^2(t_2) = x^3(t_2).$$

Optimal trajectories are then determined for $x_1^2(t)$ and $x_2^2(t)$.
    Now define x^1 so that

$$x_1^1 = \text{volume of water in reservoir } A,$$

$$x_2^1 = \text{volume of water in reservoir } B,$$

$$x_3^1 = \text{volume of water in reservoir } C,$$

$$x_4^1 = \text{volume of water in reservoir } D.$$

The vector $x^1(t_1)$ has to be determined from the known value $x^2(t_1)$ such that

$$x_1^1(t_1) + x_2^1(t_1) = x_1^2(t_1),$$

$$x_3^1(t_1) + x_4^1(t_1) = x_2^2(t_1).$$

Optimal trajectories are then determined from the four known initial conditions $x_i^1(t_0)$, $i = 1,\ldots,4$ to the four required final conditions.

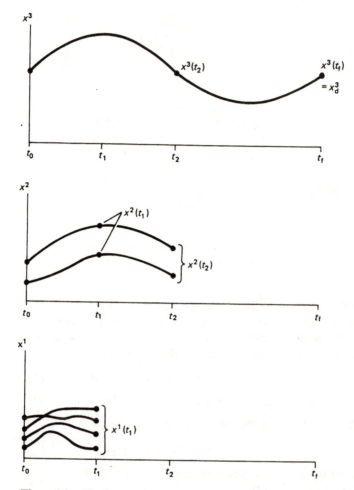

Figure 9.9   The four-reservoir problem: outline of the optimal
multilayer solution.

Notice that the times t_1, t_2, t_f are left to the user to decide. It is clear that this multi-horizon optimization makes good sense since it provides detailed day-to-day control that is completely consistent with the aggregated long-term plan. Figure 9.9 shows sketches of the optimal solutions.

In the real situation, suitable values for the times t_1 and t_2 might be

$$t_1 = t_f/100,$$

$$t_2 = t_f/10.$$

Once time t_1 has been reached and the state $x^1(t_1)$ has been measured, it would be normal strategy to repeat the predictions at the three levels, starting from the measured rather than the predicted values.

Any systematic errors between measured and predicted states, at any of the levels, would call for some measure of model adaptation with the aim of improving the accuracy of predictions.

9.6 Digraphs as a tool in the modelling of complex systems

A *digraph* is a set of subsystems that are interconnected by directed flows. (A signal-flow diagram is an example of a digraph.) A closed loop within a digraph is called a *cycle*. As a simple example, Figure 9.10 shows a digraph containing a cycle.

The *transition matrix* Φ of a digraph is a square matrix that completely characterizes the digraph. Let the subsystems in the digraph be numbered from 1 to n; then

$$\Phi = \{\phi_{ij}\} \tag{9.10}$$

where $\phi_{ij} = 1$ if there is a flow from subsystem i to subsystem j,

    $\phi_{ij} = 0$ otherwise.

The transition matrix for a given digraph can be written down by inspection. For instance, the transition matrix for the digraph of Figure 9.10 is shown in Table 9.1.

The transition matrix is a Boolean matrix. It can be raised to powers using the usual Boolean rules, i.e. that multiplication is the same as logical 'and' and addition is the same as logical 'or'.

Cycles within a digraph can be detected by examining Φ^k for different integral values of $k > 1$. Define $(\phi_{ij})_k$ as an element in the matrix Φ^k; then it is clear that whenever $(\phi_{ii})_k = 1$ this implies that the ith subsystem is within a k-step cycle. In other words, diagonal elements in the matrix indicate the presence of k-step cycles. On the other hand, a digraph that contains no cycles at all will

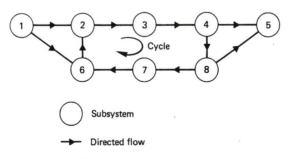

Figure 9.10   A digraph.

Table 9.1

$$
\Phi =
\begin{array}{cccccccc}
0 & 1 & 0 & 0 & 0 & 1 & 0 & 0 \\
0 & 0 & 1 & 0 & 0 & 0 & 0 & 0 \\
0 & 0 & 0 & 1 & 0 & 0 & 0 & 0 \\
0 & 0 & 0 & 0 & 1 & 0 & 0 & 1 \\
0 & 0 & 0 & 0 & 0 & 0 & 0 & 0 \\
0 & 1 & 0 & 0 & 0 & 0 & 0 & 0 \\
0 & 0 & 0 & 0 & 0 & 1 & 0 & 0 \\
0 & 0 & 0 & 0 & 1 & 0 & 1 & 0 \\
\end{array}
$$

satisfy $\Phi^k = 0$ for all k sufficiently large. Some of these points can be appreciated better by examining the entries in the matrix for different values of k for the example of Figure 9.10.

Powers $\Phi^k, k = 2,\ldots, 6$ should be examined in conjunction with the digraph, Figure 9.10. The matrices clearly allow a systematic representation of how the subsystems are linked by $2,\ldots, 6$ step routes in the digraph.

The matrix Φ^6 has 5 elements on its leading diagonal. The interpretation is that the subsystems corresponding with the diagonal elements are the starting and finishing points of 6 step cycles. Consistent with these facts, it is seen that $\Phi^7 = \Phi$, i.e. the transition matrix for the system is cyclic with period 6.

Tables 9.2–9.7 show the transition matrices Φ^2–Φ^7.

The *reachability matrix* R relating to a particular digraph is defined by

$$
R = \sum_{k=1}^{\infty} \Phi^k. \tag{9.11}
$$

In fact R reaches a steady value for some finite value of k, so that its computation is simple. The reachability matrix for the example of Figure 9.11 has been calculated and is shown in Table 9.8.

The intersection of R with its own transpose $(= R \cap R^T)$ has the property that it identifies those subsystems that are linked within a cycle. For our example $R \cap R^T$ is shown in Table 9.9. This matrix confirms that the subsystems 2, 3, 4, 6, 7 are within a cycle.

An alternative system description uses vector methods. Let u_i, y_i be the

Table 9.2

$$\Phi^2 = \begin{bmatrix} 0 & 1 & 1 & 0 & 0 & 0 & 0 & 0 \\ 0 & 0 & 0 & 1 & 0 & 0 & 0 & 0 \\ 0 & 0 & 0 & 0 & 1 & 0 & 0 & 1 \\ 0 & 0 & 0 & 0 & 1 & 0 & 1 & 0 \\ 0 & 0 & 0 & 0 & 0 & 0 & 0 & 0 \\ 0 & 0 & 1 & 0 & 0 & 0 & 0 & 0 \\ 0 & 1 & 0 & 0 & 0 & 0 & 0 & 0 \\ 0 & 0 & 0 & 0 & 0 & 1 & 0 & 0 \end{bmatrix}$$

Table 9.3

$$\Phi^3 = \begin{bmatrix} 0 & 0 & 1 & 1 & 0 & 0 & 0 & 0 \\ 0 & 0 & 0 & 0 & 1 & 0 & 0 & 1 \\ 0 & 0 & 0 & 0 & 1 & 0 & 1 & 0 \\ 0 & 0 & 0 & 0 & 0 & 1 & 0 & 0 \\ 0 & 0 & 0 & 0 & 0 & 0 & 0 & 0 \\ 0 & 0 & 0 & 1 & 0 & 0 & 0 & 0 \\ 0 & 0 & 1 & 0 & 0 & 0 & 0 & 0 \\ 0 & 1 & 0 & 0 & 0 & 0 & 0 & 0 \end{bmatrix}$$

vectors of input and output flows relating to subsystem i. Define

$$u = \begin{bmatrix} u_1 \\ \vdots \\ u_n \end{bmatrix}, \qquad y = \begin{bmatrix} y_1 \\ \vdots \\ y_n \end{bmatrix};$$

then a system of n subsystems can be described by the equation

$$u = Hy, \tag{9.12}$$

where H is defined as the *interconnection matrix* of the system.

Table 9.4

$\Phi^4 =$

0	0	0	1	1	0	0	1
0	0	0	0	1	0	1	0
0	0	0	0	0	1	0	0
0	1	0	0	0	0	0	0
0	0	0	0	0	0	0	0
0	0	0	0	1	0	0	1
0	0	0	1	0	0	0	0
0	0	1	0	0	0	0	0

Table 9.5

$\Phi^5 =$

0	0	0	0	1	0	1	1
0	0	0	0	0	1	0	0
0	1	0	0	0	0	0	0
0	0	1	0	0	0	0	0
0	0	0	0	0	0	0	0
0	0	0	0	1	0	1	0
0	0	0	0	1	0	0	1
0	0	0	1	0	0	0	0

The interconnection matrix can be obtained by the following method:

(a) replace each of the p unity elements in row i of the matrix Φ by elements y_{ij}, $j = 1, \ldots, p$ to form a matrix Y;

(b) replace each of the q unity elements in column j of the matrix Φ by elements u_{ji}, $i = 1, \ldots, q$ to form matrix U;

(c) compare the result of operations (a) and (b)—the matrix H must be chosen such that equality exists between the matrices Y and U.

The following example makes the method clear.

Table 9.6

$$\Phi^6 = \begin{pmatrix} 0 & 0 & 0 & 0 & 1 & 1 & 1 & 0 \\ 0 & 1 & 0 & 0 & 0 & 0 & 0 & 0 \\ 0 & 0 & 1 & 0 & 0 & 0 & 0 & 0 \\ 0 & 0 & 0 & 1 & 0 & 0 & 0 & 0 \\ 0 & 0 & 0 & 0 & 0 & 0 & 0 & 0 \\ 0 & 0 & 0 & 0 & 0 & 1 & 0 & 0 \\ 0 & 0 & 0 & 0 & 1 & 0 & 1 & 0 \\ 0 & 0 & 0 & 0 & 1 & 0 & 0 & 1 \end{pmatrix}$$

Example Construction of the interconnection matrix H from the transition matrix Φ for the example of Figure 9.10.

(a) Replace unity elements by elements y_{ij}:

$$Y = \begin{pmatrix} 0 & y_{11} & 0 & 0 & 0 & y_{12} & 0 & 0 \\ 0 & 0 & y_{21} & 0 & 0 & 0 & 0 & 0 \\ 0 & 0 & 0 & y_{31} & 0 & 0 & 0 & 0 \\ 0 & 0 & 0 & 0 & y_{41} & 0 & 0 & y_{42} \\ 0 & 0 & 0 & 0 & 0 & 0 & 0 & 0 \\ 0 & y_{61} & 0 & 0 & 0 & 0 & 0 & 0 \\ 0 & 0 & 0 & 0 & 0 & y_{71} & 0 & 0 \\ 0 & 0 & 0 & 0 & y_{81} & 0 & y_{82} & 0 \end{pmatrix}$$

(b) Replace unity elements by elements u_{ji}:

$$U = \begin{pmatrix} 0 & u_{21} & 0 & 0 & 0 & u_{61} & 0 & 0 \\ 0 & 0 & u_{31} & 0 & 0 & 0 & 0 & 0 \\ 0 & 0 & 0 & u_{41} & 0 & 0 & 0 & 0 \\ 0 & 0 & 0 & 0 & u_{51} & 0 & 0 & u_{81} \\ 0 & 0 & 0 & 0 & 0 & 0 & 0 & 0 \\ 0 & u_{22} & 0 & 0 & 0 & 0 & 0 & 0 \\ 0 & 0 & 0 & 0 & 0 & u_{62} & 0 & 0 \\ 0 & 0 & 0 & 0 & u_{52} & 0 & u_{71} & 0 \end{pmatrix}$$

Table 9.7

$$\Phi^7 = \begin{pmatrix} 0 & 1 & 0 & 0 & 0 & 1 & 0 & 0 \\ 0 & 0 & 1 & 0 & 0 & 0 & 0 & 0 \\ 0 & 0 & 0 & 1 & 0 & 0 & 0 & 0 \\ 0 & 0 & 0 & 0 & 1 & 0 & 0 & 1 \\ 0 & 0 & 0 & 0 & 0 & 0 & 0 & 0 \\ 0 & 1 & 0 & 0 & 0 & 0 & 0 & 0 \\ 0 & 0 & 0 & 0 & 0 & 1 & 0 & 0 \\ 0 & 0 & 0 & 0 & 1 & 0 & 1 & 0 \end{pmatrix}$$

The interconnection matrix H can then be written down by inspection to ensure equality of U and Y:

$$\begin{pmatrix} u_{21} \\ u_{22} \\ u_{31} \\ u_{41} \\ u_{51} \\ u_{52} \\ u_{61} \\ u_{62} \\ u_{71} \\ u_{81} \end{pmatrix} = \begin{pmatrix} 1 & 0 & 0 & 0 & 0 & 0 & 0 & 0 & 0 & 0 \\ 0 & 0 & 0 & 0 & 0 & 0 & 1 & 0 & 0 & 0 \\ 0 & 0 & 1 & 0 & 0 & 0 & 0 & 0 & 0 & 0 \\ 0 & 0 & 0 & 1 & 0 & 0 & 0 & 0 & 0 & 0 \\ 0 & 0 & 0 & 0 & 1 & 0 & 0 & 0 & 0 & 0 \\ 0 & 0 & 0 & 0 & 0 & 0 & 0 & 0 & 1 & 0 \\ 0 & 1 & 0 & 0 & 0 & 0 & 0 & 0 & 0 & 0 \\ 0 & 0 & 0 & 0 & 0 & 0 & 0 & 1 & 0 & 0 \\ 0 & 0 & 0 & 0 & 0 & 0 & 0 & 0 & 0 & 1 \\ 0 & 0 & 0 & 0 & 0 & 1 & 0 & 0 & 0 & 0 \end{pmatrix} \begin{pmatrix} y_{11} \\ y_{12} \\ y_{21} \\ y_{31} \\ y_{41} \\ y_{42} \\ y_{61} \\ y_{71} \\ y_{81} \\ y_{82} \end{pmatrix}$$

In this type of systems description an element u_{ij} represents the jth input to the ith subsystem. Similarly, y_{ij} represents the jth output from the ith subsystem.

The digraph, together with its associated transition matrix and interconnection matrix, is a useful systematic tool for the investigation and characterization of large systems.

THE TRANSITION MATRIX Φ—FURTHER EXAMPLE

Figure 9.11 shows an arbitrary interconnection of subsystems. The transition

Table 9.8

$$
R = \begin{bmatrix}
0 & 1 & 1 & 1 & 1 & 1 & 1 & 1 \\
0 & 1 & 1 & 1 & 1 & 1 & 1 & 1 \\
0 & 1 & 1 & 1 & 1 & 1 & 1 & 1 \\
0 & 1 & 1 & 1 & 1 & 1 & 1 & 1 \\
0 & 0 & 0 & 0 & 0 & 0 & 0 & 0 \\
0 & 1 & 1 & 1 & 1 & 1 & 1 & 1 \\
0 & 1 & 1 & 1 & 1 & 1 & 1 & 1 \\
0 & 1 & 1 & 1 & 1 & 1 & 1 & 1
\end{bmatrix}
$$

Table 9.9

$$
R \cap R^T = \begin{bmatrix}
0 & 0 & 0 & 0 & 0 & 0 & 0 & 0 \\
0 & 1 & 1 & 1 & 0 & 1 & 1 & 1 \\
0 & 1 & 1 & 1 & 0 & 1 & 1 & 1 \\
0 & 1 & 1 & 1 & 0 & 1 & 1 & 1 \\
0 & 0 & 0 & 0 & 0 & 0 & 0 & 0 \\
0 & 1 & 1 & 1 & 0 & 1 & 1 & 1 \\
0 & 1 & 1 & 1 & 0 & 1 & 1 & 1 \\
0 & 1 & 1 & 1 & 0 & 1 & 1 & 1
\end{bmatrix}
$$

matrix Φ for the system can be seen to be given by

$$
\Phi = \begin{bmatrix}
0 & 1 & 1 & 0 & 0 & 0 \\
0 & 0 & 1 & 0 & 0 & 0 \\
0 & 0 & 0 & 1 & 0 & 0 \\
0 & 0 & 0 & 0 & 1 & 0 \\
0 & 0 & 0 & 1 & 0 & 1 \\
0 & 0 & 1 & 0 & 0 & 0
\end{bmatrix}.
$$

The system can be represented alternatively as in Figure 9.12 where H represents the interconnection matrix.

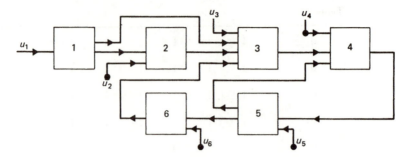

Figure 9.11   The digraph for the example.

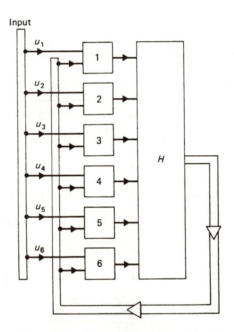

Figure 9.12   Figure 9.11 redrawn using the
interconnection matrix H.

9.7 Multilevel control

The characteristic of multilevel control is that at the lowest level are *local controllers* whose actions are coordinated by layers of *coordinating controllers*. Figure 9.13 illustrates the concept.

Each local controller can be considered to have its own local targets. It is the task of the coordinators to reconcile the interactions and conflicts of goals to obtain the best possible overall performance.

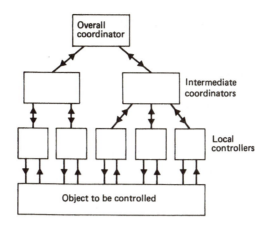

Figure 9.13   Multilevel control.

9.8 The coordination problem in multilevel control

The advantages of breaking a large system Σ into n subsystems Σ_i, each of manageable size, such that

$$\Sigma_1 + \cdots + \Sigma_n = \Sigma$$

have been outlined earlier. A distributed control system for such a system envisages n local controllers interlinked by some coordinating function. The aim of coordination is to ensure that, so far as possible, the individual subsystems Σ_i are controlled in such a way that the overall performance of the combined system is maximized.

Theoretically, ideal coordination could be achieved by treating the problem as an exercise in dynamic optimization. This would involve defining a cost function J for the complete system Σ and then insisting that the coordination must ensure that each of the n local controllers operated in such a way that J was minimized to achieve a true optimum.

As Findeisen (F2) points out, a global cost function J may exist that can be subdivided into convenient J_i to allow distributed control of a large system. Alternatively, the individual J_i may exist for individual subsystems and an artificial global cost function J can be defined to allow coordination. This approach leads to very complex solutions that are computable and justifiable only in a small minority of applications.

A less ambitious but more realistic approach is to neglect the process dynamics and to aim for steady-state optimization of the overall system.

9.9 The steady-state coordination problem in multilevel control

Suppose that a large system Σ is decomposed into n subsystems Σ_i, $i = 1, \ldots, n$ (see Figure 9.14).

Each subsystem has a state x_i governed by the equation

$$\dot{x}_i = f_i(x_i, m_i, u_i, w_i) \tag{9.13}$$

where m_i is the manipulable input to the subsystem; u_i is the input from other subsystems; w_i is the disturbance input.

Each subsystem has the output y_i governed by the equation

$$y_i = g_i(x_i, m_i, u_i, w_i). \tag{9.14}$$

A generalized control vector c is defined by the equation

$$c_i = h_i(x_i, m_i, u_i). \tag{9.15}$$

A local controller enforces the condition

$$c_i = c_{d_i} \tag{9.16}$$

(where c_{d_i} is a desired value for c_i emanating from a supervisory controller) by manipulation of m_i.

If the process dynamics can be neglected, so that the state $x_i(t)$ follows perfectly any trajectory specified by the local controller and provided also that

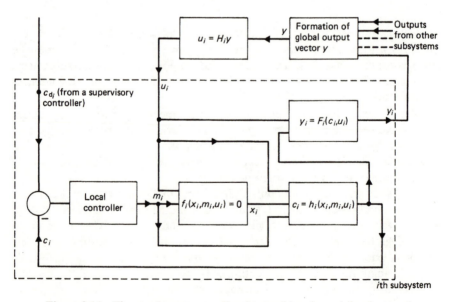

Figure 9.14   The steady-state coordination problem in multilevel control.

the functions h_i have been specified such that c_i and u_i uniquely specify m_i and x_i, then we can write

$$y_i(t) = F_1(c_1(t), u_i(t), w_i(t)),\qquad (9.17)$$

which is a relation between instantaneous values.

In the sequel, disturbances will be neglected so that equation (9.17) becomes

$$y_i(t) = F_i(c_i(t), u_i(t)).\qquad (9.18)$$

The input u_i is given by

$$u_i = H_i y\qquad (9.19)$$

where H_i is the appropriate submatrix of the interconnection matrix H. A local cost function

$$J_i = J_i(c_i, u_i)\qquad (9.20)$$

is assumed given.

A global cost function

$$J = \psi(J_1, \ldots, J_n)\qquad (9.21)$$

is given, where the function ψ is assumed to be monotonic to avoid anomalous situations arising.

The problem definition is completed by the imposition of various constraints

$$(c_i, u_i) \in K_i,\qquad (9.22)$$

i.e. the pairs (c_i, u_i) must belong to some admissible set

$$\sum_i^n r_i(c_i, u_i) \leqslant R.\qquad (9.23)$$

This last equation represents a global resource constraint. The constraint R is specified as part of the problem formulation.

9.10 The direct method of coordination

In this method the coordinating controller fixes a desired output vector y_d for the whole system and in general allocates a local resource constraint r_{d_i} to each of the i subsystems.

Thus, the problem, as seen by the ith subsystem, is

maximize $J_i(c_i, u_i)$,

subject to $u_i = H_i y_d$.   (9.24)

(Note that u_i depends on the whole of the vector y_d and not only on y_{d_i}.)

$$y_{d_i} = F_i(c_i, u_i), \tag{9.25}$$

$$r_i(c_i, u_i) < r_{d_i}. \tag{9.26}$$

The local results depend on y_d and r_{d_i}. Since the input u_i is fixed by y_d, the only free local variable is c_i.

We denote the resultant c_i obtained at the ith subsystem by $\hat{c}_i(y_d, r_{d_i})$ and the resultant objective function by

$$J_i(\hat{c}_i(y_d, r_{d_i}), u_i) = J_i(\hat{c}_i(y_d, r_{d_i}), H_i y_d) \tag{9.27}$$

which we agree to denote by

$$\hat{J}(\hat{c}_i(y_d, r_{d_i})).$$

The coordination problem for the complete system can then be defined as:

Choose y_d, r_d so as to maximize

$$J = \psi(\hat{J}_1(y_d, r_{d_1}), \dots, \hat{J}_n(y_d, r_{d_n})) \tag{9.28}$$

subject to the constraint

$$\sum_{i=1}^{n} r_{d_i} \leqslant R. \tag{9.29}$$

9.11 Implementation of the direct method using penalty functions

The coordinating controller does not usually have a detailed knowledge of the local constraints existing at each of the i subsystems.

A workable method for implementation of the direct method of coordination under these conditions uses penalty functions.

At the ith subsystem, the maximization of Section 9.10 is modified to become

maximize $J'_i = J_i(c_i, u_i) - P_i(y_i - y_{d_i}) \tag{9.30}$

(in which P is a weighting matrix) subject to the same constraints as before.

Since $y_i = f_i(c_i, u_i)$, this local maximization will now try to achieve $y_i = y_{d_i}$ by choice of c_i, insofar as constraints allow this, in addition to maximizing the local objective function J_i as before.

The coordinating controller is required to maximize the modified global objective function

$$\max_{y_d, r_d} J' = \psi(\hat{J}_1(y_d, r_d) - P_1(\hat{y}_1 - y_{d_1}), \dots, \hat{J}_n(y_d, r_d) - P_n(\hat{y}_n - y_{d_n})). \tag{9.31}$$

The coordinating controller will now, with this modified objective function, tend to move the y_{d_i} towards the corresponding y_i.

The penalty terms in $(\hat{y}_i - y_{d_i})$ are reduced by adjustment of $\hat{y}_i$ at the local level and by adjustment of y_{d_i} at the coordinating level. It can be shown (Reference F2) that, if iteration at the two levels is continued until the penalty terms disappear, then the solution so obtained is optimal within the constraints.

9.12 The interaction balance method

This method is also called the price method, since it has many features in common with coordination of a free economy based on price regulators.

We recall that the subsystem outputs y_i are given by the equation

$$y_i = F(c_i, u_i). \tag{9.32}$$

Each of the local problems has the modified cost function

$$J_i(c_i, u_i) + \langle \lambda_i, u_i \rangle - \langle \mu_i F_i(c_i, u_i) \rangle \tag{9.33}$$

subject, as before, to the constraint $(u_i, c_i) \in K_i$.

The vector λ sets a price for every input in the overall system. The vector μ_i is defined by the equation

$$\mu_i = \sum_{j=1}^{n} H_{ji}^T \lambda_j. \tag{9.34}$$

Thus μ_i sets the prices on the outputs of the ith subsystem.

Maximization of the cost function produces the values

$$\hat{c}_i(\lambda), \hat{u}_i(\lambda), \hat{y}_i(\lambda) = F_i(\hat{c}_i(\lambda), \hat{u}_i(\lambda)). \tag{9.35}$$

The equilibrium prices $\hat{\lambda}$ are defined to satisfy

$$\hat{u}(\hat{\lambda}) - H\hat{y}(\hat{\lambda}) = 0. \tag{9.36}$$

The supervisory controller varies λ, monitors $u(\lambda)$ and $Hy(\lambda)$ and attempts to fix λ such that equation (9.36) is satisfied.

Once the local minimizations have been performed, the supervisory controller attempts to maximize the function ϕ. Note that this is a modified sum of local cost functions and is less general than the cost functions possible with the direct method.

$$\phi(\lambda) = \sum_{i=1}^{n} J_i(\hat{c}_i(\lambda), \hat{u}_i(\lambda) + \langle \lambda, \hat{u}(\lambda) - HF(\hat{c}(\lambda), \hat{u}(\lambda)) \rangle. \tag{9.37}$$

A number of conditions need to be satisfied if maximization of $\phi(\lambda)$ is to provide a well-posed algorithm to determine $\hat{\lambda}$. Provided that ϕ satisfies continuity conditions, then a gradient method can be used to determine $\hat{\lambda}$.

$$\nabla \phi = \hat{u}(\lambda) - HF(\hat{c}(\lambda), \hat{u}(\lambda)). \tag{9.38}$$

This gradient can be seen to depend on the difference between input and output that the supervisory controller has to bring to zero by choice of λ.

Note that the method is a model-based open-loop algorithm. In a physically existing system, the interconnection requirement is bound to be always satisfied, so that the method cannot be used for feedback control.

9.13 Coordination with feedback for on-line control

The coordination methods that we have been discussing up to this point made no use of feedback from process signals. In this sense they can be considered as open-loop model-based methods whose accuracy is necessarily dependent on the accuracy of the available models. In practice, models are inaccurate and disturbance signals cause deviations from the values predicted by determinate equations. It is therefore natural to seek to include feedback from measured signals to improve the performance of a coordinating control system.

FEEDBACK TO THE SUPERVISORY CONTROLLER
One approach is to use feedback to the supervisory controller while leaving the local control problems as before. For the interaction balance method, Figure 9.15 shows how feedback to the supervisory coordinator might be realized.

At the supervisory level, λ is fixed such that

$$\hat{u}(\lambda) - u_m(\hat{c}(\lambda)) = 0, \tag{9.39}$$

where u_m is the measured value of the u vector.

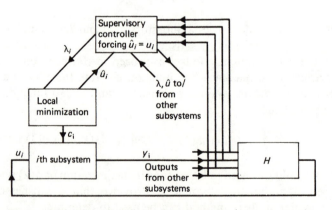

Figure 9.15 Coordination with feedback control to the supervisory controller.

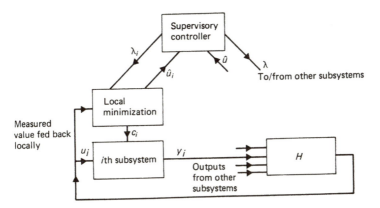

Figure 9.16   Coordination with feedback to the local controller.

FEEDBACK TO THE LOCAL CONTROLLERS

If the interaction balance method (Section 9.12) is modified so that the local controllers use measured values instead of model based values, the system of Figure 9.16 results.

The problems of convergence and stability need to be considered in a structure of this type and the reader is referred to Reference F2 for a discussion of these points.

Exercises

9.1 Discuss the advantages and disadvantages of decomposing one large problem into several smaller problems. Outline the alternative strategies to problem decomposition, illustrating each with an example. For a large decomposed control problem, what are the implications in terms of computing power and information sharing when an implementation is to be achieved?

9.2 Describe the concept of multilayer control, indicating typical activities that may take place at the individual layers.

Explain how optimization techniques may be used to achieve efficient coordination of a complex multilayer system.

9.3 Figure 9.17 shows a digraph representing a system. State the system transition matrix and use powers of this matrix to confirm the presence of closed cycles. From the transition matrix derive:

(a) the reachability matrix R (go on to calculate $R \cap R^{T}$ and state its significance);

(b) the interconnection matrix of the system.

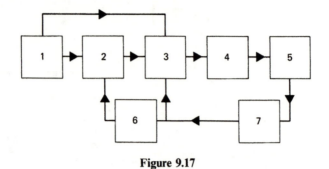

Figure 9.17

9.4 Compare and contrast the direct method with the interaction balance method for coordination of multilevel systems.

10

Control System Implementation
and Integration

10.1 Introduction

In this chapter, we consider the methodology of control system implementation through various choices of computer configuration. The chapter ends with a discussion of integration topics—that is the integration of control systems into management and production systems and the interaction between control, IT and software engineering.

10.2 Implementation—some initial points

In control system implementation, it is necessary to consider:

1. How a design will be converted into reliable, maintainable real-time software.
2. The choice of supporting systems software and hardware base.

For medium to large applications, there exists a good range of proprietary distributed control systems supported by implementation expertise, as described for instance in Chapter 11. For most end users, one or other of these commercially available systems will almost always provide the best solution.

For small- to medium-sized applications, there are many choices. At the top of the range, it may be viable to use small versions or subsets of large distributed control systems. At the lower end of the range, a PC-based solution may be attractive.

For some applications, for instance where high-definition graphics are required, a workstation solution may offer advantages.

Complicating the choices are:

the increasing moves towards networked systems (so that even a small system cannot be chosen in isolation);

the rapid moves in open systems, requiring users not to be left behind by emerging standards;

the industrial suitability of operating systems (UNIX, VMS)—with the first being popular with academics and the second with industrialists;

the relative difficulty or ease with which real-time software may be developed—either directly on the target machine that will later run the programmes or on a separate development machine;

the availability of software packages and expert systems shells—most of these run only specific host machines.

10.3 Computers for control design, for real-time software development and validation and for implementation

Comprehensive, user-friendly control design software to run on personal computers or workstations is now, after successive generations of improvement, almost universally accepted as the tool for control systems design.

However, one important point that must not be overlooked is the following. The modelling tools (to some extent) and the control design tools (to a very large extent) that are provided so conveniently to work on PCs are linear system tools. The very convenience of the tools tends to push an inexperienced user to early and unwarranted linearity assumptions.

The proper sequence for use of the kit of modelling, design and simulation tools will usually be:

1. Obtain process data.
2. Develop a nonlinear model based on the process data.
3. Produce a linear approximation to the nonlinear model.
4. Use essentially linear design tools to design a control system for the linear model.
5. Use simulation to check/modify the behavior of the control system on the original linear model.

Implementation of medium to large systems is routinely achieved through a commercial configuration consisting of semi-autonomous local controllers linked in a hierarchical manner through data highways to a central supervisory command point.

Referring to Figure 10.1, the areas described above complete only two of the requirements that are needed for the total control task.

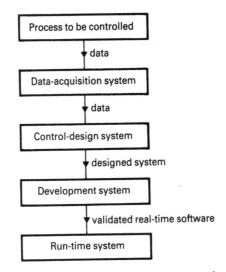

Figure 10.1 The control-design and implementation sequence.

The two areas for which the solutions are not so obvious are:

1. The system for data acquisition.
2. The system for development of validated real-time software.

On the first, there seems to be no clear choice of system for data acquisition, data treatment, data storage and onward transmission—despite the practical importance of these topics in the overall scheme of things. However, some of the well-known control design packages have facilities for data acquisition and filtering provided as front ends.

On the second, professional software houses tend to develop systems on the same target computers that will eventually be used to run the systems. Producing dependable, high-quality, well-documented software is a major task that needs to be undertaken with considerable discipline. CASE tools can assist in providing a framework within which such a disciplined development of software can take place.

10.4 Computers for control application

10.4.1 VMEbus modules for control applications

For real-time control applications, the necessary computing power can be provided on one or two boards, plugging into a standard card cage. A typical current device, suitable for industrial applications, will operate at 33 MHz, have 16 MBytes of RAM and incorporate a high-resolution graphics controller and serial, parallel and keyboard interfaces.

10.4.2 Personal computers (PCs)

Ruggedized personal computers with plug-in interfaces are increasingly presenting themselves for serious consideration in industrial control applications. Often a PC-based system configured for industrial control will be similar in many essential reports to the VMEbus modules discussed above.

10.4.3 Workstations as control elements

Workstations offer great computing power and high-definition graphics combined with an ability to be integrated with real-time databases. A workstation equipped with powerful software is a formidable tool. For instance, given a SUN workstation equipped with the software XMath and suitable interface cards it is nominally possible to:

1. Log data from an unknown process.
2. Model the process.
3. Design a controller for the process, using the model from (2).
4. Simulate the controller from (3) in action on the process model.
5. Allow the controller from (3) to control the real process directly through interfaces.

It would be simplistic to expect this design sequence to happen just as described, but the potential is clear.

10.4.4 Programmable logic controllers (PLCs)

Programmable logic controllers were originally introduced to replace banks of relays and switches for logistical control of sequences of events. They have become quite sophisticated, with interfaces to other control devices. Often PLCs will be delegated, cost-effectively, to the control of the start-up and start-up and shut-down sequences required by batch control.

10.5 Distributed control systems (DCS) for control of interlinked processes

10.5.1 Introduction to distributed control

The true potential of computers in control applications only begins to be realized when their information processing and display possibilities are applied to the supervisory control and coordination of complex systems with many variables.

Control theory offers large-scale systems theory and multivariable theory as aids towards the design of supervisory and coordinating control systems. Nominally at least, such systems have a high degree of modularity so that design, for a particular application, consists of (a) choosing a suitable configuration and (b) 'parameterizing' the configuration. A chief advantage of the approach is that it potentially avoids the need for expensive one-off system development that characterized previous approaches to computer control.

In practice, true 'openness' is still well over the horizon and considerable software still needs writing for each individual project simply to overcome incompatibilities between subsystems.

10.5.2 Distributed computer networks

By a distributed computer network we mean a set of geographically separate, partially autonomous computing devices having plant input/output capability and being interconnected through digital communication links. Control systems realized through distributed computer networks have two significant advantages over control systems based on a single centralized computer:

(a) Practicable methods for the design of control systems for large complex plant generally produce (as we have seen in Chapters 8 and 9) functional solutions in either hierarchical or modular form.

   Such solutions are ideally implemented through distributed computer networks with physical elements in the networks corresponding with modules in the function solution.
(b) The resulting systems have all the advantages associated with local control and most of the advantages associated with centralized coordinated control.

   Failure of a piece of equipment has either a local affect only or else part of the coordinating activity is lost—no equipment failure causes the loss of the complete system operation.

COMPONENT PARTS OF THE NETWORK

A distributed computer network consists of:

(a) several *local controllers*, each able to handle several control loops simultaneously;

(b) interconnecting digital *data links*, together with organizing protocols;
(c) at least one *coordinating controller*;
(d) a central *information display unit*.

Figure 10.2 shows how the elements are typically interconnected.

THE LOCAL CONTROLLER

A typical local controller is shown in Figure 10.3. It can take in sixteen analog and sixteen digital signals and, under the control of a program, can process the signals sequentially through algorithms called down from memory until finally the required output signals are produced to close the control loops. Figure 10.4 makes clear the main functions of a local controller.

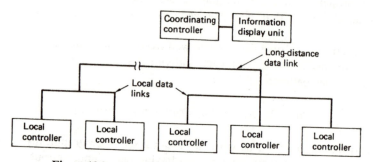

Figure 10.2 A typical distributed computer network.

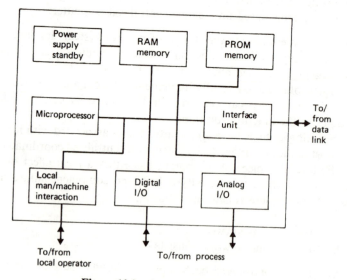

Figure 10.3 A typical local controller.

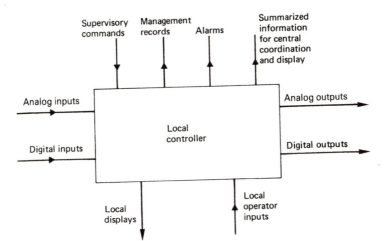

Figure 10.4 The main functions of a local controller.

MAIN PARAMETERS OF A LOCAL CONTROLLER

In designing or specifying a local controller, some of the most important parameters to be fixed are:

the maximum number of control loops that can be closed;

the maximum possible sampling rate;

types and levels of input/output signals;

degree of local autonomy required;

envisaged method of local user interaction;

degree of self-diagnosis, redundancy or self-repair capability;

degree of intelligibility to a non-expert user (a high degre of intelligibility would probably allow a user to modify and extend his own system);

degree of standardization/modularity of both hardware and software, within a controller and between different controllers.

The data links   Physically, the data links consist of coaxial cables or optical fibers, several kilometers long, capable of carrying high-speed digital data transmitted serially.

The *configuration* of the data links and the *protocols for data transfer* determine the character of the network.

The configuration   Figure 10.5 shows a number of possible data-link configurations. Often, as will be discussed later, there is an advantageous correspondence between physical and functional configurations.

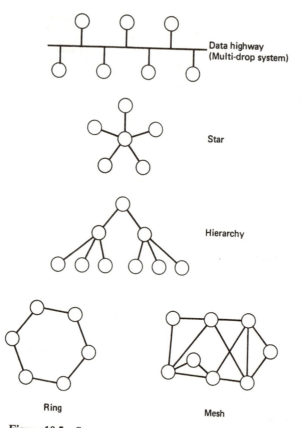

Figure 10.5 Some possible data-link configurations.

The protocols Where many devices need to intercommunicate along a shared data link, some system is needed to organize the communication as efficiently as possible, taking into account priority of messages and the conflicting requirements of speed and accuracy. The set of rules for communication is called the *protocol*. It may be implemented by a centralized explicit device or it may be implemented implicitly by ensuring that every device in the network is programmed to communicate according to the rules.

The coordinating controller A coordinating controller operates at a higher level than a local controller, in a coordinating and supervisory role. It has communication via data links with local controllers, with displays and other man/machine interfaces and possibly with a large conventional computer.

The central information display unit Most networks for computer control have a central information display unit, consisting of a set of visual display units, probably all visible from one location, with the formats of the displays being variable on-line according to user requirements.

10.5.3 Outline of the approach to applications

Manufacturers are beginning to be able to supply systems that can be configured to reflect some of the alternative broad strategies described in Chapter 9. See Figures 10.6 and 10.7, which show how one commercially available system has been developed with alternative process subdivisions in mind.

The most common applications practice is to divide the plant to be controlled into naturally occurring geographically distinct processes or

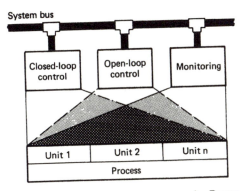

Figure 10.6 The Hartmann and Braun 'Contronic P' system can be configured on a functional basis.

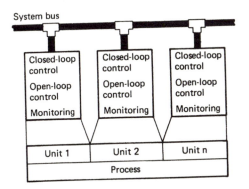

Figure 10.7 The Hartmann and Braun 'Contronic P' system can be configured on a geographical basis.

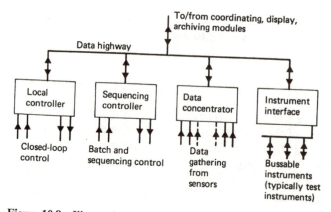

Figure 10.8   Illustration of specialist low-level control and data devices that can be selected to match the needs of an application.

subprocesses of manageable size and to allocate each to the care of a local controller that carries out closed-loop control under supervision by higher-order devices in the network. Other requirements—for instance, the interconnection of intelligent test instruments, logic and sequencing control or intensive data logging—are met by hanging appropriate specialist hardware devices on the data highway (Figure 10.8). Some of the currently available devices will be encountered in Chapter 11.

10.5.4 Provisions for enhancing availability

INTRODUCTION

The mean time between failures for a typical microcomputer-based local controller may be eight years or less. In a configuration employing many such controllers, reliability aspects are always important and often, rightly, dominate discussions on approaches and configurations.

REDUNDANCY WITHIN A LOCAL CONTROLLER UNIT

In this approach, two controllers are run side-by-side with one controller operational and the other on standby duty. Changeover occurs if diagnostic checks show that the operational controller is not performing correctly.

These diagnostic checks are carried out automatically at frequent intervals. Perhaps the most difficult part of the design is to devise a set of reliable and comprehensive self-diagnostic checks. Figure 10.9 illustrates the concept.

ONE SPARE LOCAL CONTROLLER AS STANDBY FOR SEVERAL OPERATIONAL LOCAL CONTROLLERS

$n + 1$ controllers can be allocated to n tasks with the spare controller on standby for the n duties. The system requires less hardware than the method of

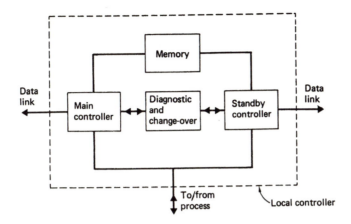

Figure 10.9 Redundancy within a local controller.

Figure 10.9 but the cost is increased complexity and reduced reliability. The spare controller needs to compute concurrently with all those controllers for which it acts as standby. Consequently, it needs to be larger than a standard controller, unless the other controllers are to be deliberately under-utilized. Figure 10.10 illustrates the concept.

DYNAMIC REDUNDANCY
A development of what has been put forward above envisages so-called dynamic redundancy. In this system, all plant signals are available, through data links, to all local controllers. A coordinating controller allocates tasks to individual local controllers and, under fault conditions, re-allocates ('dynamic re-allocation') according to a set of stored priorities. Figure 10.11 illustrates the

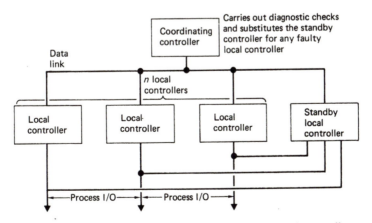

Figure 10.10 One standby controller for n operational controllers.

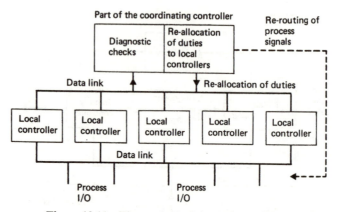

Figure 10.11   The method of dynamic re-allocation.

principle. It is clear that in return for considerably increased complexity, the reliability of high-priority base tasks can be made very high indeed.

This system presupposes that the data links are of inherently high reliability—in practice, they would seem to be as likely to fail as the local controllers.

ENHANCING THE RELIABILITY OF THE DATA LINKS

Since data links consist of cables of one type or another, they are vulnerable to accidental rupture. In an industrial environment, worthwhile redundancy can be obtained only by laying standby links physically separated from the operational links with automatic failure detection and transfer to the standby cable under fault conditions (Figure 10.12).

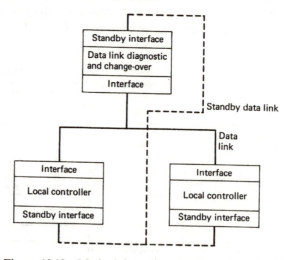

Figure 10.12   Method for enhancing the reliability of data links.

Finally, attention needs to be given to the mundane topic of connectors: they are the least reliable parts of many systems.

10.5.5 Some general aspects

FURTHER DISCUSSION OF CONFIGURATIONS AND DATA-TRANSFER PROTOCOLS
In general, messages are transmitted within a network by:

(a) *The broadcast system*, in which every node in the network receives every message that is transmitted.
(b) *The routed block system* (packet switching system), in which blocks of information are routed to specific receiving points.

The most common configuration for distributed control systems is the *data highway*, often referred to as the *multi-drop system* (Figure 10.5). This is a broadcast system and at different times many different devices need to transmit, yet only one device may transmit at any one time. Efficient operation of a system based on data highways depends on the existence of an efficient decision-making procedure that allocates the right to transmit devices within the network. A heavily loaded system may incur delays while devices queue for the right to transmit.

Each device on the network is connected by a tee into the highway. The highway carries high-speed serial data packages each containing an address word and error-checking bits as well as the information content. An intelligent interface at every tee node reads the address words and identifies those packages intended for its own device.

Conversely, the same interface acts as a source when its own device is the sender. The failure of any particular device does not interrupt the flow of data in the main highway. Physically the data highway can be a coaxial cable, a twisted-pair cable or a fiber-optic link.

Some progress has been made in standardization, for instance, the Ethernet specification lays down recommendations that cover integrity of the data links. All systems include error-checking procedures in the form of redundant repetitions, and check words and/or byte counts.

Designing a user-friendly, high-reliability distributed system that has short response times and that can be flexibly reconfigured is a daunting task. Trends are towards the use of ever more intelligent interfaces with direct memory access.

Other configurations that may be met are sketched in Figure 10.5.

The disadvantage of the star configuration is that all communication must pass through the central node unless additional links are added to form an augmented star.

The ring configuration (Figure 10.5) operates by one device taking on the role of sender and transmitting an address-containing message to its neighbor in the ring. The neighbor re-transmits the message to its neighbor until the device defined by the address is reached. Clearly, failure of any device in the ring will disrupt communication although redundancy techniques are available to main communication despite failures, using for instance double redundant rings.

STANDARD INTERFACES AND COMPATIBILITY

It would be a great advantage if different makes and types of microcomputers, user peripherals and process interface devices were compatible and easily interconnectable to form working systems. At present this is far from the case, and in many projects considerable effort has to be devoted to time-consuming one-off interfacing activities.

Standardization would involve enforcing standards in respect of:

interconnection logic;

data-transfer protocols;

signal coding, signal levels and tolerances;

impedance matching.

Progress on standardization has been disappointingly slow, but improvements in this direction are promised and expected. The Fieldbus standard, see Chapter 11, is one welcome development in this direction.

A great deal is being expected from open systems, often based around the UNIX operating system. In theory, open systems allow arbitrary equipment to be interconnected successfully, removing a whole host of compatibility problems. In practice, there is still resistance, scepticism and some way to go before the open systems concept becomes reality.

REAL-TIME LANGUAGES

Early real-time programs were written in machine code to obtain the benefits of great flexibility, speed and running efficiency. Some low-level applications are still implemented using assembler code usually with the help of development support such as a cross-assembler that runs on a larger machine to produce the necessary code.

Increasingly, however, real-time programs are being written in high-level languages, perhaps with blocks of machine code inserted for certain real-time tasks that are not within the scope of the high-level language.

Real-time programs need to be highly reliable and to this end they need to contain checking, diagnostic, fail-safe and redundancy features.

In a typical application, a wide variety of devices is connected to the computer through I/O interfaces and the program needs to control the operation of the data transfers through these interfaces.

Within the overall operation of the system software, many different programs of different importance and different immediacy need to interleave according to the designers' perceived priorities. An interrupt facility is needed to give the designer control over the priorities with which shared facilities are allocated.

Many vendor-provided control packages provide algorithm libraries that need only to be parameterized and interconnected by software links, so that programming consists of little more than choosing a configuration with the aid of a menu and then inserting numerical values.

10.5.6 Criteria for selection of distributed control systems

Compared with earlier control systems based around a single process computer, distributed control systems have many advantages. However, a new problem arises—that of specifying the best control system to suit a particular project.

Clearly, the profile of the control system should, in some sense, match the profile of the process to which it is to be applied. A bewildering choice of parameters exists that might be used to establish these profiles and hence assist the decision making.

The list below gives some broad initial guidelines. It lists aspects that may be important or critical in the process profile and, for each, discusses the implications for the control system.

Geographical layout of plant | Most of the available systems can cover, as standard, plant extending over a few kilometers. Certain systems have been applied particularly to utilities such as gas piplines, extending over 100 km or more.
It is worth considering whether the plant layout leads to a preference for a particular type of layout for the control system (ring, star, hierarchy, etc.).

Degree of coupling between sub-processes | The degree of autonomy of the local controller is an important parameter. Highly autonomous local controllers depend little on the data link—this is a reliability advantage. A low degree of autonomy in the local controller increases dependence on the data links but allows for easier coordination from the center. Close coupling between sub-processes implies that low autonomy is needed in the local controllers and conversely.

Process environment | The system must be adequately constructed to match environmental problems of dust, vibration, electrical or magnetic interference and must satisfy special requirements imposed in hazardous areas.

Sequencing needs | Certain systems have comprehensive sequencing facilities provided in the local controllers. Other systems allow the connection of a programmable logic controller alongside a local controller. Ease of programming the sequences should be checked.

Operator needs

Where local operators exist, they need local information displays and keyboards to make their inputs. Processes that are predominantly centrally controlled naturally concentrate their operator interaction centrally—this has implications on the data network.

Flexible, ergonomically attractive displays are a key aspect of distributed systems. Most systems offer mimic diagrams, historical trend displays and lists of out-of-tolerance variables (alarms). Some questions to ask: Can the mimic diagrams be easily user-modified? Are the fast trend data held on disk ready for call or do they only start to be built when a request is made? Can any process variable be singled out for detailed central inspection? Can an arbitrary set of process variables be displayed together (group display possibilities)? Are colors used to best advantage in the displays? Is the total display requirement satisfactorily met?

Reliability requirements

High reliability is generally ensured by duplication of processors, interfaces and data links with self-diagnosis and automatic changeover in the event of failures. The strategy used to achieve such diagnosis and changeover should be examined carefully.

Many systems will ride-over a short power failure of 0.1 second, but for longer power failures, a reserve power supply with automatic changeover is needed.

A highly modular hardware structure with easily understood diagnosis and good manufacturer back-up will be sought.

Batch/continuous nature of processors

Most systems began life in response to a particular market need. Most still betray their origins and still have an orientation towards a particular type of application. Predominantly continuous processes have little need for the comprehensive sequencing facilities that some of the systems provide.

Types of sensors and actuators

Particular I/O cards may be available to provide for the special needs of sensors (thermocouples, for example, need break protection and cold-junction compensation; other sensors need excitation).

High-speed/high-accuracy requirements

Process characteristics combined with control-loop specifications fix closed-loop bandwidths—these in turn fix minimum sampling rates. Most systems on the market are rather slow. They sample, perhaps between $\frac{1}{2}$ and 4 times per second. Certain control loops (for instance, those containing electromechanical drives) cannot be controlled at these modest rates and the situation needs to be assessed very carefully and quantitatively.

Where there are only few such loops, special provision can probably be made.

High-accuracy requirements will require good-quality instrumentation amplifiers in the signal lines and high-quality A/D and D/A converters with a sufficient number of bits to meet the accuracy specification.

Need for complex supervisory control	When considerable central computation has to be carried out, a system must be selected that provides or at least allows for the use of such a facility. Where a truly hierarchical multilayer or multilevel system is needed for coordination purposes, this needs to be reflected in the disposal of computing power and displays and in the configuration of the data links.
Number of closed loops at each location and in total, number of variables to be logged	A straightforward factual check of manufacturers' literature reveals the capability of the systems in these respects.
The intention to extend the system in the future	Is the manufacturer prepared to guarantee that, for some specified period, any new system that he produces will be compatible with the system that he currently offers? Are hooks and interfaces provided to allow addition of higher levels of control in the future?
The need to interface the new system with existing earlier generation equipment	Certain manufacturers can offer complete compatibility with (their own) earlier generation systems.
The need to significantly modify the system frequently	A system that needs to be modified significantly, quite frequently, implies the need for user-friendly programming and easy documentation.
The intention to make plant-based personnel responsible for the commissioning, operation and maintenance of the system	The availability of a suitable set of vendor-provided algorithms should be checked. The programming method and its suitability for use by plant-based personnel needs to be assessed. (Some systems offer a block-diagram oriented programming approach, which may be preferred.) Different systems can best be compared for ease of programming by actually documenting the program for a typical benchmark task (that is representative of the project) as it would be under the alternative systems.
The existence of large groups of sensors or actuators situated remotely	Some systems offer what is virtually a data concentrator. It may take in (say) 100 analog inputs and transmit the whole of their data serially along a single twisted pair or equivalent. Analogous equipment exists for the converse problem of driving a large group of actuators situated remotely.
Coordination software	Vendors may have available packages for common coordination problems such as energy management.
Links into management system	A distributed control system that allows a direct link to an EDP system lays the way open for the future development of an integrated process and business management system.

The table should help to obtain a broad assessment of which particular systems would suit a particular project. It is, of course, necessary to go into fine detail before making a final decision.

Two very important aspects could not be covered here: cost and manufacturers' capability. Manufacturers' capability includes the topics of: experience in the particular application area, willingness to train operators and maintenance engineers, response to call-out and general reputation on project management.

POINTS FOR THE FUTURE
Rapid developments are to be expected, and some welcome extensions will be in:

Availability of worthwhile coordinating/supervisory algorithms for use at higher levels of the system hierarchy. These would implement routine supervisory policies and leave a human supervisor free to concentrate on non-routine events.

Self-documenting systems. System documentation is important, difficult and expensive. A system that is largely self-documenting will represent a significant step forward.

High-speed local controllers. A system that offers the option of loops with (say) 1 ms sampling rates allows applications to spread over a wider range of industrial needs.

10.5.7 Implementation by distributed computer networks as a natural follow-on from modern control design techniques

Large, complex control problems are in practice rarely analyzed monolithically. Rather, they are subdivided according to the methods of Chapter 9, into manageable sub-problems whose solutions are then combined to yield an overall solution. Such solutions are implemented by software subroutines that can be identified with the subdivisions used in the earlier system analysis.

Using distributed computer networks, it becomes possible and even natural to identify physical elements in the network with specific tasks that were allocated during the control-system design phase. This has many advantages. For instance, a large system can be commissioned piecewise; failure of a device in general has a known limited effect on the process and process operators can understand the function of each piece of equipment in the network.

10.6 Integration of process control systems into overall systems of plant management

Historically, process-control systems, and currently, even state-of-the-art sophisticated distributed process-control systems, are operated more or less in isolation from business systems relating to the same plant. Separate departments are involved, with process control being considered an engineering function and business systems being linked into the administrative departments of a company.

As the potential benefits of integration have become clear, there have been attempts to link distributed process-control systems (DCS) with business control systems (BCS). Very crudely the attempts at integration have taken two forms.

1. Physical interlinking of a process control computer with a business computer using some form of networking.

   It often seems to have been assumed that interconnection of the two subsystems would rapidly lead to benefits but in practice the software to achieve efficient coordination has often been lacking and the nature of the data exchanged has been restricted to that enabling a narrow set of objectives to be attained with little possibility of complete integration.
2. Hierarchical process control with business objectives being fed in at the upper layers.

   These systems have had some success but they can at best be fairly primitive milestones on the path towards complete integration.

Conventional hierarchical systems with their reliance on sending summarized data upwards are necessarily restricted in their evolution by the original design assumptions on the bandwidths to be used in data transmission upwards and by the pre-judging of the types of summarization that would be satisfactory. Both design choices (bandwidth, nature of scope of data summarization) could well be regarded as limiters of system/flexibility and hence life-span.

PRACTICAL SOLUTIONS TO THE TOTAL INTEGRATION OF PROCESS CONTROL AND BUSINESS SYSTEMS

We may regard a factory as consisting of a set of interlinked processes fed by raw materials and producing products. The factory interacts with the outside world, receiving orders, quoting delivery dates, scheduling internally, monitoring progress, ensuring quality and despatching products.

One promising approach to total integration (Reference B10) is as follows.

The first level is that the interlinked processes are controlled by one or more distributed process-control systems (DCS).

The second level is an operational support level. This level contains in real-time all of the information in the first level. Many users access this level—none access the lowest level. High speed and little flexibility are the aims at this level.

The third level is a decision support level. This level operates off-line, responding to complex unanticipated enquiries.

10.7 Safety-critical and high-integrity designs

High reliability is of course demanded in all industrial applications. This need is met in commercial distributed control systems by providing optional doubling up of communication links and system elements, linked to diagnostics and automatic changeover strategies.

Certain applications require such high standards of reliability that they can properly be described as high-integrity applications. Safety-critical systems, where lives will be lost in the event of failure, are in this category.

To illustrate current approaches to the design of high-integrity systems, we examine the design of the attitude control system of an aircraft. Attitude control is concerned with the orientation (i.e. angular position) of the aircraft with respect to three axes (normally called yaw, pitch and role axes). See Figure 10.13.

To control orientation the pilot moves control surfaces on the wings, tail plane and tail fin and, through its nonlinear dynamics, the aircraft responds by rotating about three axes passing through its centre of gravity.

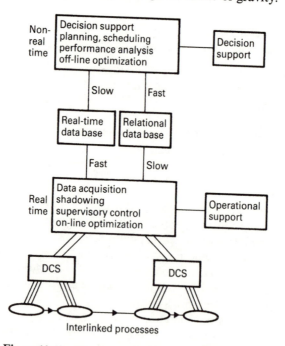

Figure 10.13   The three-level CIM System proposed by Brooks.

Some possible modes of failure for this system are:

One or more sensors may fail   Redundant sensors may be fitted but then there is a need to pick out healthy signals by some method. In very critical cases, three sensors all measuring the same quantity can be fitted, allowing a two-out-of-three voting strategy to separate healthy from unhealthy signals.

The communication links may fail   This can be overcome to some extent by duplication of channels. However, if original and duplicated channels follow the same route (as is to some extent inevitable with tail-end to cockpit channels in an aircraft), both channels may be destroyed together in an accident.

One or more actuators may fail, partially or completely, to move the corresponding control surface   Since there are usually seven control surfaces, but only three control dimensions, there is clearly some built-in redundancy that can be exploited by intelligent control. Furthermore, there is interaction between the three axes {as can be seen from the nine main nonlinear differential equations governing orientation, (see Reference M3)} and this may be exploited to allow complete orientation control to be achieved even though one set of control surfaces may have become immovable.

The obvious approach to designing the necessary intelligent control implied above is to build a system that:

(a) recognises that there is an actuator problem and that then
(b) reconfigures the control system to perform optimally with the 'as-is' set of damaged actuators.

Such an approach is called *reconfigurable control.*

The one small disadvantage of reconfigurable control is that the aircraft may have crashed before the diagnostic and reconfiguration phases have been successfully completed.

If there are fears that such a situation may arise, it will be preferable to

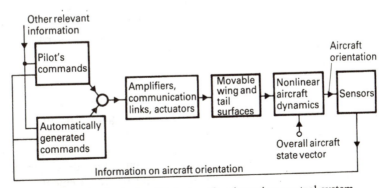

Figure 10.14   A simplified aircraft orientation control system.

design a *fault-tolerant control system* that makes use of actuator interactions at all times so that if and when an actuator fault occurs, no reconfiguration will be necessary since the system has been designed to work with partial actuator sets from the beginning.

The pilot may issue an erroneous command or the automatic system (see Figure 10.13) may generate an erroneous command   Gates to limit the magnitudes and rates of change of command signals may be built-in as a crude check on these types of error—in other words, the system may be required to stay within a safe operating envelope.

The overall aircraft state vector may move into an area outside the expected envelope   This could arise if for instance the aircraft lost engine power and forward speed, if it was physically damaged—as would be likely with a military aircraft or if it came into freak weather conditions.

These changes to the state vector can be considered as severe input disturbances to the orientation control system. To counteract all such foreseen and unseen disturbances, it is clear that the orientation control system needs to have maximum robustness to disturbances, including disturbances to the parameters of the aircraft dynamics. Unfortunately, the requirement for maximum robustness conflicts with the requirement for rapid response to the controls. The final control design requires detailed quantitative studies to fix the boundaries of compromise between performance and robustness.

The computer system may fail   Process control systems will, in general, be based upon programmable general purpose computers.

These are generally highly reliable, compared with physical devices, particularly connectors and actuators, and compared with specially built electronic systems. Even avionic systems, as in this example, will usually be based upon proven computers—albeit perhaps specially packaged micro-processors.

The risks and maintenance problems attached to the development of a special-purpose computer for a specific application generally outweigh any performance advantages—which are usually short-lived given the rapid evolution of general purpose equipment and devices.

Similarly, the software—the programming of the computers—can usually be relied upon when it is based upon configuration and parameterization, by the control engineer, of standard, well-proven, packages for the particular type of application. There is still, clearly, a need to take great care to ensure that the packages are robust, and that their 'programming' is done correctly.

The greatest risk to the computing system is likely to arise from any newly-written special-purpose software. This often entails development work by software specialists who may not understand the requirements of control engineering in general, or of the specific application in particular.

In the sense that it does not 'wear out', software is never 'unreliable'. However, it is generally so complex, and the failure modes so diverse, that it can never be completely tested. So software can contain latent design errors which remain undetected for years, but which may surface under just those rare critical conditions when correct operation is most important.

Such design errors are often loosely divided into two categories— 'completeness' and 'correctness'. Errors of completeness are usually associated with specification errors, where the specification failed to mention some input conditions—such as what the system should do with out-of-range data from a sensor. Correctness errors are usually associated with mistakes in the design process. There is some evidence that specification errors are the most prevalent.

There are many widely varying techniques for software development which are intended to address one or both of these problems. They range from the use of mathematical logic in programming to software quality management, which is based for the most part on requirements to follow a collection of 'best practice' procedures, including configuration management, the use of design codes of practice, and design reviews.

The respective merits of software development techniques generate much debate in the software industry, without, as yet, any conclusion or consensus. References A4, A5 contain discussions of some of the issues.

10.8 The interaction of control with information technology (IT)

As societies and industries mature, their typical problems move from the separable, well-delineated, well-defined to the interconnected, interacting, less well-defined and more organizational. Thus, it is true to say that civil aviation is more critically affected by the organizational problems of air traffic control than by the more purely engineering problems of aircraft flight control.

It can be argued that control problems are increasingly to be found embedded within larger organizational problems whose solution requires an appropriate multidisciplinary approach that typically involves control and IT including software engineering.

This multidisciplinary interaction may be required through all the stages of a project. For instance, in problem analysis in the drawing up of functional requirements in the design stages, in the choice of approaches to implementation and in verification. For instance, in a typical industrial project a control specialist will perhaps meet CASE (computer-aided software engineering) tools, architecture issues, preferences for particular operating systems, preferences for particular languages and preferences for particular forms of human–computer interaction.

As progress in IT continues, there will increasingly be a need to consider parallel computer implementations, architectures containing neural networks and other machine-learning or artificially-intelligent artefacts. The control subject needs not only to make use of current and future advances in electronics, IT and software engineering to design even better control systems, it also, and very importantly, needs to interact flexibly with other disciplines to solve a whole range of emerging complex problems. Control engineering will best serve the community by working with other disciplines to provide the hybrid skills that such complex organizational problems call for.

Exercises

10.1 True optimization of a plant that is controlled by a set of semi-autonomous interconnected local controllers is impossible. Is this statement true? Is it of practical significance? Are there in fact any theoretical (as opposed to practical) advantages of distributed control over centralized control?

10.2 Describe, with diagrams, a configuration for the overall control of a linked group of processes through distributed computing power.

Describe the layout of a typical local controller.

Suggest criteria by which a complete system and the individual elements in the system (local controller, data highway, central facility) may be characterized and hence assessed for fitness for a particular purpose.

To what extent are these types of control systems universal and to what extent applicant-dependent?

10.3 Much effort often needs to be expended on interfacing because there are no universally agreed standards on device compatibility. Find out from the literature as much as you can on progress in standardization for both serial and parallel satisfying. Try to list the main features that a good interfacing standard needs to embrace. Some of the bodies concerned with standardization are the International Electrotechnical Commissions (IEC), the International Standards Organization (ISO) and the Institute of Electronic and Electrical Engineers (IEEE) and you should seek out their latest definitive publications.

10.4 Flow-measuring devices need to be carefully matched to the particular characteristics of their application. This may mean a great deal of expenditure of manpower in selection, installation, re-scaling and re-calibration. An attractive alternative is to specify a single universal programmable flowmeter and to set it up and keep it calibrated by signals from the control computer system. Such a flowmeter is an example of a *smart sensor*. Find out more about smart sensors and summarize this information in a report that marshals the arguments for and against smart sensors.

11

Commercially Available Distributed Control Systems and Their Industrial Application

11.1 Introduction

Distributed control systems (DCS) have evolved into very powerful tools that are able to cater for the technical requirements of a complete plant and to link into business control systems.

Some of the established manufacturers of distributed control systems include:

ABB Process Automation

Ferranti

Fischer and Porter

Fisher Controls

Foxboro

Hartmann and Braun

Honeywell

Siemens

Valmet

Yokogawa

Here we review a selection of systems and then describe case histories that together span a wide range of application areas.

It will be seen that all the DCS systems are basically similar. However, each detailed description that follows has a different emphasis and a different viewpoint. By the end of the chapter the reader should have a good understanding of the state-of-the-art in distributed control systems and their application.

11.2 Computer integrated manufacturing through the Fisher PROVOXplus system

11.2.1 System overview

The PROVOXplus system architecture can form a foundation for computer-integrated manufacturing (CIM). CIM is the integration of process, plant and business operations (from order entry and scheduling through production, quality control, maintenance, shipping and accounting), made possible through computerized information networking. The aim of CIM is to ensure that process and business information is available at the proper place and time for better decision making at all levels.

Figure 11.1 illustrates the hierarchy of production and management functions in a typical process plant. Each of the five levels of the pyramid

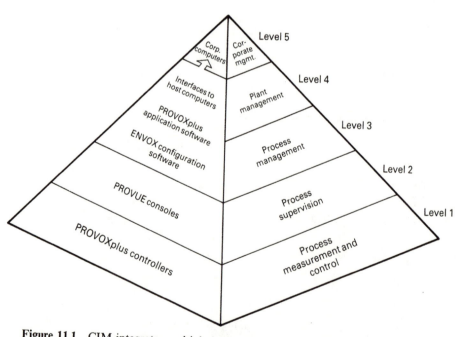

Figure 11.1   CIM integrates multiple levels of process equipment and information.

represents a different process management function, each with its own decision-support requirements.

The first level of the pyramid (Level 1) is process measurement and control. Here controllers directly adjust final control elements.

At Level 2 (process supervision), the operations console allows operators to monitor and control the process via the Level 1 controllers.

At Level 3, computer-based plant automation functions are integrated into the process control system. For example, a process host computer might generate management reports, calculate optimization strategies and communicate instructions (or recommendations) to the operations console.

Levels 4 and 5 are plant and corporate management functions. Higher-level computer functions, such as cost accounting, inventory control and order processing/scheduling, are performed at these levels. Being on-line with the lower levels of the pyramid allows these higher-level computers to provide up-to-the-minute information for better management decision support.

PROVOXplus controllers and PROVUE operations consoles are the key elements of Levels 1 and 2. At Level 3, PROVOXplus application software, computer interfaces and ENVOX configuration software help manage activities in the lower levels as well as providing process information in usable form for Levels 4 and 5.

11.2.2 System functions and features

CONTROL AND MONITORING

The most basic interfacing function is the handling of input and output (I/O) signals between the process and the control system. PROVOXplus I/O devices can provide the necessary signal conversions for a wide variety of measurement devices and control elements. Some devices allow the I/O subsystem to be located almost a mile from the controller and allow one I/O device to serve as a back-up for as many as eight others.

Continuous control functions provide control capabilities from the simplest pressure, temperature, flow, or level loops to complex strategies involving inter-related loops and advanced calculations. Most of these functions are 'pre-engineered' and only require parameterization. To add further control and computation power, other specialized functions are available to allow the design of advanced, customized control algorithms and complex calculations not catered for in the pre-engineered library.

For applications that require logic, interlocking and simple sequencing in addition to continuous control, logic-control functions provide capabilities for interlocking, permissives, conditionals and step-by-step sequences.

Advanced sequencing and batch control functions provide the control and interface capabilities required for batch processing, from single-product/single-stream processes to multi-product/multi-stream applications. These capabilities include:

    Advanced logic and sequencing operations
    Powerful continuous control functions
    Integration of discrete and continuous functions
    Equipment coordination and allocation
    Recipe management
    Batch scheduling and reporting
    Operator interfacing

These control and monitoring functions are available in a variety of devices, each offering a different combination of functions and capacity. By selecting the appropriate controller for each control loop (or group of loops) in the process, a customized system results.

OPERATIONS INTERFACE

The operator's console gives operations personnel the ability to monitor and control activities across a broad area of the method that permits operations on local highways to proceed simultaneously. The result is a very high effective rate of information transfer. The network traffic director gives each local traffic director and device attached to the network highway a 'turn' at using the highway. It thus controls communication between devices on different local highways as well as between devices on the network highway.

MAXIMIZING DATA THROUGHPUT

The PROVOXplus communication system maximizes data throughput by reducing the communication load on each device, freeing it to do its control and/or reporting functions in a timely manner. The communication system does this by using both solicited and unsolicited data transfer.

Solicited or *request–response* data transfer includes supervisory control changes from a host computer, operator-initiated changes, diagnostics, displays of tuning parameters, and end-of-process report information. In such cases, a device on the highway solicits an action or data from another device. The response of the requested device completes the request–response data transfer cycle.

Routine operating data, however, can be reported using unsolicited data transfer. The techniques used include reporting data periodically, periodically by exception and on change of state.

In *periodic* reporting, a device reports operating data at a user-configured, fixed-interval sample rate. For example, a controller may automatically report a tank level to the console every 10 seconds. This reporting method assures the operator of up-to-date information without going through the request–response process. It also reduces the data-transfer burden on the controller and the console by about one-half, as well as reducing data highway loading.

In *periodic-by-exception* reporting, the sending device compares the present value with the last-sent value. If the difference between the two values is greater than a specified dead zone, the information is reported at the next periodic interval.

Using the example above, the controller could be configured to report the tank level to the console only if the level changes by more than, say, 1%. The operator would still be notified of any significant change, but the console would not be updated unnecessarily. This further reduces the data-transfer load.

It is also possible to send data from one device to another on *change of state*. Using this method, the data is reported only when a point goes into or out of alarm or—for discrete values—if the alarm state or on/off value changes.

If the exact level in the example tank were relatively unimportant, this method could be used to notify the operator only when a high-level switch indicates that the tank is nearly full.

These reporting methods maximize data throughput and free PROVOXplus devices to perform control, alarming, diagnostics, operator interface and other functions.

By comparison, a highway that uses primarily request–response communications or puts all plant communications on a single local-area network would require many times the bandwidth of the PROVOXplus data highway.

DATA SECURITY

The assurance that a digital message is correct is essential to the successful operation of a distributed control system. The PROVOXplus communication system uses special security techniques to make sure information moves safely over the data highway and that any differences between transmitted and received data are detected and counteracted.

PROVOXplus systems use three error-checking methods to detect any errors that occur during message transmission:

Bit checking

Length checking

Mathematical checking

Messages that are received without errors are acknowledged by a 'received-okay' message from the communication interface assembly (CIA) of the destination device to the source device's CIA. If a source CIA does not receive this acknowledgement, it will re-transmit the message to the destination device a number of times. If this is not successful, the message will be sent via the redundant data highway, if present. If the source CIA still does not receive acknowledgement, an alarm alerts operators that a communication problem has occurred and that data from this device is suspect.

SYSTEM INTEGRITY

Continuous diagnostic functions check communication integrity between devices and the internal integrity of each device.

The CIA for each PROVOXplus device continually checks the communication integrity of the device and of the data highway and reports any problems to the traffic directors. In addition, each device performs internal diagnostic tests to confirm its own integrity.

The local and network traffic directors keep track of devices that report internal problems and of devices that do not respond to messages. This information is organized into an integrity report that is broadcast periodically to every device on the data highway.

The CIA for each device stores this report in its memory and, before sending a message, checks to make sure the message is being addressed to a device that

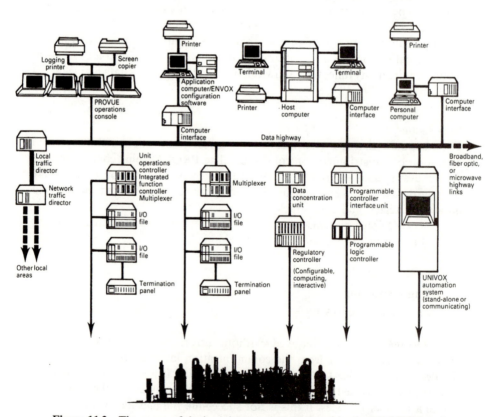

Figure 11.2 The range of devices that can be connected in the PROVOX system.

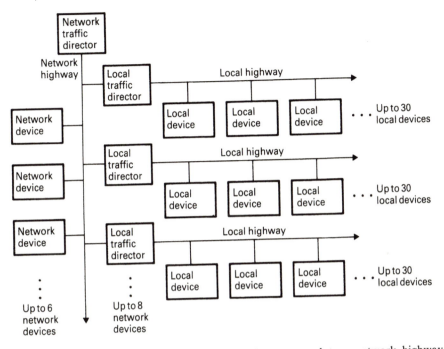

Figure 11.3 Up to eight local highways can be connected to a network highway, integrating as many as 246 devices in a single network that can span more than 4 miles (6 km).

will be able to receive it. In addition, the integrity broadcast can generate console displays that alert operators to communication failures.

By using these integrity reports to eliminate undeliverable messages and by having local data highways operate continuously and independently, the PROVOXplus communication system is able to give rapid highway access to all devices on an equal basis.

REDUNDANCY

For further security, communication redundancy can be provided by installing redundant local data highways, each with its own traffic director. Each highway device is connected to both highways through individual CIAs. The primary and secondary highways are electrically isolated from each other so that no single failure will affect both highways.

The secondary data highway is maintained in hot standby condition. If the communication path to the primary highway is functional for all devices, then the data is transmitted via the primary highway and the secondary highway receives an indication that no communication is required.

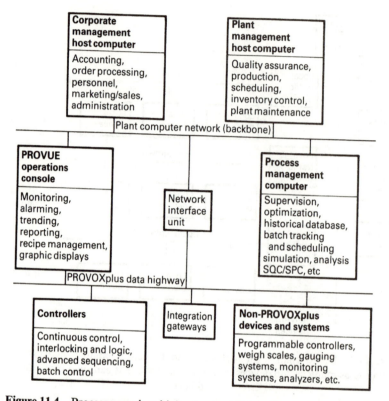

Figure 11.4 Process-area data highways can be linked into plant computer networks for plantwide communication.

If a communication failure occurs between a device and the primary highway, only that device switches to the secondary highway. If total failure occurs on the primary highway, however, all communications switch to the secondary highway. The loss of communication is reported in the diagnostics displays, or a special display can be configured to relay this information to the operators.

Redundancy is also available to increase the reliability of network communication. As shown in Figure 11.8, additional CIA is installed in each device and additional traffic directors are installed to control the operation of these redundant highways.

INTERFACING

To enhance integration of computers, subsystems and other devices, the PROVOXplus system has the flexibility to interface with a number of broadband

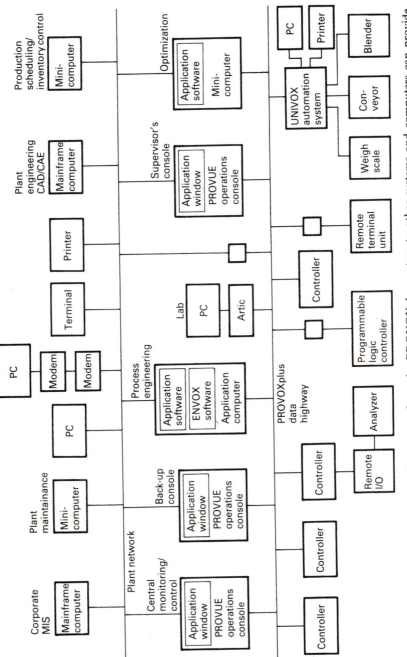

Figure 11.5 Extensive interfaces and gateways from the PROVOXplus system to other systems and computers can provide true plantwide information exchange for computer integrated manufacturing (CIM).

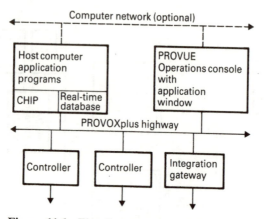

Figure 11.6 The Computer/Highway Interface Package (CHIP) gives computer programs access to process data.

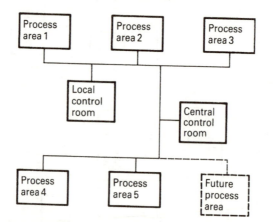

Figure 11.7 The geographic distribution to PROVOXplus system components allows several plant areas to be connected and provides for simple, cost-effective future expansions.

networks, including DECnet, Token Ring, PC NET, 802.3/Ethernet, TCP/IP, IEEE-488 and MAP.

11.2.3 Computer-Integrated Manufacturing (CIM)

Computer-integrated manufacturing (CIM) is usually defined as the integration of process, plant and business operations made possible through computerized

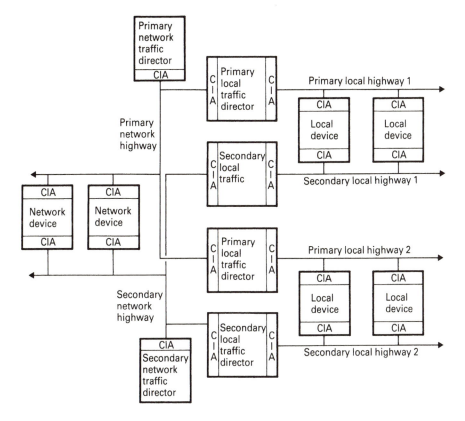

Figure 11.8 Network as well as local highways can be redundant.

information networking. The technology that makes this networking possible is based primarily on communication and interfacing.

Because the PROVOXplus communication system has extensive, powerful interface capabilities, it can play a vital part in plans for a plantwide information system. A global, relational database is used for configuration of the system, providing the basis for a plantwide database. Application software products, such as process supervision and plant management packages, optimize the effectiveness of plantwide automation systems.

Eight figures show a selection of the features of the PROVOX system. Figure 11.2 shows the range of devices that can be interconnected in the PROVOX system. Figure 11.3 shows how local highways in the PROVOX system can be interconnected to a network highway. Figure 11.4 shows the arrangement for plantwide communication. Figure 11.5 shows the interfacing and gateway arrangements for the network to serve a computer-integrated

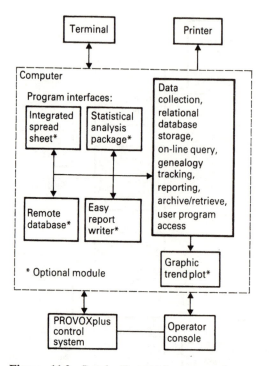

Figure 11.9 Batch Data Manager software provides capabilities for batch histories, flexible report generation, material tracking, and plantwide batch data collection.

manufacturing (CIM) system. Figure 11.6 shows the arrangement by which computer programs access process data. Figure 11.7 shows the possible geographic distribution of PROVOX system components. Figure 11.8 shows the redundancy that can be built into the network. Figure 11.9 shows the facilities of the batch data manager.

11.3 The Bristol Babcock system

Figure 11.10 shows the extensive communication facilities provided by the 'Enterprise' Open Architecture of Bristol Babcock. Both Ethernet and Token Ring provide links to the process control system. The diagram summarizes many other features offering powerful company-wide integrating possibilities.

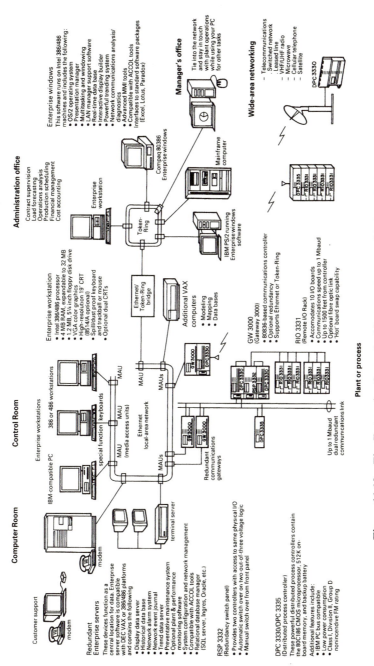

Figure 11.10 The Enterprise Open Architecture system of Bristol Babcock.

11.4 The TCS 6000 integrated system for distributed process control

NETWORK 6000 is a process instrumentation and control system for both continuous and batch processes. Its architecture blends the benefits of distributed and supervisory control with open access at all levels—a few loops to thousands of points. Its physically distributed units can 'stand alone', or be networked to give single-loop or redundant multi-loop integrity. Application-specific packages allow TCS to implement turnkey systems.

MAXI-VIS WORKSTATION SYSTEMS (FIGURE 11.11)
Maxi-Vis IV, based on the DEC VAXserver operating under VMS, offers a range of hardware and software tools for medium/large control systems. Its DEC hardware allows great flexibility in redundant processor and distributed workstation configurations—whilst TCS standard software provides data acquisition, display, logging and powerful batch/sequence control. Maxi-Vis IV

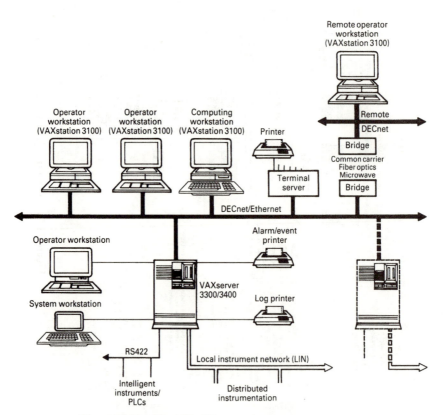

Figure 11.11   Maxi-Vis IV supervisory computer system.

integrates Tactician products, S6000 Instrumentation and foreign devices into a large system architecture.

TACTICIAN T2001 WORKSTATION

T2001 is a 386 PC-based control system offering powerful display and control facilities for small/medium applications. T100 I/O front-end units acquire data with S6000 instruments distributing the control. Icon-based configuration toolkits, SPC and reporting facilities, drivers library for most programmable controllers and Ethernet/Novell and RS485 Networks are available.

TACTICIAN T1000/T100 UNIT CONTROLLER

Tactician T1000, featuring an IP66/NEMA 4 operator panel with wipe-pad pointer and electroluminescent display, gives graphics plus computer flexibility— right at the process. Multi-loop PID and batch/sequence control can be distributed amongst T1000s and T100s communicating via the instrument network in redundant configuration if necessary, while S6000 instruments can provide single-loop integrity. Historical/configuration data can be passed between T1000s and to a PC for analysis.

INTELLIGENT INSTRUMENTS

These microprocessor-based instruments offer in-built diagnostics, block-structured configuration and high-level programming. The range includes multi-loop and specialized single-loop controllers, signal processors, powerful flow computers and advanced block-structured PID controllers.

Figure 11.12 shows a typical Maxi-Vis supervisory computer system incorporating networked workstations.

The implementation of an open network architecture, together with the powerful hierarchical database structure of Maxi-Vis, makes system data accessible to all computers and workstations on the network.

For applications where full workstation facilities are not required at the central computer location and for non-networked systems, a directly connected lower-cost operator workstation may be used.

DISPLAYS

Pre-formatted displays   A hierarchy of overview, group and point displays is provided which can be configured automatically and then edited to suit individual requirements.

Customized mimic displays   Mimic displays are fully user-configurable via advanced picture creation and editing facilities.

Using pixel graphics and up to 64 foreground and background color combinations, customized graphic symbols and process plant diagrams can be created.

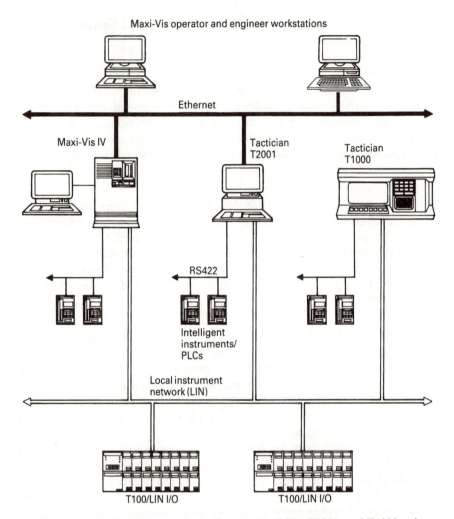

Figure 11.12   A TCS system including the Maxi-Vis, T2001, and T1000 units.

All database values can be accessed allowing the creation of flexible and fully dynamic screen displays.

Trend   Trend facilities provide full color graphical displays of real-time and historical process data on a common time axis, together with associated instrument fascias.

Axis pan and zoom features are provided for the detailed examination of data by moving the trends backwards and forwards in time and altering their magnification.

11.5 The Contronic P. System of Hartmann and Braun (Figure 11.13)

The system is based on process stations for the automation functions of loop control, logic control and monitoring. The process stations contain a modularly extendable process interface which is connected to transmitters, signal transducers and process-correcting equipment.

Apart from the standard loop functions, the digital processing functions include, for example, dedicated blocks for sequence control, dosing circuit, state correction, quantity registration and feedforward control.

Variable cycle times permit optimum matching to the process dynamics.

Different kinds of digital measuring systems with manufacturer-specific data transmission protocols can be coupled directly to the process stations via a freely programmable interface.

Machines and individual plant units which are equipped with PLC devices can also be coupled and thus integrated into the central process management.

THE UNIVERSAL STATION

The universal station is a functional unit with integrated video display and operator control for stand-alone operation. It combines the functions of process and operator stations and so provides a solution for small-process automation. At the same time it offers all the advantages of the Contronic P system and uses the same cards and the same operator control and monitoring functions as in the large system. Thus the universal station is fully upward-compatible when it is connected to the system bus.

THE SERIAL SYSTEM BUS

The serial system bus is used for communication between the decentralized process stations and the central stations. The bus system operates with a data transmission rate of 1 MBd.

Up to 127 stations can be connected. The system bus can have a length of up to 2 km without, or up to 12 km with, bus refreshers.

THE OPERATOR STATIONS

The operator stations in Contronic P are used for central operator control and monitoring of the process events in the decentralized process stations.

Up to three color VDUs can be connected to one central operator station for visual display of the process. Operator keyboard and light pens are used to select process points, setpoint values, manipulated variables or to change alarm values, to start log functions and to acknowledge messages. Recipe disposition also takes place at the central operator station.

For on-screen curve display and for archiving, the time characteristic of measured variables is recorded and stored with high resolution.

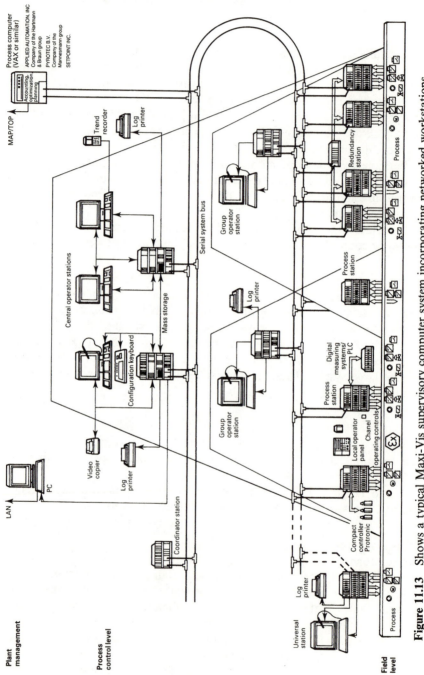

Figure 11.13 Shows a typical Maxi-Vis supervisory computer system incorporating networked workstations.

Past-history process data as well as the configuration data are available for quick access.

A configuration keyboard can be connected for configuration of all system functions at a central point.

The group operator station is installed with a minimum use of space in a trolley container and can be used efficiently in different places for operator control of individual process sections. It is a compactly constructed operator station with the same scope of functions and the same cards as in the central operator station.

THE COORDINATOR STATION

The coordinator station is used in larger systems for implementing higher-level automation strategies such as start-up and shut-down control. Apart from dedicated functions, it also provides free arithmetic for implementing special automation solutions such as optimizing calculations, special logs and special dialogs.

It also provides a fast and flexible interface for coupling higher-level computers.

FUNCTIONS OF THE PLANT MANAGEMENT LEVEL

Recipe processing On the basis of the NAMUR concept of basic operations, Contronic P provides integrated recipe processing for automation of batch processes with changing recipes.

'Recipe integration' in Contronic P means:

Central management in the operator station—decentralized execution in the process stations.

Standard operator and configuration interface for recipe and loop functions.

Batch-related logging.

Data archiving and evaluation with the PC Personal computers can be coupled to Contronic P for archiving and evaluation of extensive process and production data. The coupling permits direct access to real-time data and evaluation with customary PC software.

Complex optimization with process computers Complex tasks on the production management level such as optimization calculations are assigned to process computers. Marketable user-packages of well-known consulting organizations can be used, e.g. the OPTROL 7000 system from Applied Automation, Inc., which has proven its ability in numerous applications, for example when optimizing product quality, exploitation of energy and materials and reducing operating and maintenance costs in process engineering plants. The computers have direct access via the system bus to all process data.

Computers from all leading manufacturers can be connected via the powerful interface.

Structure-conditioned dependability   The functionally decentralized structure and modular design inherently lead to very high dependability, e.g.

The decentralized process stations operate automatically and independently of central functions.

Process interface cards such as analog input and controller output are organized according to channels.

Project redundancies   Establishment of redundancies for certain functions increases the availability of the process-control system due to the fact that in case of a fault these functions are fully preserved.

Thus an extremely high availability is achieved. According to actual process requirements, the availability can be projected differently on various levels of the process control system. Depending on the requirements a redundancy of 1:4, 1:2 or 1:1 is possible for complete process stations. The central operator stations and mass storages can also be designed with redundancy. This is also true for the power supply. The system bus is installed redundantly as a standard feature.

Self-monitoring   If a fault occurs in spite of the measures described above, the fault will be detected by the self-monitoring function and reported to the central fault diagnostic system.

The central fault diagnostic ensures quick fault recognition and leads, together with the modular system structure, to rapid error correction.

In addition, service programs are available for rapid fault tracing in on-line as well as off-line diagnostic mode.

11.6 Analog devices: DCS systems

11.6.1 Complete measurement and control system: the Micromac 5000 (Figure 11.14)

FEATURES

A single-board measurement and control system designed to interface to hundreds of real-time I/O points. The board is fully programmable and can be used in a distributed control application or as a local front end.

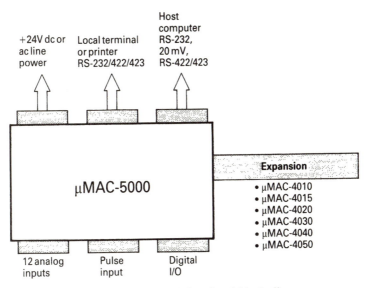

Figure 11.14 μMAC-5000 functional block diagram.

Completely Integrated Measurement and Control Systems on a Single Board.

Wide selection of analog and digital I/O

Fully programmable in μMACBASIC

Serial communications

Rugged industrial chassis

Sensors: thermocouples, RTDs, strain gauges, load cells

Millivolt, voltage, current source

Current outputs

1000 V Channel-to-channel isolation for most input types

Solid-State Digital Interface for Direct Connection to Switches, Lights, Tachometers, Solenoids, Stepper Motors.

a.c. or d.c. inputs or outputs, from 5 to 280 Volts

Counter or pulse inputs for frequency or event counting

μMACBASIC Programs developed and run on board

Provides links to I/O commands, user-written procedures and functions and interrupt support

Program storage in PROM or battery-backed RAM

Supports IBM PC and other hosts for program development

Powerful Communications Capability

> Supports RS-232C, 20 mA, RS-422/423 up to 19.2 kbaud
> Asynchronous communications
> Protocol emulation easily accomplished with μMACBASIC
> Fully buffered input and output characters

Rugged Industrial Design

> 0–60°C Operating temperature range
> Rack mountable
> Hardware watchdog timer
> Battery-backed memory

APPLICATIONS

> Energy management
>
> Process monitoring and control
>
> Facilities monitoring
>
> Furnace/boiler control
>
> Pot line control
>
> SCADA

The μMAC-5000 Measurement and Control System is a complete solution designed for a broad range of industrial measurement and control applications. Because of its many communications modes and I/O interfaces, it can be used in applications requiring distributed intelligence or a local front end. The μMAC-5000 can grow with the application; from its single-board configuration of 28 I/O points to the μMAC-4000 expander boards capable of handling over 100 mixed I/O points. Up to 16 such clusters are supported in a multi-drop configuration. A μMAC-5000 must reside in each cluster of boards to coordinate all I/O functionality.

The μMAC-5000 board is driven by the 8088 microprocessor with 80 K of ROM and 32 K of battery-packed RAM (expandable). Programs can be run from battery-packed RAM or PROM (jumper selectable). The board can be powered from either a.c. line voltage or from +24 V d.c. power or both.

The μMAC-5000 has two serial communication ports—local and remote—that can be used with virtually any host computer. The local serial port supports RS-232C or RS-422/423 and is used primarily for program development or as a local printer/terminal. The remote serial supports RS-232C, RS-422/423 or 20 mA and is primarily used for host system communication.

The μMAC-5000 is designed to meet the reliability needs of industrial applications. It is capable of withstanding significant variations in temperature (0–60°C) and is resistant to shock and vibration. Battery-backed RAM and PROM provide reliable storage media for program code as well as application data. A watchdog timer monitors the system and notifies the user when the system has failed. Single-board enclosures, card cages and rack mount units are available.

11.6.2 Analog devices: Industrial controller

THE MICROMAC 6000 (FIGURE 11.15)
The system has been designed for high-performance measurement and control for a wide variety of industrial applications. The unit's modular design allows for application specific configuration—the exact number of analog and digital signal conditioning modules. Expansion via the I/O processor extends I/O capability without sacrificing system performance.

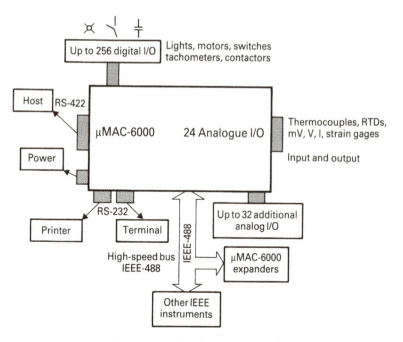

Figure 11.15 μMAC-6000 functional block diagram.

FEATURES

Complete Industrial Computer with Real-World Signal Interface and Communications to Higher Level Computers.

High-quality signal conditioning

Powerful CPU with optional co-processor

Serial or parallel communications

Modular design

Signal modules

Sensors: thermocouples, RTDs, strain gauges, load cells

Millivolt, voltage and current source

Current outputs

1500 V r.m.s. of isolation

Guaranteed 0.05% accuracy, no-field calibration

Low-cost expansion panels available

Solid-state digital interface for direct connection to switches, lights, tachometers, solenoids, stepper motors

a.c. or d.c. inputs or outputs, from 5 to 280 Volts

High-speed card/frequency inputs

Pulse outputs

8 MHz, 8088 Microprocessor with optional 8087 co-processor

Fully programmable in μMACBASIC or C

256 KBytes battery-backed RAM

Battery-backed calender clock

μMAC-6000 Expander available for extra I/O processing

Communications flexibility allows connection to any computer, peripheral or other instrument.

Isolated RS-422/RS-485 high-speed serial port

Dual RS-232 serial communication ports

IEEE-488 Parallel bus provides links to expander CPUs

Modular design provides the greatest possible flexibility for an integrated system.

Analog signal conditioning on a per channel basis

Field-replaceable CPU

Analog and digital I/O expansion available through optional backplanes

Expander CPU can be added to increase processing power

The μMAC-6000 is a modular I/O processor that integrates real-world I/O into any industrial automation strategy. Because it offers powerful computing capability, communications and input and output interfaces, it can be used as a stand-alone system, in concert with a host, or as part of a distributed control system. A μMAC-6000 can also grow with the application, from an initial 24 analog and 48 digital I/O points to over 200 analog and 1000 digital I/O points per cluster. Its stand-alone capability allows it to continue to operate in the event that the central host computer goes down.

The μMAC-6000 is a modular system in that it is configured by combining the necessary components needed in a particular application—the exact number of analog and digital signal conditioning modules. The CPU, a metal enclosed two-board set, is also modular in that it can be replaced in the field without disturbing the field wiring. Programming modularity is also provided since the system may be programmed in either μMACBASIC or C by adding the proper routines to the application library as needed.

The μMAC-6000 can operate at a number of different levels—monitoring, analyzing, or controlling. The base system is driven by a 16-bit processor (8088) with the option to add a co-processor for faster computation. For large-point-count applications, additional I/O processor CPUs can be added—over the local high-speed bus—so that the additional I/O channels do not consume the bulk of the base system's processing power. The μMAC-6000 system (including expansion CPUs) can also be part of a multi-drop configuration containing up to eight systems linked to a single host.

Communication between the μMAC-6000 system and other devices is performed through either the three serial communication ports or the IEEE-488 (GPIB) bus. Two of the serial ports are RS-232; the third is an isolated RS-422/RS-485 (up to 19.2 kbaud) for direct or multi-drop connection to a host computer.

11.7 The Siemens Teleperm M System

The Siemens Teleperm M System is a decentralized process control system which, in addition to the Simatic S5 automation system and the Siemens 300 Systems process computers, offers solutions in the field of process automation. It consists of a series of different components with differing performance levels, and easy combination of these components allows individual adaption of the system to processes of different scales and levels of complexity. All system configurations are based on the automation subsystems (AS) (local controllers) which, due to their design, can also be used as stand-alone systems in small- and medium-scale automation plants and thus successfully close the gap between decentralized large-scale systems and specific task-oriented systems.

FEATURES AND PERFORMANCE LEVELS OF THE AUTOMATION SUBSYSTEMS

The AS220 and AS230 automation subsystems of the Teleperm M system can be used according to their performance levels for solving process control tasks including any combination of closed-loop control, computation, open-loop control, monitoring, signalling, logging and operator communication and observation. The AS231 automation subsystem is a version of the AS230 designed solely for signalling and logging tasks.

All three systems—the AS220, AS230 and AS231—are stored-program automation systems. Figure 11.16 shows a block diagram of the AS220 subsystem, which applies in principle to the other two subsystems.

The basic unit contains the power supply, the processor and the different interface modules and the remaining slots can be used for input/output modules whith represent the interface to the process.

In addition to the basic unit, one or more expansion units can be used, each with a maximum of 14 I/O modules.

In the case of a fault in the central unit, these modules automatically take over control of the corresponding control loops and the operator interacts with the system either from the front panel of the modules or from special manual/automatic control stations.

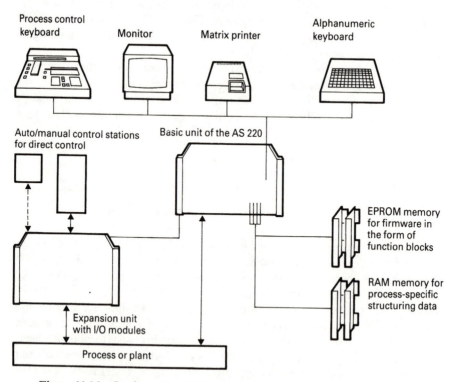

Figure 11.16 Configuration of the Siemens AS220 automation subsystem.

The memory is subdivided into two areas. The EPROM memory contains the system firmware which contains the function blocks. More than fifty standard blocks for closed-loop control, computation, open-loop control, monitoring, logging and indication as well as for operator communication and monitoring are available. These blocks can be configured using the alphanumeric keyboard, i.e. they can be called up as often as is required and supplied with the corresponding interconnections and parameters. This application-oriented structuring data is stored in RAM memory. The individual control loops and open-loop control functions can be monitored using the process control keyboard.

The stand-alone use of the AS220, AS230 and AS231 subsystems does not exclude the interfacing of these subsystems with each other or with supervisory central operating and monitoring systems.

AS220 AUTOMATION SUBSYSTEM

This system is based on a microprogrammed central processor with a 16-bit internal processing width predominantly for analog values. The system has a total memory of 64 K words, of which 22 K words in RAM are required for storing application-oriented structuring data. The basic cycle time is 250 ms. A maximum of 47 I/O modules can be connected to the system and this results in an application area for small- to medium-scale automation tasks with approximately the following performance level:

10 to 40 control loops,

20 to 80 additional analog values for further processing,

3 to 6 sequence controls,

20 to 100 binary logic functions,

2 to 4 black and white mosaic diagrams for graphical display of process subsections and standard listings.

AS230 AUTOMATION SUBSYSTEM

This more powerful automation subsystem is based on a microprogrammed central processor with a 16- or 32-bit processing width. Three basic cycles are available (125 ms, 1 and 8 s). A total of 94 I/O modules can be connected. Compared with the AS220 system, the AS230 contains an extended range of firmware function blocks.

For special purposes, the TML (Teleperm M language) process control language is available in addition to the firmware blocks. This allows special function blocks to be built.

For operator communication and monitoring two color monitors can be connected. The maximum memory capacity of 256 K words matches the high performance level and of this the user has a maximum of 240 K words at his disposal.

Medium-scale automation tasks can be implemented with AS230. Taking into account that it is possible to combine differing tasks, this results in an application area with approximately the following performance level:

30 to 50 control loops,

50 to 120 additional analog values,

100 to 250 binary logic functions,

5 to 20 color mosaic diagrams and individual listings.

OPERATOR COMMUNICATION AND MONITORING

Operator communication and monitoring subsystems, graded according to the problem, are available in addition to the automation subsystems. They extend from conventional operating controls to systems with color monitors and light pens. Parallel operator communication and monitoring means the simultaneous display of all the information and associated intervention.

A cable-saving easily planned serial bus subsystem is used for the communication between the individual automation components and the components in the process control room. It transmits the data between the individual components and coordinates and controls the data flow.

CS275 BUS SUBSYSTEM

Data are exchanged between individual TELEPERM M systems, including Siemens process computers, via the CS275 bus subsystem. The main requirement here is a coupling between several automation subsystems of one and the same plant for the transmission of control information as well as a coupling to the subsystems provided for operator communication, signalling and logging.

The CS275 bus subsystem consists of a local bus and a remote bus for the transmission of data.

Information of various types, compiled into messages, is transmitted between the individual subsystems via the bus subsystem in serial mode. The data paths can also be designed with redundancy where high requirements for availability of data transmission exist.

Each subsystem is coupled to the bus subsystem via the autonomous bus interface module. This module decouples the exchange of data, in time and function, between its subsystem and the bus subsystem. Every bus interface module can control the exchange of data, i.e. it can take control of the bus.

The master function can be transferred to another bus interface module of the same autonomous bus subsystem (master transfer). A higher availability than with a central bus control is achieved in this way. The local and remote buses are connected together through a bus coupler. With a bus coupler, each bus connected to the coupler remains autonomous, i.e. data transfer is possible on a local bus and a remote bus independently of one another.

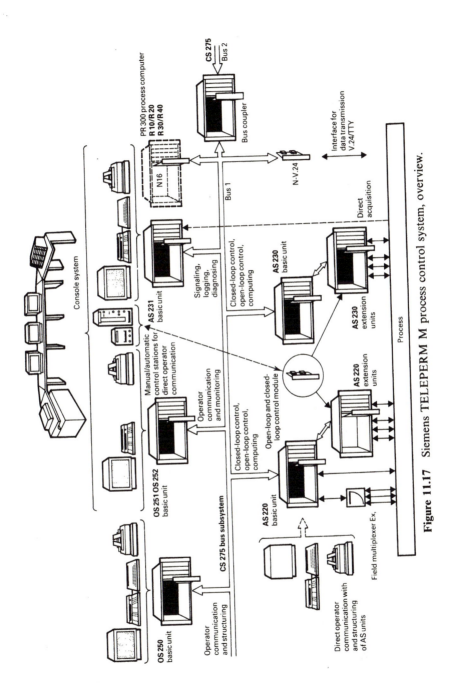

Figure 11.17 Siemens TELEPERM M process control system, overview.

OPERATOR COMMUNICATION AND MONITORING STRATEGY

Direct operator communication and monitoring The necessary elements or devices are directly coupled to the AS automation subsystem. The front panel elements on the open-loop and closed-loop control modules permit the direct operation and monitoring of motors, actuators, solenoid valves and controllers. The manual/automatic control stations can be used to remotely control motors, actuators, solenoid valves and controllers if enabled by the process and if no protection commands are present.

Central operator communication and monitoring The technical devices and equipment for central operator communication and monitoring are coupled to AS automation subsystems via the CS275 bus subsystem. Monitoring is carried out on black and white or color monitors, operator communication with a process communication keyboard or light pen.

Figure 11.17 shows an overview of the TELEPERM M system.

11.8 The Honeywell TDC 3000 System

11.8.1 General aspect

The Honeywell TDC 2000 system (totally distributed control) was announced in 1975—as such it was one of the first distributed systems to become commercially available. The Honeywell TDC 3000 system, into which it has now evolved, will be described. (Figure 11.18 shows TDC functionality. Figure 11.19 shows TDC hardware features.)

The four key elements of TDC 3000 are:

1. Distributed process control and data acquisition devices linked by universal control networks and data highways.
2. Communication links consisting of gateways, local control networks (LCNs), universal control networks (UCNs) and data highways for total system communications.
3. Distributed history, application, computing and communication modules for enhanced control operations.
4. The universal station a single Expert Window™ to all plant operations.

GATEWAYS FOR EFFICIENT INFORMATION INTERCHANGE

TDC 3000 gateways provide the data conversion, buffering and processing necessary for efficient information interchange between the local control network and devices with different communication protocols and speeds. Built-in conversion routines handle data base translations and minimize configuration requirements.

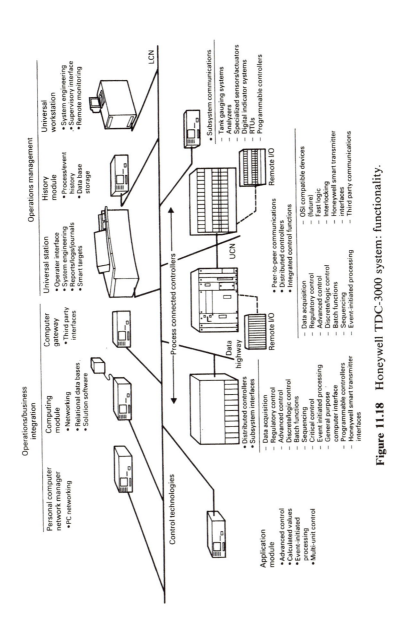

Figure 11.18 Honeywell TDC-3000 system: functionality.

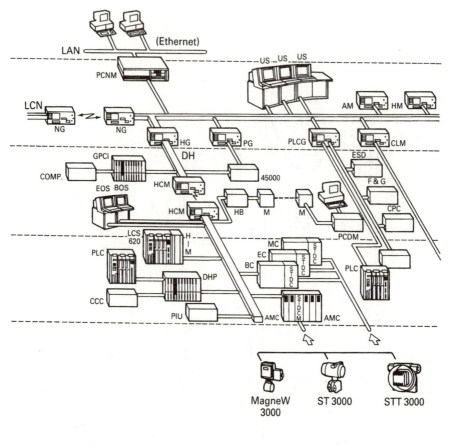

MagneW
3000   ST 3000   STT 3000

AM	= Application Module	EOS	= Enhanced Operator Station
AMC	= Advanced Multifunction Controller	ESD	= Emergency Shut Down
		F&G	= Fire & Gas
BC	= Basic Controller	GPCI	= General Purpose Computer Interface
BOS	= Basic Operator Station		
CCC	= Compressor Control Corporation	HB	= Hiway Bridge
		HCM	= Hiway Coupling Module
CPC	= Critical Process Controller (TRICONEX)	HG	= Hiway Gateway
		HM	= History Module
CG	= Computer Gateway	HIM	= Hiway Interface Module
CLM	= Communication Link Module	HTG	= Hydrostatic Tank Gauging
		LAN	= Local Area Network
CM50N	= Computing Module 50 N	LCS620	= Honeywell PLC
DH	= Data Hiway	LCN	= Local Control Network
DHP	= Data Hiway Port	LCNE	= Local Control Network Extender
EC	= Extended Controller		

Figure 11.19 Honeywell TDC-3000 system: hardware.

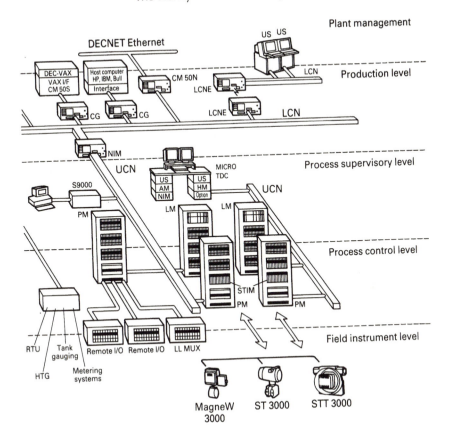

LLMUX	= Low Level Multiplexer		ST3000	= Smart Transmitter 3000
LM	= Logic Manager		STDC	= Smart Transmitter Digital Communication
M	= Modem			
MC	= Multifunction Controller		STDCM	= Smart Transmitter Digital Communication Module
NG	= Network Gateway			
NIM	= Network Interface Module		STIM	= Smart Transmitter Interface Module
PC	= Personal Computer			
PCNM	= Personal Computer Network Module		STT3000	= Smart Temperature Transmitter 3000
PCDM	= Personal Computer Data Manager		UCN	= Universal Control Network
			US	= Universal Station
PG	= Processor Gateway			
PLC	= Programmable Logic Controller			
PM	= Process Manager			
RTU	= Remote Terminal Unit			

A variety of gateways are available, exploiting the features in the TDC 3000 system.

The Personal Computer Network Manager (PCNM) Allows a network of IBM-compatible personal computers and other LAN users to interface directly to the local control network (LCN) providing a distributed information system.

The Programmable Logic Controller Gateway (PLCG) Provides a high-performance LCN interface to programmable controllers and third-party subsystems.

The Communication Link Module Integrates various speciality subsystems such as analyzers, remote terminal units, tank-gauging systems and paper machines directly to the LCN.

The Computer Gateway (CG) Allows two-way communication between computers and the LCN.

The Data Hiway Port (DHP) Interfaces programmable controllers and devices that emulate their protocol to the data highway.

11.8.2 Individual elements in the control loop

PROCESS MANAGER

The MAP-based TDC 3000 Process Manager (PM) regulatory, sequential, logic, data acquisition and computational control strategies. Peer-to-peer communications allow distributed multi-unit control strategies.

Process Manager features include:

Fiber-optic remote I/O

Selectable processing rates

32-point low-level analog inputs

Pulse inputs (up to 20 KHz)

Control strategy integrity

Controller and I/O redundancy

Key-lock capabilities to control user access

Large memory capacity for user-created programs

The process manager uses Motorola 68000 CPU elements and its outline architecture is as shown in Figure 11.20.

Referring to Figure 11.20, each of the interface units I/O autonomously brings in, verifies and conditions data before passing it via the 'communication'

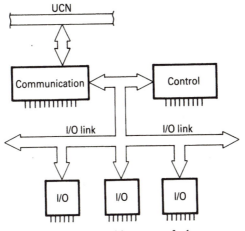

Figure 11.20 The architecture of the process manager.

block to the universal control network. The 'control' block undertakes whatever local control is required and undertakes peer-to-peer operations with other process manager modules.

THE BASIC CONTROLLER (BC)

The basic controller contains the General Instruments CP1600 16-bit microprocessor and is designed for continuous control of conventional analog feedback loops. The controller takes in up to 16 analog signals, processes the signals through any of the 28 ready-made algorithms stored in PROM, and produces 8 analog outputs. Each input variable is scanned and each output variable is updated at a fixed rate of 3 times per second. All plant input/output signals are in the form of 4–20 mA currents. The algorithms are selected and parameters set from a basic operator station on the data highway. The basic controller is powered by a 24 V supply with battery back-up.

THE EXTENDED CONTROLLER

The extended controller is capable of handling both analog and digital input/output data. Otherwise it serves a similar purpose to the basic controller.

THE MULTIFUNCTION CONTROLLER

The multifunction controller is a module of the control system which is used for controlling discontinuous or nonstandard processes. It contains the hardware necessary to receive analog, digital and counter inputs from field devices, perform control computations according to the assigned algorithm and control sequences, and provide the resulting analog and digital outputs to the field devices. In addition to interfacing with field devices, it can also interface to a PC through a communications link which enables plant personnel to enter control strategies and monitor/modify process operation.

The multifunction controller has the capability of handling up to 32 analog inputs, up to 72 analog outputs, up to 256 digital inputs, up to 128 digital outputs, and up to 64 counter inputs.

The multifunction controller contains PROM-resident diagnostic programs. The diagnostics test various controller functions and report the results to the operator. The results of the diagnostic tests are also displayed on light-emitting diodes in the controller.

The multifunction contoller has a data highway interface which allows an operator at a multifunction operator station to monitor/modify process operations at the multifunction controller. In addition, this feature also allows a host computer on the data highway to provide setpoint and output values to the batch controller for certain algorithms and for output points, and to change the status of digital output points.

A case history, showing how the multifunction controller has been applied to a thermoset-resin plant, is given in Section 11.16.

UNIVERSAL STATION

The universal station provides a comprehensive window to the process together with control facilities for both continuous and batch processes and data retrieval. A keyboard and touchscreen allow rapid and flexible operator interaction with the system. The displays operate in three modes: 'normal', 'process malfunction' and 'control system malfunction', and in each of these modes a hierarchy of displays can be run through until the required level of detail is reached.

The universal station can be switched between an 'operator mode', an 'engineer mode' and a 'maintenance mode'. The first mode allows normal process operation, the second allows control strategies, the system data base and the displays to be built or modified, the third mode allows diagnosis and in this mode a telecommunications link to the vendor's technical center may be utilized.

HISTORY MODULE

The history module serves as a store for historical process data, software images for use in setting up or re-starting the system and a library of display-generating software.

APPLICATION MODULE

Each application module can handle up to 1500 loops and process these at 100 points per second. Back-up support can be supplied by additional redundant application modules. The module contains over twenty ready-written algorithms and has the provision for users to write their own algorithms in a special control-oriented high-level language.

COMPUTING INTERFACES

A range of computing interfaces can be geographically distributed to provide computing power wherever it is needed within the TDC network. Proven interfaces exist for VAX, IBM, HP, etc. machines.

GATEWAYS

Various 'gateways' allow plug-in interfacing between other vendors subsystems and the TDC communication network.

THE RESERVE CONTROLLER DIRECTOR (RDC)

This is the central unit of the uninterrupted automatic control system concept (UAC). The RDC detects the failure of any of eight controller files, informs the operator of the failure and switches in a reserve file with correct parameters—all within one second of the failure occurring.

PROCESS INTERFACE UNITS (PIUs)

They allow:

(a) Efficient data acquisition from plant sensors into the data highway.
(b) Switch-type signals originating from a supervisory controller on the data highway, to be output to the plant. (Recall that the basic controller cannot undertake discontinuous duties so this facility is needed for logic operations during start-up, etc.)

DATA HIGHWAY (DH)

Three branches, each up to 1500 m (5000 ft) long and consisting of standard coaxial cable, connect into a single HTD (see below) to interconnect up to 63 devices (up to 28 to a branch). Data is transmitted serially as 31-bit words at 250 000 bits per second. Each device is 'hung on' to the highway via a standard interface. Each branch of the highway is duplicated, to give a redundant data highway (RDH), with the HTD performing switchover between primary and back-up cables. The primary and back-up cables are usually separated physically (one overhead and one underground perhaps) to improve reliability. A number of accuracy checks are built into the communication protocol, including the requirement for a device to echo each signal that it receives to allow its accuracy to be checked at source.

HIGHWAY TRAFFIC DIRECTOR (HTD)

The HTD controls the protocol of intercommunication along the data highway. It contains duplicate logic to cover both primary and back-up operation with switchover to a back-up data highway cable.

DATA HIGHWAY PORT TO PROGRAMMABLE CONTROLLERS

The port matches the communications protocol of the system data highway to that of the programmable controller. The port allows both reading and writing to the data tables of the programmable controller. It scans the data tables of connected programmable controllers at a selectable rate from 1 to 15 seconds and updates its data base which is then available to the data highway. A single port can support up to eight programmable controllers.

BASIC OPERATOR STATION (BOS)

This is a 47 cm (19 inch) color VDU with keyboard connected into the data highway. In a typical arrangement three identical BOS are sited adjacently so that all three can be seen by the same operator. One is usually showing a plant overview, one is switched to group displays and one is dedicated to alarm duties. A hard-copy unit is normally arranged to record all alarm conditions that occur. Flexible, comprehensive monitoring of up to 6000 points and up to 1200 alarms at up to 200 points per second is a feature of the system. The operator interacts with the system (calling up different types of displays of changing parameters or configuration in the basic control loops) by means of push buttons.

ENHANCED OPERATOR STATION (EOS)

The enhanced operator station is a microprocessor-based console with keyboard, diskette drive and 47 cm video screen. It connects to the data highway as a fully integrated part of the system to provide communication and display functions and act as an interface for all configuration, programming and operating of the system.

Compared with the basic operator station, with which it is compatible, the enhanced operator station provides for:

supervision of up to 1500 process variables from a single operation station;

extensive graphic display capabilities including custom display building;

dot trending features for display of past performance;

trend displays updated in real time to enable tuning for optimum performance;

automatic free-format hard-copy reporting;

simplified and extended sequence generation and batch recipe handling;

a flexible calculated point capability.

Through the enhanced operator station, the operator can communicate with as many as four clusters of batch controllers, each containing up to eight devices.

Sequence programs are generated off-line using the standard keyboard and the sequence-oriented procedural language. They are brought on-line as required.

The generation procedure is similar to that using the stand-alone batch controller but is simplified by the use of diskettes. Multiple-sequence programs can be compiled at one time and be stored on a single diskette. Up to 36 recipes and associated sequences can be merged into one recipe diskette and there is no limit to the number of diskettes.

THE UNIVERSAL CONTROL NETWORK (UCN)

The UCN is a high-performance, peer-to-peer, real-time, MAP-based process control network. It features advanced technologies that provide the performance, reliability and security required by today's process industries. The UCN provides a platform for open systems interconnection capability at the process-control level in the TDC 3000 system hierarchy. The UCN is a five-megabit-per-second carrier-band, token-passing network based on IEEE 802.2 and 802.4 standards. It is a high-performance, real-time process-control network with enhancements to provide the redundancy, reliability and security required by the process industries.

LOGIC MANAGER

The logic manager is a sequential fast logic process on the TDC 3000 MAP-based universal control network. Motor controls, for example, are easily implemented by function block configuration. Specifically designed for applications requiring fast program execution, extensive digital and interlock logic, the logic manager teams with the process manager to provide full function process and programmable logic control devices in one unified architecture.

PERSONAL COMPUTER NETWORK MANAGER

The PCNM is a PC-based database server which interfaces a network of personal computers to the TDC 3000 LCN. It delivers easily configured, real-time data to all users of personal computers for trending, reporting or statistical analysis.

PC DATA MANAGER

This UNIX-based, IBM 386-based system allows remote supervision of small to medium local control systems. It interfaces to TDC 3000 Hiway devices and remote terminal unit (RTU) networks.

SERIES 9000

This is a low-cost, easy-to-use hybrid type controller that replaces and integrates control hardware such as single-loop controllers, multi-point recorders, setpoint programmers, multiplexers, PLCs and auxiliary control devices. It combines loop control, logic control and data acquisition functionality in one efficient package.

11.8.3 Fieldbus standard

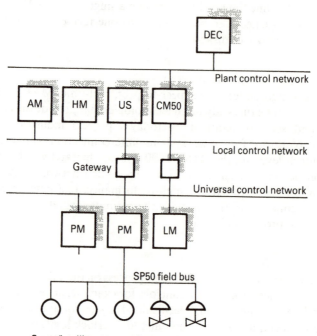

Figure 11.21   Honeywell TDC 3000 uses SP50 field bus for special bi-directional communication with field-mounted intelligent sensors.

11.8.4 Application software for the TDC 3000

TDC 3000 EXPERT

This expert system runs on a Microvax under the VMS operating system connected with the TDC system via a computer gateway. The system uses a frame-based DEC/write language and an object-oriented description of the process. The aims of the system include the early detection of fault conditions and operation guidance on evolutionary improvements to the process.

HPC PREDICTIVE CONTROL

HPC is a multi-input single-output model-based package that resides in the TDC application module and replaces any PID controller. Identification is by limit cycling and/or user-selected perturbation followed by least squares fitting—the process is kept under control during identification. The system accommodates lead-times of up to 50 sampling periods and can deal with open-loop unstable processes.

PROCESS MODELLING AND OPTIMIZATION

This system operates on a DEC VAX platform connected to the TDC 3000 system. The software is divided into: equation manager and solver, real-time optimizer, Fortran support.

The Honeywell equation manager and solver (HEMS) modelling package allows users to easily enter process modelling equations in their usual engineering form. These equations can be linear or nonlinear, simple or complex, and can be entered in any order. In addition, HEMS automatically orders equations and converges simultaneous equations.

Models consist of the following three parts applications:

Model equations

Specifications of external data

Definition of either 'square' or optimization problems

Models can be used to:

Converge a material balance and report the results

Optimize nonlinear process operations by including a definition of optimization variables and their bounds in the model

In addition, the system:

invokes the RTO (real-time optimizer) to perform optimization calculations;

automatically handles all set-up and interfacing procedures with RTO.

A report writer produces a report of the model solution, including details of the optimization variables (and constraints if the optimizer is invoked).

The real-time optimizer (RTO) uses the powerful MINOS optimization solver to determine optimum conditions for the process. RTO is based on MINOS—a Fortran-based application design to solve large-scale optimization problems. (The proven MINOS program is installed at over 200 sites worldwide, and has an active users' group.)

RTO features include:

Solutions for large, nonlinear problems

An efficient linear programming module

Automatic scaling of linear constraints

Automatic estimation of some or all gradients

Sparse matrix techniques

DATA BASE

The data base for a PMO application allows the application to communicate with the process and with process operators and engineers. A large application may access several hundred or even thousands of items of data. The TDC 3000 data base serves as the main data base for use by PMO applications.

STATISTICAL CONTROL

Real-time statistical process control, an on-line statistical process-control software package improves product uniformity.

Statistical control allows process problems to be separated into two types of variations. One type is referred to as *Special* causes and can best be handled by people close to the process such as the local supervisor and the operator. The second is referred to as *Common* causes, which represent approximately 85% of process variations and usually requires higher management action. Having the ability to separate these two types of variation can focus more attention on achieving:

Higher effective capacity

Lower unit costs

Higher uniform quality

Process and quality control engineers do not usually share the same line management and their immediate objectives are sometimes different. Their approach to controlling the process is not always the same. One is concerned with 'is the process working', the other is concerned with the 'probability of the process working'. One is concerned with reaction to a process alarm, the other with the interpretation of statistical control limits, for example. Real-time $SP_Q C$ helps bridge these two job functions so as to share in a better understanding of an immediate process problem and the long-range problem of product uniformity.

A wide variety of user-selectable statistical alarms are available to help predict process-variable drift along with variations on the standard statistical frequency distribution and sigma calculations. X-bar and R charts and histogram displays are provided to substantially decrease time in manually preparing the control charts.

Operators are responsible for the process operation and the prevention of upsets. All too often setpoints need to be adjusted for a variety of causes, some known and some suspected. Real-time $SP_Q C$ can assist substantially in determining process variability and its direction drift. The probability of which setpoint to adjust, which variable is the real culprit, or whether a sensor is drifting out of calibration, are other examples. The graphical aids give the operator a mental picture of the process history to help interpret satisfactory product-quality level. Displays are provided for manual entry of data, such as from the laboratory.

Real-time SP_QC automatically collects subgroups of process data observations on a regular and selectable time base, calculates the mean (average) and plots the X-bar value of each subgroup on X-bar control charts. The range of the subgroup observations is calculated and plotted on R charts. Selectable statistical methods of continuously re-calculating the upper and lower control limits are provided.

Five types of statistical routines help to identify non-random variations in the process and to quantify causes and sources of variations.

The aim of the package is to help the operation to achieve better control and in particular better uniformity.

Figure 11.22 shows schematically how successful statistical process control may narrow the variability band of a process.

AUTOMATIC LOOP-TUNING
LOOPTUNE-II is an off-the-shelf software product that provides three methods of tuning PID control algorithms.

LOOPTUNE-II operates in the TDC 3000 application module (AM), and offers a variety of features, including:

Tuning capabilities for loops in all controllers including process manager, application module, multifunction controller, extended controller, or basic controller.

Pre-configured configuration, tuning and status displays for operation from the universal station.

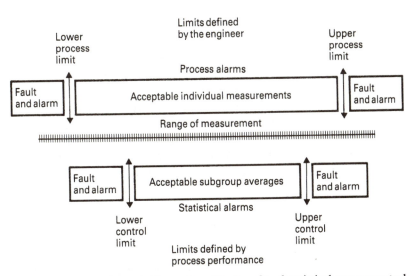

Figure 11.22 A schematic illustration of the benefits of statistical process control.

TUNING ALGORITHMS

Search algorithm

optimizes absolute error and weighted average output changes. Tightness of control adjustment minimizes absolute error or output changes (trade-off);

gathers information from natural loop disturbances without perturbing the process;

handles loop dynamics (order, dead-time, response speed) without user input;

field-proven in over 100 installations.

Fast algorithm

determines Ziegler–Nichols tuning values: proportional, integral and derivative;

provides tuning response adjustment (over-shoot adjustment);

offers on-line limit cycle (relay) control while tuning;

features user-adjustable disturbance limits (typically 3–5%).

Tuning constant scheduling   This algorithm features four sets of tuning constant (three scheduled sets and one override set), and offers:

automatic scheduling based on any system value;

operator override from a universal station target;

easy interface to user-written scheduler functions.

11.8.5 MAS/C for automation of manufacture

MAS/C is a specialized controller designed to provide supervisory control in finishing, converting, formatting, packaging, shipping and receiving operations. The manufacturing automation system controller platform features a single virtual system architecture that is transparently distributed and easily scalable. It can be used as a stand-alone node or in a network of nodes. Regardless of configuration, the physical distribution of processors and resources is transparent to the application program and appears to application engineers and operators as a single, seamless system. Applications are intended to be configured by production engineers, reducing the need for systems engineers and programmers. MAS/C gives an overview of operations for any system terminal—production floor or remote office. IBM-compatible computers can be used as graphics terminals. Graphic displays offer a pictorial view of the

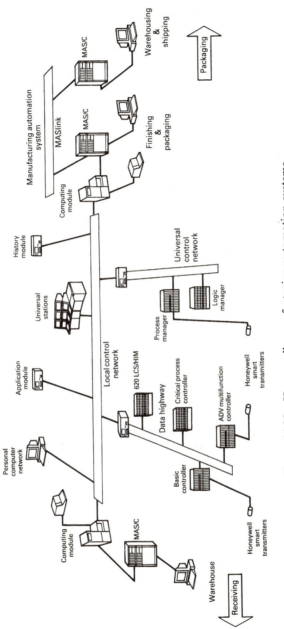

Figure 11.23   Honeywell manufacturing automation systems.

process with timely data—providing early detection and correction of production problems.

Real-time production monitoring and control allows the user to:

relate finished product quality to raw materials and the process that transformed the raw materials;

create a flexible manufacturing environment that helps to:
- minimize downtime
  increase productivity
  shorten lead times
  improve quality
  decrease scrap
  improve yields
  increase equipment utilization
  decrease operating costs

11.9 The Foxboro System

Two of the case histories that follow in this chapter are applications of the Foxboro 'Intelligent Automation' distributed control system.

Figure 11.24 shows the architecture of the Foxboro system. Refer to Section 11.12 for a detailed description (within a case history) of the system.

11.10 Case history C: Control of a 1000 tonne press

The control system for a 1000 tonne Fielding & Platt hydraulic forging press, recently installed in the press shop of Thomas Wild Ltd in Sheffield, utilizes a Quarndon Electronics VME microcomputer system operating with a 1000 mm analog absolute position transducer, to accurately control the position and motion of the main ram.

The system comprises:

XVME-600/1 68000/68010 processor module with SRAM/EPROM sockets

XVME-201 48 channel I/O module

XVME-500 16SE/8DI channel input module with 10 μs A/D converter

3U component unit—5 channels of analog signal conditioning and one analog output.

The CVME-201 handles 7 channels of digital input and 13 channels of digital output, all via optical isolators, and includes 11 channels of addressed

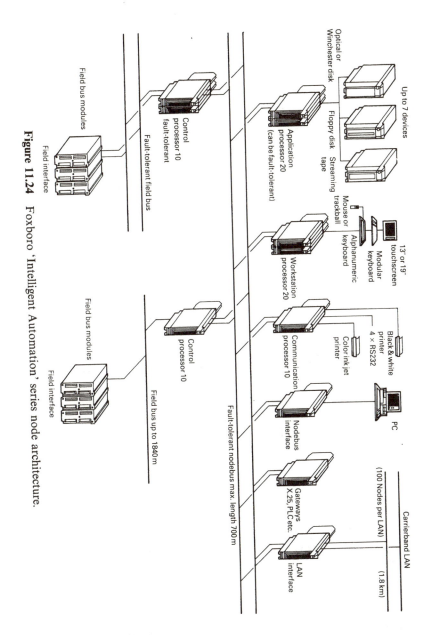

Figure 11.24 Foxboro 'Intelligent Automation' series node architecture.

The following labels appear in the figure:

Field bus modules

Field interface

Control processor 10 fault-tolerant

Fault-tolerant field bus

Application processor 20 (can be fault-tolerant)

Optical or Winchester disk

Floppy disk

Streaming tape

Up to 7 devices

Mouse or trackball

Alphanumeric keyboard

Modular keyboard

13" or 19" touchscreen

Workstation processor 20

Communication processor 10

Black & white printer

Color ink jet printer

4 × RS232

Nodebus interface

PC

Gateways X.25, PLC etc.

LAN interface

Carrierband LAN

(100 Nodes per LAN)

(1.8 km)

Fault-tolerant nodebus max. length 700m

Field bus up to 1840m

Control processor 10

Field bus modules

Field interface

information. A further 6 channels are used for pump control via the D/A converter on the component board.

The XVME-500 handles five channels of conditioned S/E analog input from the component unit.

AUTO FORGE CYCLE

During the approach sequence the ram speed is monitored, such that when the speed is less than 10 mm/s, it is assumed that the work has been contacted. At this point, the ram position is stored and the press forge cycle initiated. The forging speed is controlled by a variable delivery pump. The speeds are:

Fast approach 150 mm/s.

Slow approach 35 mm/s.

Forge speed 25 mm/s.

The system software runs in a multi-tasking PDos operating system loaded into VME Prom 68000 cards.

11.11 *Case history D: VMEbus control of a fully automatic woodworking machine*

15 minutes of manual set-up time has been reduced to 38 seconds by the utilization of a VMEbus microcomputer system (couples to an industrial PC) to Wadkin's OMNISET™ fully automated C.N.C. controlled wood moulder machine.

Wadkin has applied C.N.C. control to planing and moulding machines that reduces machine down-time and stock levels with improved management techniques and better tool practice.

Before the introduction of the electronic C.N.C. setting system, manual adjustment could take as long as 15 *minutes*. A typical setting with OMNISET, excluding tool change, is 30 *seconds* for a typical machine of 19 axes.

OMNISET automatically controls position, chipbreaker setting, pressure shoes, side pressures and feed speed. The accuracy is such that the first component produced will be perfectly to size, reducing wastage and the need for a trial price.

All axes are moved simultaneously, by geared a.c. brake motors and have uni-directional approach for absolute accuracy.

ELECTRONIC CONTROL

The Quardon microcomputer control equipment comprises an industrialized 19" VME rack and industrial PC/AT system.

Depending on the number of axes to be controlled, the 19" rack has the appropriate mix of four-channel motion control, two-channel relay and 68000 CPU cards plugged in.

The 68000 processor-based VMEbus motion control and relay boards provide signals to accurately position the motor-driven spindles.

A 68000 CPU board carries out supervisory functions, such as monitoring control software programs.

SOFTWARE

Quandtron's PDOS[R] multi-tasking real-time operating system software loaded in VMEPROM[R] is embodied on all 68000 processor cards.

The 19" VME rack is mounted inside an IP65 control cubicle having a front cut-out into which the industrial PC is mounted.

11.12 Case history E: Control of a paper-mill complex

SYSTEM DESCRIPTION

The Foxboro I/A system (refer to Section 11.9) has three levels of communication. The lowest level exists as the Fieldbus, shielded-pair cable that can be up to 4000 ft (1220 m) in length. A control processor (CP) communicated with up to 48 input/output (I/O) cards called 'Fieldbus modules' (FBMs) via the Fieldbus. Analog FBMs provide D/A (digital/analog) and A/D conversions only. All scaling, filtering and analog control algorithms are executed in the CP.

The second level of communication is called the Nodebus. The Nodebus is a coaxial cable that can be up to 100 ft (30 m) in length. The individual control processors communicate with each other over the Nodebus and with all the other processor types described below. Each CRT, with its touch screen, keyboard and trackball, is serviced by an individual workstation process (WP). The applications processor (AP) manages all the bulk memory devices, such as hard disks, floppy drives and streaming tapes. The AP hard drives contain all the system software, user control configuration, graphics and memory images for all other cards in the system.

The communication processor (CommP) serves four serial devices, which are typically three printers and a system CRT. The foreign device gateway is used to interface the Foxboro I/A system to other computer systems and can be used with specific software to interface to programmable logic controllers (PLCs). The processor cards interconnected with a nodebus and the control processors' associated FBMs are collectively called a 'node'. The smallest stand-alone I/A system is a node.

The third level of communication is the carrierband local area network (LAN). The carrierband interface (CBI) connects the nodebus to the carrierband. This communication level is used by processor cards on two separate nodes that need to exchange data.

The I/A system can be configured to employ a fault-tolerant design in which two pieces of hardware can be used to process the same information.

11.13 Case history F: Computer integrated manufacturing at a photographic film plant

3M Italia's Ferrania photographic film plant comprises a series of batch and continuous processes, each requiring precise control to achieve very high levels

Southeast Paper's Foxboro I/A system consists of 12 nodes (11 are used for control). The 12 nodes contain 12 pairs of fault-tolerant application processors and 24 80 megabyte hard drives configured as 12 mirrored drives. The 33 workstation processors control 33 touch-screen CRTs. The 66 pairs of fault-tolerant CPs have a total of 475 digital FBMs and 396 analog FBMs. These analog FBMs provide a total of about 7600 discrete I/O points and 3200 analog I/O points. Four foreign device gateways provided interfacing to another 500 points. There are 14 printers. The hardware is divided almost equally among the three operational areas of power, pulp and paper.

The design of the process graphics incorporates the advantages of the touch-screen CRTs to minimize the use of keyboards and trackballs. The designs stress ease of operation and uniformity of controls. Most operators learned to operate the I/A system in less than one hour.

The design and the levels change both graphically and numerically on displays. For example, the pumps change color when started or stopped and the levels change both graphically and numerically on displays. Operators can see the results of their actions on the same graphic. For example, the pumps change color when started or stopped and the levels change both graphically and numerically on displays.

Another part of Southeast Paper's control philosophy is the use of dynamic process graphics as the operator's interface to the process. These graphics represent the process equipment, flows, conditions and control elements. From the graphics, the operator can control the process by touching appropriate areas on a screen. A two-touch approach was implemented to prevent accidental changes. The first touch selects the parameter to be changed and the second touch implements the change. Operators can see the results of their actions on the same graphic. For example, the pumps change color when started or stopped and the levels change both graphically and numerically on displays.

CONTROL PHILOSOPHY

Where practical, Southeast Paper integrated as many controls as possible into the I/A system. In many cases, when a vendor-furnished control system or PLC would normally be used, Southeast Paper bought logic drawings instead. The logic was then implemented in an I/A system. This approach for control decreased the number of systems and spare parts maintained, reduced the amount of operator and maintenance training required, enhanced the flexibility of the controls and lowered the number of system interfaces for control and data collections.

The two outputs are compared. If they are unequal, an internal diagnostic check is run and the faulty hardware is automatically taken off-line. The good hardware continues to function normally without disturbing the process. When the faulty hardware is replaced, the new card is automatically downloaded and placed in service.

of quality and productivity. By basing its process control strategy on distributed control systems, the company has been able not only to integrate a wide range of third-party control and measurement subsystems, but to bridge the gap to higher-level supervisory and plant management computer systems.

In each production area, 3M has installed control systems to minimize wastage, enhance product quality and, wherever possible, remove the need for personnel to enter the production area. This last requirement arises not just because many of the processes involve fire and explosion hazards, but because most of them must be conducted in total darkness.

Film production commences with the manufacture of the triacetate base. Flakes of triacetate, suspended in a solvent, are cast on a continuous moving stainless steel belt, which then forms a web over one meter wide. This is then passed through a succession of drying stages before reaching the final take-up reel.

The process requires precise control of temperature and web speed to ensure a uniform thickness, and freedom from surface blemishes and scratches. 3M installed a PROVOX distributed control system (DCS), from Fisher Controls, to control temperature throughout the process, and a Modicon 584 programmable logic controller (PLC) to control the complex drive arrangements for the caster's belts, rollers and fans. Web thickness is continually monitored by a Measurex scanning nucleonic gauging system.

The DCS and PLC come into their own at start-up, when the former's trending facilities, and the latter's ability to control the complex sequence of clutch operations, are used to achieve on-specification product in the shortest possible time.

EMULSION MAKING

While the production of triacetate base is essentially a continuous process, emulsion production is an archetypal advanced batch operation. Ferrania's recently commissioned new production facilities manufacture emulsion for the company's color film, graphic arts and X-ray film.

One PROVUE console is dedicated to each of the three main production activities—solution preparation, emulsion making and digestion—and communicates with its respective controllers over the common PROVOX data highway. Overseeing the entire operation is a digital MicroVAC II-based data management systems, which interacts with the DCS via Fisher's Computer Highway Interface Package (CHIP). CHIP maintains and continually updates a copy of the process-control database.

In essence, the functions of the data management system are to conduct batch validation prior to initiation, to provide continuous acquisition and logging of data from the DCS (while the batch is running), and to generate full batch reporting once it is complete. While control of the actual batch processes is handled by PROVOX, precise weighing of raw materials is performed by an automated weighing system, supervised by a digital PDP 11/73 computer. A second PDP 11/73 controls the automated storage and retrieval of the finished emulsion. Integration of these functions with the DCS is provided via the data management MicroVAX, which communicates with the PDP 11s over an Ethernet Local Area Network (LAN).

System configuration is handled on a Fisher engineer's workstation comprising a DECPro 380 computer. This is interfaced to the DCS highway to allow access to the PROVOX database, and permit downloading of new configuration data to the PROVUE consoles without interfacing with the normal operation of the system. The actual setting up of individual batch strategies is handled by the operator, by selecting from the appropriate procedure and operations libraries within the system.

A feature of the processes involved in emulsion making is the rapid change in key parameters that occurs during the first few seconds of some of the batch operations. Control of these interactions is achieved through the controllers' function sequence table facility, which allows sophisticated algorithms to be developed, while providing a capability to handle up to 256 simultaneous events. Precise measurement of key parameters is ensured by interfacing the DCS with high-accuracy Coriolis effect mass flowmeters and analytical sensors, for such parameters as silver content and pH.

The PROVOX system is supervised by a department host computer, again a digital MicroVAX II, which interacts with the control system through CHIP.

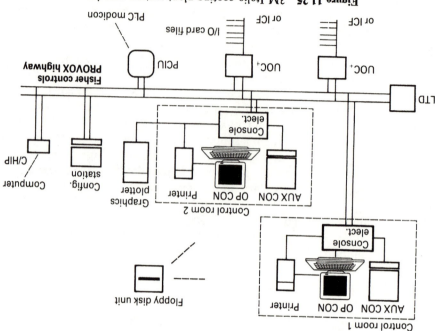

Figure 11.25   3M Italia coating plant system overview.

Ethernet

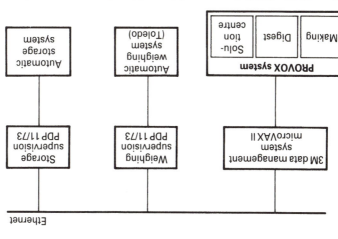

Figure 11.26   3M Italia new emulsion making plant system overview.

INTEGRATION

In the coating department, 3M has taken its integration strategy a stage further by establishing a link with the Ferrania plant's IBM mainframe, thus providing for direct transfer of data, such as production reports, materials handling and inventory information and planning schedules.

Production proceeds continually and the principal task of the PROVOX system is to control the flow of emulsion from the kettles on the basis of measurements of silver concentration. Interaction between the PROVOX system and the Modicon PLC allows drive status and speed information together with alarms to be displayed at the PROVUE consoles, while PROVOX itself determines the deviation from setpoint of the web taper tension, derived from a load cell measurement, and passes the result to the PLC for speed-control purposes.

The PROVOX DCS has integrated controls with third-party equipment, such as PLCs, process computers and specialist measurement systems, including sophisticated supervisory computer systems. This strength will allow progressive integration over the next decade. (Figures 11.25, 11.26 provide a system overview.)

11.14 Case history G: Computer integrated manufacture at a chemical manufacturing plant

Monsanto's initial approach was to ignore systems, machines, hardware and software and instead to collect and analyze the data and information needed to maximize business results.

The factors to be taken into account will always be unique to a particular organization but all companies at the corporate level can be expected to be interested in manufacturing strategy, production forecasting, materials planning, order billing, distribution, inventory control, sales and service and corporate accounting.

What information is needed at site management level? Some examples could be that of manufacturing performance, quality performance, accounting analysis, decision support information and time attendance.

Different data are needed to carry out the actual manufacturing process. These can be split broadly into process management, laboratory management, warehouse management, utilities management and process control data. Clearly, each of these can, if required, be broken down further.

Finally, a bank of information classified as manufacturing support could cover data related to maintenance, design and construction, office automation, access control and other areas.

This completes a general overall framework, but a stair-stepping approach can be used to break down data definition even further. For example, process management could consist of a further subset of data such as charge calculation, usage monitoring and production scheduling as opposed to production planning, quality control, performance monitoring and efficiency monitoring.

Similarly, laboratory management requires such functions as sample log-in, recording of results, quality assurance, on-line data collection, calibration, test scheduling and laboratory efficiency analysis.

THE REALITY

At Newport, there is in essence a three-tiered network with transparent links between the different levels. These levels are designated as *dedicated* (to a specific plant), *group* (a group of plants or a department) and *site*.

At the dedicated level is a site-wide Fisher PROVOX highway with, currently, five production units linked to it and operating under differing degrees of PROVOX control and a separate training and development unit. This uses both Hewlett-Packard and Digital Equipment Corporation hardware.

At the group level is a number of machines running various applications and utilizing differing techniques to ensure that the data can be transferred across the various boundaries. Three process management machines run batch and continuous data historians and SPC, a laboratory management system, access control, engineering, personnel and office automation. Ethernet, SNA gateways and token rings are used to facilitate communications.

At the site level, an IBM mainframe computer acts as a site host, providing coordinated manufacturing management data, and acting as the switching centre on to the worldwide communications network.

Putting it all together achieves what is called the 'Newport Network 90', which is generally shown in Figure 11.27 with the three levels of 0—dedicated, 1—group and 2—site.

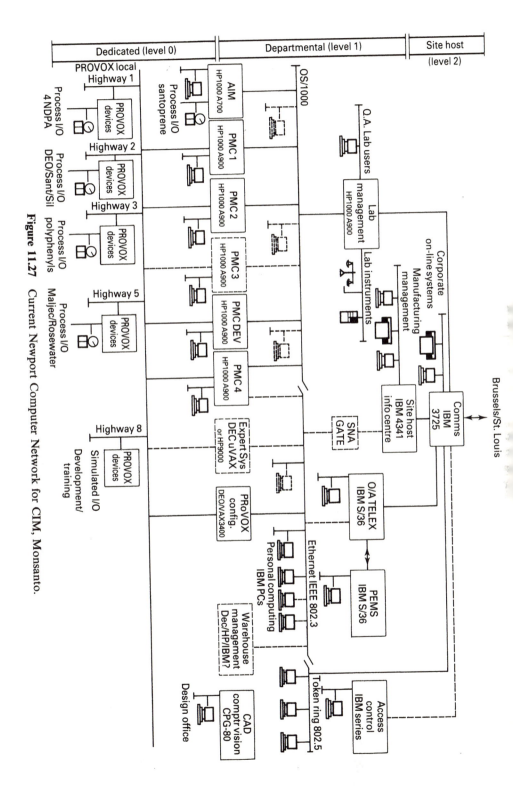

Figure 11.27 Current Newport Computer Network for CIM, Monsanto.

THE BENEFITS

It is impossible to separate out the benefits of the scheme described here from the benefits attributable to other quality and efficiency programmes that have been introduced. However, as a simple way of assessing achievement, one can relate output to input and calculate a productivity index. In terms of labour, Monsanto Newport has had a declining workforce for many years, reducing from a peak of approximately 1200 in the early 1970s to the current figure of 450, with significant reductions in the early 1980s.

Looking at finished product volume over the same period, the factory has moved from a base of 50 000 tonnes in the early 1980s steadily rising to a figure of 80 000 tonnes in the early 1990s.

Integrating the two sets of data shows that personnel productivity has risen from base 100 in 1979 to an index of 190 in 1989. 50% of this was achieved between 1982 and 1984 in the 'survival period' and the challenge has been to maintain this pace of improvement in a 'non-crisis' situation. This sounds impressive, but in the chemical industry there are other factors in addition to personnel productivity that are important in measuring overall efficiency. Monsanto, therefore, also uses a total site productivity number using an index that combines personnel, raw materials, energy and capital data. This indicator shows an average improvement of 2.5% per annum over the last ten years.

It would not be true to say that this is due totally to the introducing and use of integrated manufacturing systems. But it certainly is true to say that they have made a significant contribution.

In essence the issue is all about information. Integrated manufacturing systems are simply a powerful and essential tool for spreading this information fast. For there is no doubt that speed of response to internal and external customer needs will be a key element in determining the future success of a manufacturing organization.

11.15 Case history H: Control of a pumped storage scheme: Modcomp Minicomputer Network for control of a pumped storage scheme

The Virginia Electric and Power Company operates what is currently the world's largest (2100 MW) pumped storage scheme in Bath County (USA). Modular Computer Systems Inc. (Modcomp) has supplied the $2 million control system for regulation of the water flow from two very large reservoirs, flood control and control of turbines and electricity generation (Figure 11.28 shows the overall arrangement).

The control system is based around two Modcomp Classic 7870 'super minicomputers' that act as the hosts to eight mid-range Classic II/25 systems. The entire system is linked through advanced fiber-optic technology to form a sophisticated distributed network capable of monitoring some 33 000 process

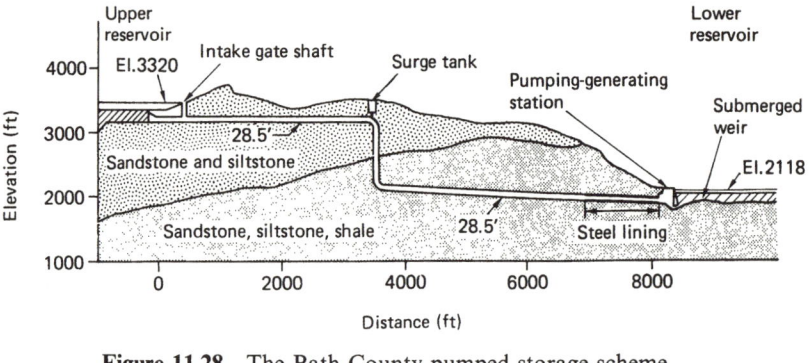

Figure 11.28   The Bath County pumped storage scheme.

variables. Main operator interaction is through five high-resolution color graphic stations.

The Classic II/25 machines used as the satellite computers have fast I/O rates (up to one megabyte per second) and memories in the range 128K–512K bytes. The host machines offer load-sharing and hot standby facilities.

Minicomputer networks such as that of Figure 11.29 provide control for large, geographically distributed problems where high availability is required.

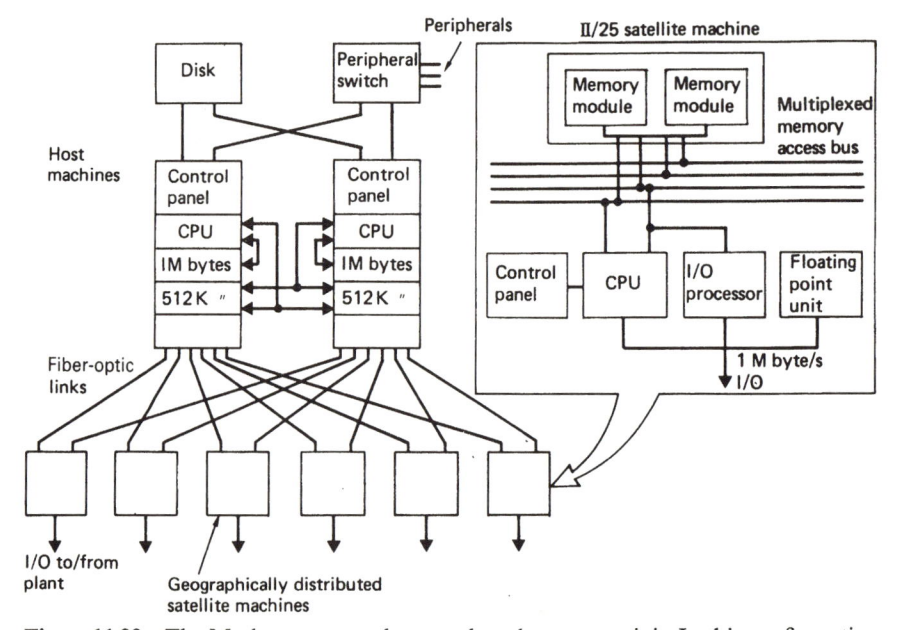

Figure 11.29   The Modcomp control system based on superminis. In this configuration, each host computer has one megabyte of private memory and 512 kilobytes of shadow memory, critical data is stored in the shadow memories. The system offers load sharing, resource sharing and instant standby possibilities.

11.16 Case history I: Control of a batch process: Application of a batch controller to a thermoset resin plant

Batch control applications require significant attention to the logic and sequencing problems associated with start-up and shut-down. Controllers intended for continuous processes have neither the sequencing capabilities nor the software structures for these tasks. The usual approach has been to design a one-off system based around a general-purpose process computer. Recently, a number of purpose-designed batch controllers have become available. They offer many savings, particularly in software design and production. An outline of the application of such a special-purpose batch controller is given below.

OUTLINE OF THE CONTROL PROBLEM

The batch starts with the transfer of a specified weight of monomers from the storage tanks A and B into mix-tank C (Figure 11.30). While mixing is taking place, the reactor D is being preheated by means of its steam jacket. Once the

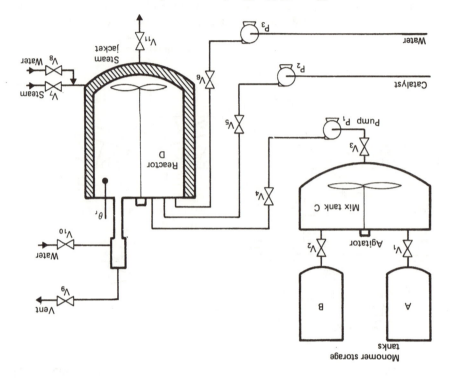

Figure 11.30 The thermoset resin process.

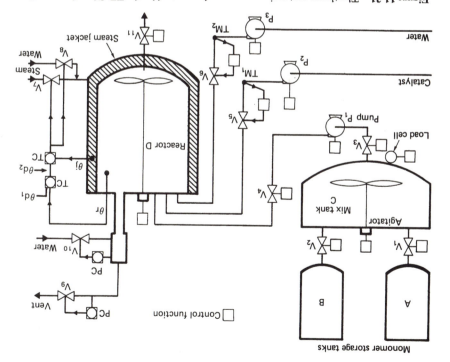

Figure 11.31 The thermoset resin process under control by the TDC batch controller.

necessary temperature is reached, the mixed monomers are transferred to the reactor. Specified volumes of catalyst and water are next charged to the reactor. The temperature of the reactor needs to be maintained at a desired value until chemical analysis of the reactor contents indicates that polymerization is complete. The reactor contents are then discharged and preparations are made for the next batch to start.

The Batch Control System achieves the requirements by implementing the following sequence (see Figure 11.31):

1. Check all valves, pumps, vessels, temperatures, etc., to ensure that they are in the correct states to allow the batch to start.
2. Open stem valve V_7 to start preheating of reactor.
3. Open valve V_1 to allow monomer from tank A into the mix-tank until the load cell indicates that the correct weight has been transferred.
4. Open valve V_2 to allow monomer from tank B into the mix-tank until the correct weight has been transferred.
5. Switch on the mix-tank agitator.
6. When reactor preheat temperature is reached, open valves V_3 and V_4 and start transfer pump P_1. When the mix-tank load cell indicates zero, close valves and shut down the transfer pump.

7. Start pump P_2, open valve V_3 until turbine meter TM_1, shows that sufficient catalyst has been added, then shut down pump and close valve.

8. Charge the required quantity of water as in Section 7 but using pump P_3 valve V_6 and turbine meter TM_2.

9. Temperature control of the reactor is achieved using cascaded loops. The temperature θ_r of the reactor contents is measured and compared with the desired (setpoint) temperature θ_{d_1}. A control algorithm operates on the error $\theta_{d_1} - \theta_r$ to produce a desired jacket temperature θ_{d_2}. The measured jacket temperature θ_j is compared with θ_{d_2} and a second control algorithm operates on the error $\theta_{d_2} - \theta_j$ to produce an output to the steam and water control valves. These nested loops can achieve tighter temperature control of the reactor contents than can be obtained through the use of a single control loop.

10. When chemical analysis of samples from the reactor show that polymerization is complete, the operator makes an input that causes valve V_{11} to open to discharge the reactor.

11. The system is prepared for the next batch by turning off the reactor agitator, cooling the reactor, etc.

12. The system prints out a log of the batch showing times, temperatures, weights transferred, etc.

The batch control system treats all the plant shown in Figure 11.30 as one *unit*. The mix-tank and the reactor each have a corresponding *sequence* of tasks. Each sequence is split into *phases*. Each phase is made up of *steps*.

A Sequence Oriented Procedural Language (SOPL) eases the implementation of the required control scheme. Each SOPL program corresponds with one sequence, defined as above. Figure 11.32 illustrates the software to control the batch reaction sequence and in particular the charge phase, consisting of three steps.

The TDC batch controller controls the stepwise execution of the process sequences by following SOPL instructions, monitors and controls process

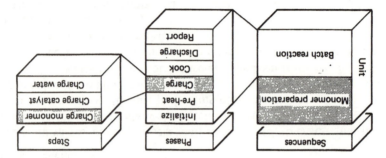

Figure 11.32 Partitioning of the process by the batch control system.

variables (temperature, flow, pressure, level, etc.) logs data on the process state and any violation of alarms.

The batch controller is also able to coordinate the operation of multiple reactors to optimize the performance of an entire thermoset resin plant. Batch controllers are monitored and manipulated by the operator through the graphic display screen and keyboard of an operation station on the TDC system data highway. The TDC 3000 batch control system is a member of the integrated family of TDC 3000 process management systems.

THE BENEFITS OF ACCURATE BATCH CONTROL

Precise weighing and charging and tight control of temperatures and pressures minimize the production of off-specification material and generally increase batch-to-batch consistency.

Better emergency actions and comprehensive management information complete the list of benefits.

The system can be interfaced to a conventional process computer to allow more comprehensive supervisory control.

HONEYWELL BATCH CONTROLLER—BRIEF SPECIFICATION

Inputs	32	analog
	256	digital
	64	from counters
Outputs	72	analog
	128	digital
Diagnostics	on PROM resident programs with reports to the operator and also displayed on LEDs. Diagnostics are performed either once per second or according to logical demands	

Optional connection on data highway

CPU	16-bit NMOS device
RAM	8K words + 8K on a spare location
PROM	36K words
A/D conversion	12-bit
D/A conversion	10-bit converting to 4–20 mA current sources

LOCAL BATCH OPERATOR STATION

23 cm (9 inch) monochrome CRT.

Communication with the batch controller is by 48 Kbits/second link called the C-Link (physically—a twisted pair with optical isolators at each end).

11.17 *Case history J: Enhanced boiler drum level control*

The aims of enhanced boiler drum level control are the following:

energy savings;

increase in process stability;

control accuracy enhancements;

more advanced calculations for control, based on reliable basic data;

operational strategy changes possible in real time;

reduction in maintenance costs of boiler and instrumentation;

ease of use and modification of control system.

Essentially, a boiler continuously evaporates feedwater into steam. Figure 11.33 illustrates its principle, as well as the equipment involved.

Boilers are installed in most industrial and power plants and because of their particular reliability and performance requirements, they present a widespread and challenging application example for control-systems engineers.

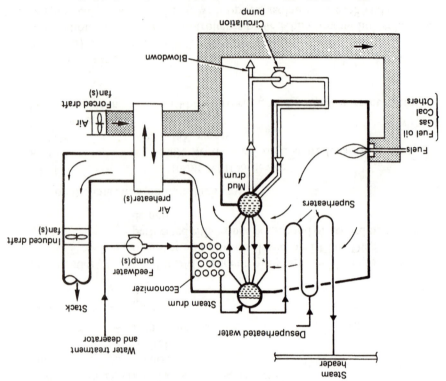

Figure 11.33 The main physical features of a typical boiler.

Traditionally, boilers have been controlled by old, well-tried methods, but the advanced system to be described below has been applied to more than 300 boilers.

The following major equipment is used:

boiler feed-water pump, turbine driven with speed governor;

boiler feed-water control valve;

boiler with economizer;

  drum

  primary superheater

  desuperheating station

  secondary superheater;

desuperheating water control valve;

blowdown control valve.

Steam quality is primarily dependent on feed-water quality. Feed-water treatment is the main tool to protect the boiler, superheater, turbine and heat-transfer equipments against corrosion and deposits.

Careful supervision of feed-water and blowdown water is therefore of great importance for the reliable operation of the boiler, turbine and other heat-transfer equipments.

Deposits in turbines and heat-transfer equipments always imply a reduction in efficiency, in heat transfer and in process output.

During boiler operation, blowdown is used to prevent build up of deposits. The mechanism involves blowing a proportion of the boiling water to waste and a proper control of this mechanism can increase efficiency.

Thus it is of a great interest to relieve only the minimum necessary boiling water to guarantee steam quality and to try to recover a maximum of energy from the blowdown water through a flash drum.

Using valves to manipulate the process results in losses of energy due to the pressure drops across them. To try to minimize these effects and still keep a satisfactory response to all process upsets is a relatively recent and new challenge to the control system.

The feed pump is a large rotating machine whose inertia prevents very rapid responses. The control scheme overcomes this problem in a way to be described below.

EXPLANATION OF THE ACTION OF THE CONTROL SYSTEM AND ITS COMPUTER IMPLEMENTATION (REFER TO FIGURE 11.34)

Primary control of drum level   Following standard boiler-control practice, primary control of drum water level is by feedback from water level measurement with feedforward from steam-flow measurement. The feedforward is particularly valuable when a step demand in steam usage occurs and the resulting pressure drop causes a misleading swell effect in the boiler drum contents. This effect

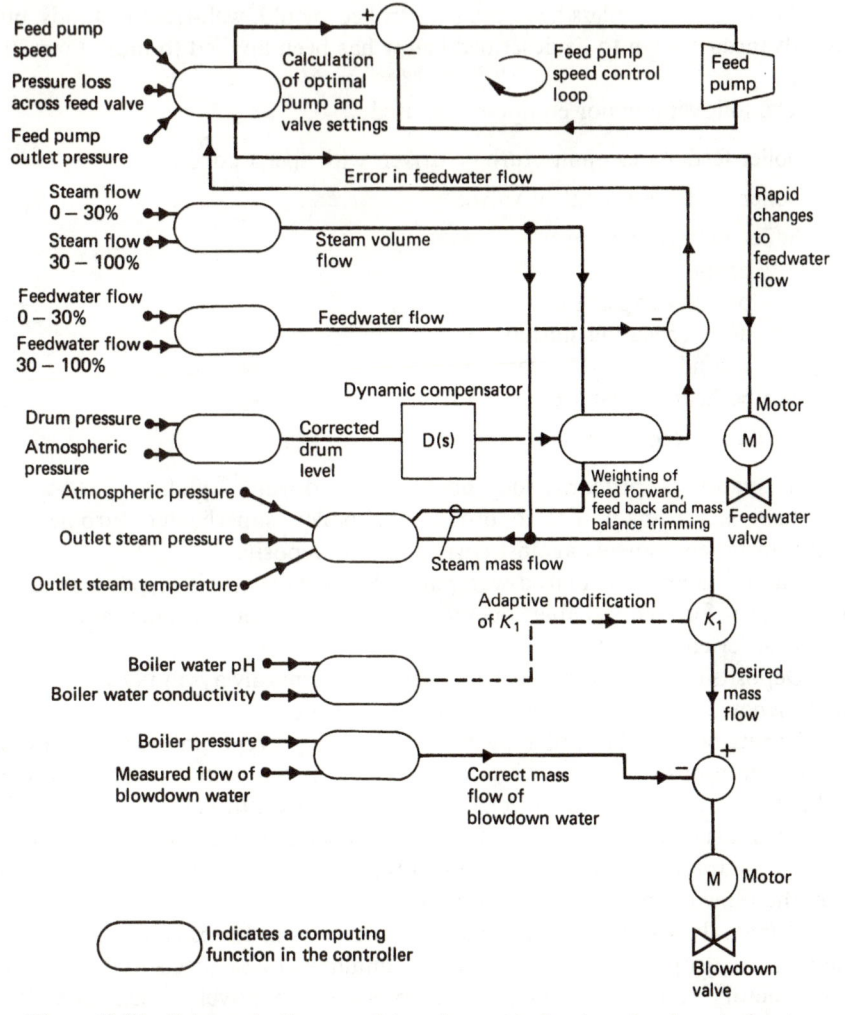

Figure 11.34 Schematic diagram of the enhanced boiler drum level-control system.

causes apparent drum level to rise and a feedwater control system based only on apparent drum level takes a temporarily incorrect corrective action.

Matched-range pairs of measuring devices It is difficult to measure flow over a wide range using standard sensors based on pressure drop across an orifice. In the application described here a 'low' sensor covers the flow range 0–30% and a 'high' sensor 30–100% of the flow range with computer changeover between the sensors and a programmed strategy to deal with failures (e.g. if the 'low' sensor fails there is automatic changeover to the (less accurate) 'high sensor'. If the 'high' sensor fails there is automatic changeover to the 'low' sensor and lowering of the boiler operating maximum to 30% of its normal maximum load). Both steam and water flow measurements are carried out by matched range pairs in this scheme.

Correct mass-flow calculations. Correct mass flow of steam is obtained by correcting the original flow measurement for temperature and pressure.

Correct mass-flow of the boiling water used in blowing down (to avoid formation of scaling deposits) involves a correction to specific gravity based on drum pressure.

Primary control of the blowdown mechanism   The system implements the rule:

mass flow of blowdown water $= K_1$ (mass flow of generated steam),

$$(11.1)$$

where K_1 is predetermined to match the expected conditions in the boiler.

Secondary control of the blowdown mechanism   Equation (11.1) represents simply an open-loop ratio control and in the long term either scaling conditions will start to be established or else the system will drift in the opposite sense and energy-wasting excessive blowdown water will be discharged from the drum.

To trim the value of K_1 in equation (11.1) and avoid the drifts outlined above, the electrical conductivity of the water in the drum is used in what can be considered as an adaptive loop. The action of the adaptive loop is itself subject to modification to ensure that at no time does the pH of the drum contents pose a corrosion risk.

Calculation of true drum level   Drum level is measured by differencing the outputs of two pressure transmitters. One transmitter is located at the bottom of the drum and the other at the top in the vapor phase. The calculation of level based on these two pressure measurements involves a number of obvious corrections to take account of the different location of the two transmitters and of second-order temperature and pressure effects. When these calculations and corrections have been completed the result can be considered to be 'apparent drum level'. A further correction to obtain true drum level requires a correction to be made to allow for the variation of the specific gravity of boiling water with pressure.

Mass-balance calculation   An accurate material balance allows adaptive correction to the feedforward action referred to above under the heading 'primary control of drum level'. Periodic calculation of the mass balance results in periodic updating of a multiplier K_2 which attenuates or amplifies the extent of the feedforward action relating boiler feedwater control with measured steam usage flow.

Compensation for swell and shrink effect Boiling water is a mixture of water and steam and the ratio between the two cannot be measured. None of the corrections described so far have made allowance for this phenomenon.

The ratio changes very rapidly when a sudden steam demand causes a drop in drum pressure. More steam forms rapidly within the water of the drum causing 'swell'. Conversely, the sudden addition to an unusually large quantity of cold feedwater causes steam to condense and the drum contents 'shrink'.

It will be seen that this interesting phenomenon results in a step response of the form shown in Figure 11.35. An approximate linear model for such behavior requires right half plane zeros in the transfer function: such a model has non-minimum phase characteristics and, not surprisingly, rapid feedback control is difficult. The simplest effective control strategy is to add a dynamic element $D(s)$ to the level-control loop to prevent large rapid responses in the opposite sense to the corrections that are required. $D(s)$ effectively ensures that the misleading swell and shrink effects are ignored by the level-control feed-back loop.

Optimum control of feedwater flow If feedwater flow is controlled only by a valve, the energy losses at the valve will be unacceptably high at low flow rates. For this reason a variable speed feedwater pump is used to regulate the feedwater flow. However, there is a problem during boiler transients when the high inertia of the rotating feedwater pump prevents a rapid response. The solution is to use a valve for making rapid changes to feedwater flow and to allow the system to return slowly to a situation where the pump speed and the valve opening are chosen to be an energy-minimizing combination.

Computer implementation The complete control strategy is implemented through a Honeywell multi-function controller of the type described in

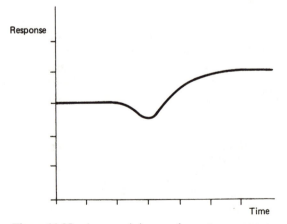

Figure 11.35 A non-minimum phase step response.

Section 11.8.2. The multifunction controller carries out the program duties indicated in Figure 11.34. It can also take care of feedwater treatment and combustion efficiency control. Through its communication by way of the system data highway, the multifunction controller allows the boiler operator strategy to be integrated into a dynamic plant-wide energy policy that will often involve such decisions as whether to use a steam to generate electricity.

11.18 Concluding remarks

Developments are proceeding rapidly in the area with which this chapter is concerned. New software tools, in association with powerful hardware and networking techniques, will surely result in great progress, particularly in the area of large integrated system design. Readers are urged to maintain an awareness of developments through conference proceedings, technical journals and manufacturers' literature.

11.19 Miniproject

By searching the literature (e.g. *Automatica, Process Engineering* (UK), *Regelungstechnische Praxis* (German), *IFAC, ECC* and *ACC Proceedings*) and using abstract journals and contacts with industry, compile your own personal file of case histories. When reading the literature, be skeptical and discard papers that describe ideas not implemented anywhere, papers based only on simulation and papers based only on pilot scale implementations. For those (few?) case histories that remain, filter out the detail, rewrite the material in your own words so as to bring out the fundamental structure of the control approach while emphasizing any novel features in either the control or the computational aspect.

──────12──────

Adaptive and Robust Control

12.1 Introduction

The field of adaptive, self-tuning and robust control is very wide and a quite large proportion of current research papers produced at international control conferences are devoted to these topics. In this chapter we begin at the most elementary level possible and then move on to the principles and practice of some of the available approaches. The adaptive section ends with a survey of some commercially available controllers.

Because on-line identification forms a central part of adaptive control, this chapter contains a section discussing some of the practical aspects of the topic. At the end of the book there is a dedicated set of references and a bibliography covering the theory, design and application of robust, adaptive and self-tuning control.

Figure 12.1 gives an overview. First we note that if process *structure* varies, as for instance under fault conditions, *fault-tolerant* or *reconfigurable control* are available. Both techniques are current research areas and they are not covered in this chapter. Where the process parameters vary predictably, the nominally simple technique of *controller scheduling* can be very effective. Where it is decided to use a variation-tolerant constant controller this amounts in the single-loop case to choosing sufficient stability margins. In the multi-loop case, it leads to *robust control*.

Adaptive control, as can be seen from Figure 12.1, is usually taken to include those cases in which an on-line mathematical model of the process is maintained and used for continuously re-designing the controller *or* where a fixed model is chosen to represent the system's desired behavior and is used on-line to aid in the synthesis of forcing signals that make the behavior of the real process agree with that of the fixed model. This latter system is called model reference adaptive control (MRAC).

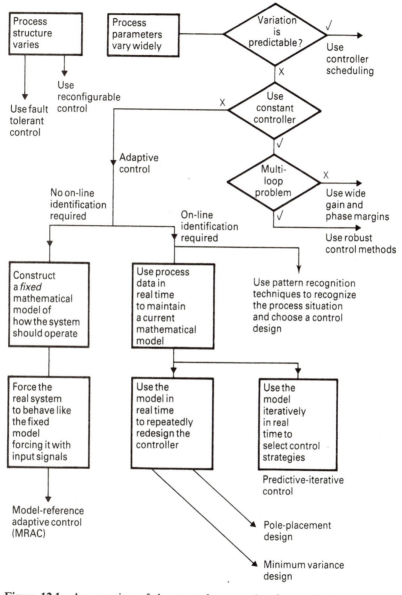

Figure 12.1 An overview of the control approaches for application to varying processes.

Finally, Figure 12.1 shows a *pattern recognition route* to adaptive control. This approach is not described in this chapter but it is incorporated in some commercially available adaptive controllers.

12.2 Some simple initial ideas on adaptive and self-tuning control

Consider the usual feedback loop of Figure 12.2 in which G is a process and D is a controller. Conventionally, controller D is designed using quantitative knowledge of G (i.e. using a process model) to meet a given design specification.

A reasonably well-designed feedback loop should still work efficiently in the presence of modest changes in the characteristics of G. Such a robustness of design is clearly essential in all practical applications since all real processes, however constant their nominal characteristics may be, will always be subject to some variability. Feedback loops do possess a considerable degree of tolerance to changes in the characteristics of the process compared with the characteristics encapsulated in the proces model G.

It is interesting to ask, for a particular loop: How far can G depart from the nominal model of G that was used in design, before the performance of the loop becomes seriously degraded?

This is an important point. Consider some examples. Suppose G is a ship that is to be steered automatically so that the control loop is that of a ship's autopilot. Ship steering is a complex subject but consider just one aspect—ship speed and its effect on rudder characteristics. When a ship is stationary or near-stationary, the effect of the rudder on ship orientation is very different from when the ship is at speed. We can consider that G is not constant but is speed-dependent and we must ask the question: If G varies significantly, can a constant controller D provide good performance over the range of variation or must D vary in some way to take account of variations in G?

Suppose, again referring to Figure 12.2, that, G is a metal rolling mill, D is a controller and the objective is to control the thickness of metal strip. It will be found that in the design of the controller D, the thickness, hardness, temperature and width of the product all have a significant effect on the model G and hence on the design process for D. Most metal rolling mills operate with a wide range of thicknesses, hardnesses, temperatures and widths of product.

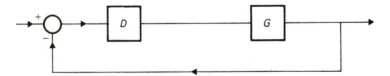

Figure 12.2   A feedback loop with process G and controller D.

The question again arises; can any constant controller D cover satisfactorily the range of operation?

Both the above illustrative examples are of the form; G is a process whose characteristics vary significantly. D is a controller that may not give adequate control over the range of variation. In case it is decided to design a best possible constant controller D that can deal with the envisaged variations in G, we have (roughly speaking) a *robust control system*. In case it is decided (by some means!) to design a variable controller D that changes characteristics in order to remain effective as G varies then we have an *adaptive control system*, i.e. an adaptive control system is one that adapts according to changes taking place in the process, to remain effective over the range of variation. An adaptive controller where the main emphasis is on the ability of the device to carry out its own setting up to control an unknown process is often called a *self-tuning controller*.

Example   A process G can be modelled by transfer functions ranging from

$$G_A(z) = \frac{0.9}{(z - 0.2)}$$

to

$$G_B(z) = \frac{1.1}{(z - 0.2)}$$

where G_A, G_B should be thought of as representing the same process under different conditions, typical of those in practical processes.

For the control of the process with model $G_A(z)$, a controller

$$D_A(z) = \frac{z - 0.2}{z}$$

yields the overall transfer function

$$\frac{G_A D_A}{1 + G_A D_A} = \frac{\dfrac{0.9}{z}}{1 - \dfrac{0.9}{z}} = \frac{0.9}{z - 0.9}.$$

However, if the same controller $D_A(z)$ is still in position when the process transfer function has become $G_B(z)$, then the overall transfer function is

$$\frac{1.1}{z - 1.1}$$

i.e. the closed loop system is now unstable. It would be necessary to use a controller

$$D_B(z) = \frac{0.9}{1.1} \frac{z - 0.2}{z}$$

to obtain the same stable overall transfer function as before. This simplistic example demonstrates that a single constant controller may be inadequate over a range of variation of a process. If a single constant controller is used then its parameters will have to be chosen to ensure stability in the worst case with consequent unnecessary degradation of operation most of the time.

In this example, the solution would appear to consist in varying the controller as conditions require, i.e. to use an adaptive controller.

Although the example is simple, the following questions that may occur to the reader are non-trivial and are relevant to real problems in which an adaptive control solution is proposed.

1. How are we to determine 'instantaneously' during operation, the model G that is current?
2. If instantaneity is impossible in 1 (as seems likely), how quickly must we determine the model G? Surely this question will be linked with the further question: how quickly can the real process change its characteristics?
3. In a non-ideal (say noisy) situation, how much data do we need to gather before we can say with confidence that a particular process model is currently valid?
4. Assuming that we have a method for determining the current model G with confidence, do we then simply adapt off-line control design methods for on-line use to determine the controller D or are there specific methods available that are specially suited to on-line applications?
5. (Thinking in detail about the simple example.) Do we have sufficient confidence to leave an automatic algorithm for adaptation of the controller D operating in a situation where it is known that some controller settings, if mismatched to some possible process conditions, will lead to a disastrous unstable condition?

These questions will be carried along in the further treatment of adaptive control. However, before proceeding, we return to the related concept of self-tuning control that was briefly mentioned above.

Refer again to Figure 12.1 in which we assume G is a process for which we know no model and that the controller D has a previously fixed structure with variable parameters that are to be set ('tuned') to fit the situation.

When the loop is first set up, it is the job of a skilled control technician to set the parameters in the controller D, i.e. to tune the controller.

As we have discussed, a mechanism whereby 'by some means' the controller D can set its own coefficients is called self-tuning and a controller D that contains such a mechanism is called a self-tuning controller. Clearly, such a self-tuning controller can be expected to be a valuable commissioning aid compared with the situation where the controller D has to be tuned manually.

It will now become clear that after successful initial commissioning of a closed loop containing a self-tuning controller, the option could remain to leave the self-tuning controller in position and allow it to continue to change the values of its parameters to match any changes that may be occurring in the process.

Thus it can be seen that the boundary between adaptive control and self-tuning control is rather blurred—a self-tuning controller operating continuously to keep tuned in line with changes in the process is a type of adaptive control.

Initial questions about the application of self-tuning controllers suggest themselves as follows:

1. To what extent can a 'universal self-tuning controller' be fitted onto an arbitrary process with expectations of success? At the other extreme, might we find that self-tuning is capable only of fine tuning and that the self-tuning controller, rather than being universal, will need to be reasonably well matched *a priori* to the particular process?
2. If a self-tuning controller is connected to a process, how does the self-tuning proceed exactly? For instance, is there a tuning phase during which no control is exercised? Will perturbations need to be injected into the process in the tuning phase? Can tuning be expected to be always successful or might there be situations where the tuning does not, in a sense, converge?
3. What effects do noise and nonlinearity in the process have on the success of self-tuning (and indeed on adaptive control in general)?
4. How far has the subject developed? For instance, can one expect to be able to design a workable adaptive optimal multivariable system? Is it instead more the case that most working applications to date are relatively simple single loops?
5. How does self-tuning proceed when the process (as in the steel mill example referred to earlier) can only exist with product present? In more detail, to self-tune a controller for strip thickness control, is it permissible to start the mill with the untuned controller and allow the tuning operation to proceed with the risk of damage to product and plant until the controller has converged to its tuned state? If this is not allowable, a self-tuning controller cannot be applied to a thickness-control loop since, with no product in the mill, the process consists only of movable rolls with air between them and with no output variables to be measured or controlled.

12.3 Approaches to adaptive control

Thinking again about the two examples (ship steering, steel mill) outlined earlier, the following approaches to adaptive control suggest themselves:

1. Calculate or determine experimentally how the process model changes as a function of known/measurable changes in operating conditions. Design, *a priori*, a range of controllers to cover the range of model variation. Implement

a system to change the controller in synchronism with changes in process operating conditions. This type of control is usually called *controller scheduling or pre-programmed control*; it depends on some modicum of measurability and consistency in the process.

Many authors would not regard controller scheduling as being a type of adaptive control—although, if applicable, it may serve the same purpose. The usually accepted approaches to adaptive control are as described in 2 or 3 below.

2. Acquire input–output data from the process over some period of time and use an algorithm, on-line in real time, to produce a currently applicable process model. *This is the identification phase of adaptive control.* Use the currently applicable model in conjunction with an on-line control design algorithm to calculate and then implement a controller. *This is the control design phase of adaptive control.* In adaptive control, the identification phase and the control design phase are cycled through alternately.

Alternatively:

3. Construct a constant real-time model G_d having the desired behavior of the actual and changing process of characteristic G. Run this so-called *reference model* G_d in parallel with the process, so that both G and G_d receive the same input signals. Compare the output of G and G_d to produce a deviation signal d. Use the signal d, to force the behavior of G to be the same as that of G_d, i.e. the aim is to make signal d be zero. This approach, to be described more fully later, is called *model reference adaptive control* (MRAC).

12.4 Approaches to self-tuning control

1. There is a one-off initial identification phase during which the process is in open loop. The control parameters are then calculated from the open-loop data and the loop is closed.

Alternatively:

2. There is an initial phase of rudimentary control during which data are generated for identification. Once identification is completed, control design takes place and the resulting controller is then implemented.

12.5 Identification methods for adaptive and self-tuning control

Adaptive and self-tuning control depend on the availability of a reliable process identification method. Table 12.1 lists some of the aspects to be considered regarding identification.

Table 12.1

Pre-programmed control	Model adaptation depends on quantitative knowledge of sensitivity to changes	Changes may not be measurable. The quantitative effect of the changes on the model may be difficult to determine
Open-loop identification	Modelling depends on numerical fitting of a pre-chosen model form to the experimentally obtained data	The process is not under control during the identification phase
Identification in closed loop	(a) perturbations are continually injected into the loop (b) no perturbations are injected	For both (a) and (b) identification is complicated by the correlation that exists between output and input signals. For (b) identification depends on normal systems operation producing sufficiently rich information to characterize the process dynamics.

12.6 Linear difference equation process models

12.6.1 The ARMAX model

A difference equation much used in adaptive control is the moving average model with auto regressive exogenous input (ARMAX model) of the form.

$$y(t) + a_1 y(t-1) + \cdots + a_n y(t-n)$$
$$= b_0 u(t-k) + \cdots + b_m u(t-k-m) + c_0 \varepsilon(t) + \cdots + c_n \varepsilon(t-n) + d(t)$$

$$(12.1)$$

in which y, u are output and input

ε	is a random variable representing stochastic effects
d	is a disturbance variable
n, m	are integers to be specified
k	is an integer representing time delay as a number of sampling instants
a_i, b_i, c_i	are the model parameters

Equation (12.1) may be expressed in terms of a backward shift operator as

$$A(q^{-1})y(t) = B(q^{-1})u(t-k) + C(q^{-1})\varepsilon(t) + d(t) \qquad (12.2)$$

where

$$A(q) = 1 + \sum_{i=1}^{n} a_i q^{-i},$$

$$B(q) = \sum_{i=0}^{n} b_i q^{-i},$$

$$C(q) = \sum_{i=0}^{n} c_i q^{-i},$$

$$q^{-1}y(t) = y(t-1).$$

In adaptive control applications, m, n, k and sampling interval T are specified in advance and A, B, C, d are determined experimentally.

The choice of m, n, k, T is an important compromise between interacting effects. Not only the modelling itself but also the type of control obtained by use of the model will be strongly influenced by these choices.

In use, the model is fitted to data by fixing the A, B, C, d parameters using an identification algorithm. Most favored is the recursive least squares approach. This aims to minimize a function

$$T = \sum_{t=1}^{\alpha} \{y(t) - \hat{y}(t)\}^2 \qquad (12.3)$$

where α is the current sampling instant and $\hat{y}$ is the estimate of y, based on an estimated parameter set

$$\hat{\theta} = [a_1, \ldots, a_n, b_0, \ldots, b_m, c_0, \ldots, c_n, d]. \qquad (12.4)$$

In what follows it is assumed that the c_i parameters are zero. Then the recursive least squares algorithm is

$$\hat{\theta}(t) = \hat{\theta}(t-1) + P(t)\psi(t-1)[y(t) - \psi^T(t-1)\hat{\theta}(t-1)] \qquad (12.5)$$

where ψ is defined by the equation

$$y(t) \triangleq \psi^T(t-1)\theta(t-1) + e(t) \qquad (12.6)$$

in which $e(t)$ is an uncorrelated error term.

P is the square (dimension $n + m + 1$) covariance matrix of the fitting error. P also needs calculation by the recursive equation.

$$P(t) = P(t-1) - P(t-1)\psi(t-1)\psi(t-1)[\psi^T(t-1)P(t-1)$$
$$\times P(t-1)\psi(t-1) + I]^{-1}\psi^T(t-1)P^T(t-1) \qquad (12.7)$$

Equations (12.6), (12.7) can be combined to yield

$$\hat{\theta}(t) = \hat{\theta}(t-1) + K(t)[y(t) - \hat{y}(t)] \qquad (12.8)$$

in which $K(t)$ is the Kalman gain, calculated from

$$K(t) = \frac{P(t-1)\psi(t)}{1 + \psi^T p(t-1)\psi(t)} \qquad (12.9)$$

(See Ljung (1983) for details).

12.6.2 Bias in the parameter estimates

If the error $(\varepsilon(t))$ in equation (12.1) has zero mean, the parameter estimate $\hat{\theta}$ is unbiased. Otherwise bias in the estimation occurs, such bias becoming severe in the presence of low signal-to-noise ratios.

In the presence of noise of non-zero mean, unbiased estimates may still be obtained by using the instrumental variable method (Young (1970)) the method of extended least squares (Clarke (1981)) or the method of pseudolinear regression (Goodwin (1984)).

12.6.3 Convergence and tracking

The way in which the recursive least squares algorithm is formulated leads to its convergence to a fixed set of model parameters. It is in the nature of the algorithm that, given experimental data from a fixed parameter process, it monotonically homes in on the best parameters for that process.

However, if the process having had fixed parameters for a long period, suddenly changes its parameters significantly, it is found that the recursive identification algorithm remains 'loyal' to the previous parameters with an inability to move off to track the new parameters.

The author has investigated the phenomenon through repetitive simulations using the well-known Bayesian relation between *a priori* and *a posteriori* probabilities,

$$P(A|B) = P(A)P(B|A)/P(B),$$

where the symbols have their usual meanings. The author's results and many others have shown the inability of the recursive least squares algorithm to track a moving target.

The solutions to this problem have been *ad hoc*, to say the least. They comprise:

1. Periodically abandoning the covariance matrix P as found recursively and rather arbitrarily substituting a new starting value $P(0)$ in the hope of rejuvenating the system's tracking ability.
2. Working with so-called 'forgetting factors' that weight process data in such a way that the importance of old data is progressively diminished.

Both approaches (covariance resetting and forgetting-factors) have been endlessly investigated but they remain as rather unsatisfactory empirical devices.

12.6.4 On-line checks on the validity of parameter estimates

Since in adaptive control the parameter vector estimate $\hat{\theta}$ will be used directly to update an on-line controller, it is important to devise automatic checks to prevent poor estimates $\hat{\theta}$ feeding through to result in poor or disastrous control. Some type of on-line confidence factor is the most obvious measure to be used. If the measure is below a pre-assigned value, the controller should not be updated but instead switched to a precalculated safe set of coefficients. Skill in this area of design will safeguard against the rogue behavior that adaptive-control systems are sometimes accused of.

12.7 Approaches to the control design phase

Once a model has been obtained in the identification phase, the method of control design can proceed based on:

1. Optimum controller synthesis, in which a cost function is minimized—briefly, *minimum variance design*; or
2. Pole placement or its equivalent, eigenvalue assignment.

It is also possible to envisage arranging the identification and control-design phases in such a way that no explicit model is ever brought to the surface and the two phases merge into a single adaptation algorithm.

12.7.2 Minimum variance design

An obvious choice for a wide class of processes is to choose a controller to minimize the variance of the output y.

Setting $m = n$ in equation (12.2) to make the problem 'square', a minimum variance solution for y is achieved by setting

$$u(t) = -\frac{F(q^{-1})}{B(q^{-1})E(q^{-1})} y(t) \qquad (12.10)$$

in which E and F can be determined by the equation

$$C(q)^{-1} \equiv A(q^{-1})E(q^{-1}) + q^{-k}F(q^{-1}). \qquad (12.11)$$

For use in adaptive control the two equations (12.10), (12.11) above are solved at each time step. The resulting control law is an optimal k steps-ahead predictor of error, combined with a compensator that attempts to bring that error to zero. In practice the algorithm is almost always used in recursive form (see Exercise 5.24 and Section 5.2.4 earlier in the book). Additionally, attention has to be paid to choice of sampling interval, sensitivity of the algorithm to errors in modelling assumptions and the possible need to incorporate integral action.

12.7.3 Control design based on pole placement

Given A, B, C from equation (12.2), the pole-placement controller is chosen to locate the closed loop poles at prespecified positions.

If we refer back to Section 8.12.2 and Exercise 8.12 in the book, it is clear that given a current $\{A, B, C\}$ description of a process, it is possible to synthesize a controller that will move the eigenvalues from their current positions to any positions chosen by the designer.

All that is nominally required to achieve an adaptive pole-placement system is to cyclically identify the current $\{A, B, C\}$ model for the process and then calculate and implement the necessary eigenvalue assignment algorithm.

Notice carefully that the placement of poles does not completely specify the performance that will be obtained. This effect occurs because the positions of the closed-loop zeros are not known in advance.

12.8 Model reference adaptive control

12.8.1 Basic Ideas

Let us first recall that in control-system design, we frequently have our aim-specification in the form of a model designated $H(z)$. Given a process, $G(z)$, we choose a feedback controller $D(z)$ so as to obtain the desired behavior using the relation

$$H(z) = \frac{KG(z)D(z)}{1 + KG(z)D(z)} \qquad (12.12)$$

Thus, in many (usually non-adaptive) control design approaches, we do have a model that we wish our system to emulate as closely as possible.

Note that an algorithm that explicitly contains a model $H(z)$ on-line is not necessarily more advanced than one containing a controller designed using equation (12.12). See Figures 12.3 (a) and (b) which are equivalent if in Figure 12.3(b) $H(z)$ satisfies equation (12.12) and K is a high gain. (The transfer function for Figure 12.3(b) is

$$\frac{G(z)(1 + KH(z))}{1 + KG(z)}$$

and for sufficiently high K, the overall transfer function approximates $H(z)$.)

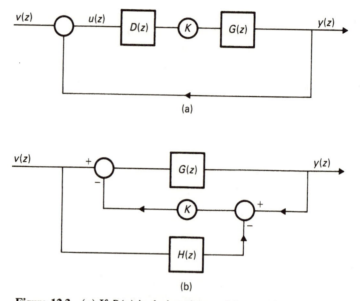

Figure 12.3 (a) If $D(z)$ is designed to satisfy equation (12.12) then the overall system transfer function is given by $H(z)$; (b) a 'model following' approach.

We have to ask: is the 'model following approach' (b) in any sense superior to the usual approach (a)?

The insensitivity to parameter changes of both approaches depends on high values of gain K being implemented.

Other questions that arise are:

1. To what extent will variations of different parameters in $G(z)$ affect the parameters within the model $H(z)$ and, more importantly, how will these changes follow through to changes in performance and stability margin in the system?

2. Can steps be taken during a static (i.e. non-adaptive) design to minimize the sensitivity of $H(z)$ to changes in $G(z)$? Is there in fact any justification for a belief that, during design, we may be able to choose between a delicate high-performance design and a more robust, somewhat lower-performance design, Figure 12.4.

The aim of the scheme is to vary the gain K to keep the process outputs $y(t)$ equal to the reference model output $y_m(t)$. The adaptive controller will of necessity operate on the integral of the signal $p(t)$, varying K in the direction to make y equal to y_m. In practice, it is by no means obvious whether variation of parameters in G can be completely compensated by variation of K. How far and how fast to vary K remains a non-trivial problem for the designer of the adaptive controller block.

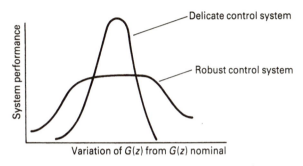

Figure 12.4   The notion of sacrificing some performance in exchange for robustness in the face of variations.

12.8.2 Algorithmic details

Given a process of model

$$A(q^{-1})y(t) = q^{-k}B(q^{-1})u(t) \qquad (12.13)$$

and a reference model described by the equation

$$E(q^{-1})y_m(t) = q^{-k}H(q^{-1})y_r(t) \qquad (12.14)$$

(see Figure 12.5) where E, H are user-specified polynomials of order p, define F, G as polynomials of order $k + 1$ and $n + 1$ respectively related to the equation

$$E(q^{-1}) = F(q^{-1})A(q^{-1}) + q^{-k}G(q^{-1}) \qquad (12.15)$$

From equations (12.13) and (12.15) we can obtain

$$E(q^{-1})y(t) = q^{-k}G(q^{-1}) + q^{-k}F(q^{-1})B(q^{-1})u(t) \qquad (12.16)$$

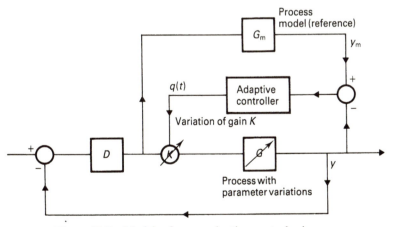

Figure 12.5   Model reference adaptive control scheme.

Equating (12.16) and (12.14) yields the expression

$$G(q^{-1})y(t) + F(q^{-1})B(q^{-1})u(t) = H(q^{-1})\dot{y}_r(t). \qquad (12.17)$$

This equation is a control law to force the process output y to track the reference model. It becomes an adaptive control law if parameter values are represented by their estimates.

To make the model reference adaptive system operational, the parameters in polynomials G, F, B in equation (12.16) need to be estimated.

An alternative procedure based on gradient methods is often used. Such a method has the advantage that convergence and stability of the adaptive loop are guaranteed.

12.9 Trajectory following control

12.9.1 Introduction

In the control of many important industrial processes, the main objective is to require certain key process variables to follow preferred trajectories—such trajectories have been found by experience or calculation to be efficient or optimal within constraints. Most of the accepted methods of adaptive control do not provide for trajectory following but a number of approaches have been developed.

A typical problem that occurs in industrial batch processes can be split into three parts.

1. Generate preferred trajectories for the system's state to follow. These trajectories will depend on: initial conditions, end of batch targets and during-batch costs and constraints.
2. Generate a tracking strategy to force the system states to follow the prespecified trajectories.
3. Implement the strategy with sub-strategies being available to deal with the non-ideal tracking that will occur in practice. For instance, if a state departs significantly from the desired trajectory part way through a batch, will it be best to try to return to the desired trajectory or should a new desired trajectory be generated from the current point onward using approach 1 for generation?

Given a nonlinear batch process described by the equation

$$\dot{y} = f(y, u, w) \qquad (12.18)$$

in which f is an arbitrary function, y is a vector of outputs, u is a vector of inputs and w is a vector of disturbances, the task is to drive the system along a preferred trajectory.

12.9.2 Optimal feedback control using the Riccati equation

Equation (12.18) is linearized about a desired trajectory, resulting in a continuous time $\{A, B, C, D\}$ representation.

$$\dot{y} = A(t)y + B(t)u + D(t)w \qquad (12.19)$$

which is then discretized to a $\{F, G, H\}$ representation

$$y(k + 1) = F(k)y(k) + G(k)u(k) + H(k)w(k). \qquad (12.20)$$

A cost function is given by

$$J = y(N)^T My(N) + \sum_{k=1}^{N-1} y(k)^T Qy(k) + \sum_{k=1}^{N} u(k)^T Ru(k). \qquad (12.21)$$

The $\{u(k)\}$ control policy that minimizes the performance index is given by the feedback law

$$u(k) = K(k)y(k) \qquad (12.22)$$

where

$$K(k) = R^{-1}G(k)^T(F(k)^T)^{-1}(P(k) - Q) \qquad (12.23)$$

$$P(k) = Q + F(k)^T P(k + 1)[I + G(k)R^{-1}G(k)^T P(k + 1)]^{-1}F(k) \qquad (12.24)$$

(this is the matrix Riccati equation) and

$$P(N) = M. \qquad (12.25)$$

The Riccati equation is solved backwards in time starting from the final condition $P(N) = M$.

The values of $P(k)$ so obtained are substituted into the equation for $K(k)$ resulting in a precomputed, time-varying, optimal trajectory-following algorithm. Note that the strategy follows the desired trajectory optimally—it does not generate the desired trajectory.

12.9.3 Dynamic Matrix Control

Dynamic Matrix Control (DMC), Cutler (1980), uses predictions based on a stored non-parametric model consisting of impulse or step responses. The method allows calculation of the necessary controls that will produce predictions that follow a given desired trajectory.

The *dynamic matrix A* stores the step responses of the system numerically, allowing for the cumulative effect of inputs applied at different times and exploiting linearities to make the task possible.

The steps in Dynamic Matrix Control are:

1. Measure the position of the output vector $y(k)$
2. Measure the changes

$$u(k) - u(k-1) \triangleq \delta u(k)$$
$$w(k) - w(k-1) \triangleq \delta w(k)$$

(12.26)

in manipulable and non-manipulable inputs that have taken place over the previous time step.

3. Use the A matrix to operate on the input changes to predict future perturbations to vector y, i.e.

$$\delta y(k+1) = A \begin{bmatrix} \delta u(k) \\ \delta w(k) \end{bmatrix}.$$

(12.27)

This allows a prediction of $y(k+1)$ using

$$\hat{y}(k+1) = y(k) + \delta y(k+1).$$

(12.28)

4. Measure $y(k+1)$ and compare with $\hat{y}(k+1)$ and use the difference vector to improve the prediction over the next step.

These four steps allow prediction based on stored step responses with provision for error correction.

The model is used backwards in time over a chosen horizon of sampling intervals to calculate the values of the manipulable inputs that will force the responses to follow the given desired trajectory.

Least squares optimization techniques can be used to calculate the necessary sequence of control signals. The user must select optimization and control horizons and weighting factors allowing emphasis to be put on particular variables.

The approach is highly practical, allowing incorporation of non-standard requirements because of its simple model, error-correcting prediction and manipulability. The approach does depend on adequate reliable process measurements or equivalent state estimates being available.

12.10 Controller scheduling

In cases where the structure of a process changes with time, a strategy that recognizes the change and then substitutes a controller of different structure may be attractive. Halme and Visala (1991) described such an approach, first illustrating the idea in terms of a water-tank process, Figure 12.6(a), (b), (c).

Halme creates a scalar-valued index, q, that is measurable on-line and that takes different integer values for each of the three states (Figure 12.7). This

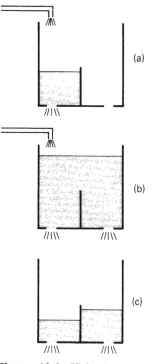

Figure 12.6 Halme's water tank in different states.

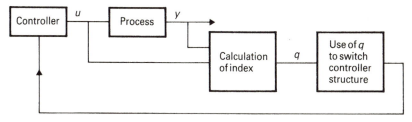

Figure 12.7 Controller scheduling activity.

index, by showing the current structure of the process, allows on-line controller scheduling.

Halme illustrated how his approach might be applied to a batch bio-technological process as in Figure 12.8.

Figure 12.8 shows the four phases typical of a batch biotechnological activity. Figure 12.7 summarizes Halme's result where the on-line-measurable index q allows the structure of the (very complex) process to be classified into one of four states.

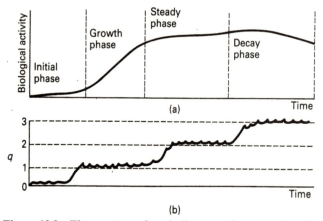

(a)

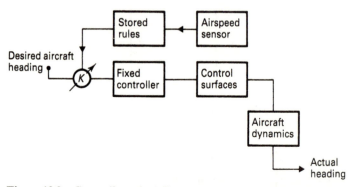

(b)

Figure 12.8 The concept of an indicator q that measures the current structure of a biotechnological process.

Figure 12.9 Controller scheduling and actual aircraft heading: the variable gain K is modified as a function of airspeed according to precalculated rules.

Most controller scheduling is less ambitious and consists of an application-dependent open-loop strategy to use an auxiliary measurement to modify control behavior as in the example in Figure 12.9. The limitations of controller scheduling are:

There is often no measurable, useful auxiliary measurement available.

The 'stored rules' may need to be quite complex and time-varying.

Despite these difficulties, controller scheduling is used with great success, particularly in the batch process industries.

Perhaps one of the most spectacular examples of controller scheduling is in electric steelmaking. For rapid melting, a short arc (high current, low voltage) is first used. For refining, a long arc (low current, high voltage) is then used.

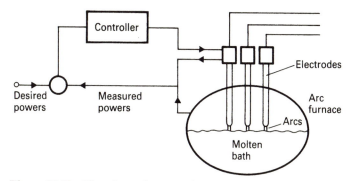

Figure 12.10 The electrode control system moves the electrodes vertically, changing arc lengths, to maintain desired powers in the three phases.

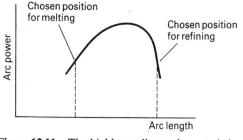

Figure 12.11 The highly nonlinear characteristic of an electric arc.

When the transition is made from melting to refining, controller scheduling consists in inserting a minus sign into the closed loop electrode positioning algorithm to allow operation on a different part of the power versus arc-length curve (see Figures 12.10 and 12.11) where the slope changes sign.

12.11 Batch-to-batch adaptation for batch processes

The problem of batch control is to choose initial conditions $x(t_0)$ and control policy $u(t)$, $t \in [t_0, t_f]$ such that the state $x(t_f)$ at the end of the batch satisfies a target requirement

$$x(t_f) = x_d \tag{12.29}$$

where typically x_d is the required specification for a product.

Noting that $x(t_f)$ is in general a function of the initial conditions x_0 and of the control policy $u(t)$, i.e.

$$x(t_f) = f(x_0, u) \tag{12.30}$$

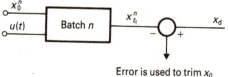

Error is used to trim x_0
and $u(t)$ that will be used
(open-loop) for batch $n+1$

Figure 12.12 Initial condition $x(t_0)$ and control policy $u(t)$ are adapted from batch to batch to minimize the error δx_{t_f}.

then by sensitivity analysis we can obtain approximate values for the partial derivatives $\partial x_{t_f}/\partial x_0$ and over a sequence of batches it is then possible to adaptively modify initial conditions and/or control policy, using relations of the form

$$\delta x_0 = \frac{1}{\dfrac{\partial x_{t_f}}{\partial x_0}} \delta x_{t_f} \qquad (12.31)$$

where $\partial x_{t_f}/\partial x_0$ is a prederived sensitivity coefficient and δx_{t_f} is a deviation, i.e. an amount by which the target x_d has been missed in the current batch, batch n, i.e.

$$\delta x_{t_f} = x_d - x_{t_f}^n \qquad (12.32)$$

and δx_0 is the adaptive change to be made to the initial conditions, compared with nominal, for the next batch, batch $n + 1$:

$$x_0^{n+1} = x_{0\,\text{nominal}} + x_0.$$

This adaptive batch-to-batch updating works well in practice: one well-known, simple example being the direction of gunners on to a distant target based on radio feedback from a gunnery observer.

12.12 Predictive–iterative control using a fast process model

Any process for which a fast on-line model exists can be controlled by the technique shown in Figure 12.13.

The process model, which should be several orders of magnitude faster than the process, receives trial control inputs, each defined over a simulated time interval T. The algorithm evaluates the consequent responses and chooses from the set of trial inputs the one that will be implemented on the actual process over the next time period T of process model time.

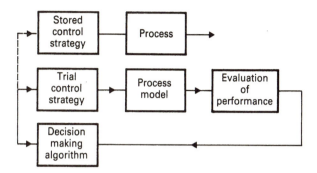

Figure 12.13   Predictive–iterative control strategy.

The advantages of the method are:

(a) No restrictions are imposed on the type of model provided that it satisfies the requirement of being suitable for fast on-line computation.
(b) A quite inaccurate model will still give good control. This is because at the beginning of each interval of process time T, the model starts off with the current initial conditions. The time T is chosen so that the prediction ability of the model is not overstretched.
(c) The difficult problem of control synthesis is avoided and no concessions need to be made to allow the synthesis problem to become tractable. Any criterion whatever can be used in the block marked 'evaluation of performance'. There is no need to be restricted to quadratic criteria.

Individual schemes will have additional features according to the process-control requirements. In particular, some means of updating the model so that it changes to match changes occurring in the process will often be required. Such additional features bring the system into the adaptive-control area. Despite such additions, the basic principle of predictive–iterative control remains as above.

12.13 Commercially available adaptive controllers

12.13.1 The Foxboro EXACT system

The EXACT system is one of the optional modules available within the Foxboro Spectrum series of controllers. The controller algorithm is PID and the self-tuning is basically a problem recognition exercise to ensure a reasonable transient response following either a setpoint change or a naturally occurring disturbance. During a transient, the pattern recognition feature recognizes the amplitude and frequency of oscillation. Stored rules then allow calculations of necessary changes to the PID coefficients to bring the performance close to the type of response that has been specified as desirable by the user.

The controller also contains a user-specifiable linearizer to counteract the effect of known and repeatable process nonlinearities. Such linearization, if effective, makes the self-tuning operation become independent of the amplitude of disturbances.

12.13.2 The ASEA NOVATUNE system

The ASEA NOVATUNE system first identifies the process from on-line data using a recursive least squares algorithm to yield a difference equation representation.

The control algorithm is then calculated to satisfy minimum variance criteria. Thus, control is not limited to PID strategies.

The system has been applied to pulp and paper processes and to strip rolling mills.

12.13.3 Leeds and Northrup ELECTROMAX V

The ELECTROMAX V is a PID controller with self-tuning capability. On receipt of a user command, self-tuning commences. The procedure is that a second-order transfer function model is fitted on-line by the instrumental variable technique (a variant of least squares fitting) to process data and that from this model, optimum PID coefficients are calculated. See Cheung (1987) for further information.

12.13.4 The Turnbull TCS 6355 self-tuning controller

The Turnbull TCS 6355 Controller uses a well-tried PID algorithm operating in a fast (36 ms interval) basic feedback loop. For initial set-up, an open loop Ziegler–Nichols type step-test allows the initial sampling period to be chosen for the adaptive loop. In the adaptive loop: (i) A recursive least squares algorithm produces a discrete time model $G(z)$ of the process, (ii) $G(z)$ is transformed by bilinear transformation into a continuous time model $G(s)$, (iii) the PID controller D is fixed so that $G(JW)D(JW)$ has a 60° phase margin, (iv) The recommended PID settings are displayed together with a confidence factor (obtained as an output from the recursive algorithm), (v) The recommended settings are implemented only if the process operator presses the 'accept' button. Figure 12.14 illustrates the scheme of adaptation.

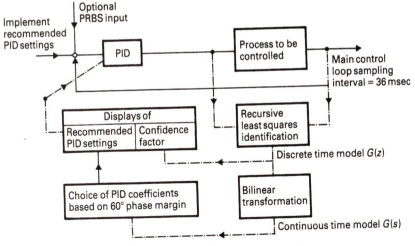

Figure 12.14 Turnbull adaptive controller.

12.13.5 The REX 1000 temperature controller of Calex Instrumentation Ltd (UK)

The REX 1000 is a dedicated temperature controller with all the normal features of a microprocessor-based three-term controller. In addition, the controller performs automatic tuning of coefficients when initiated to do so by a user-operated button.

12.13.6 The self-tuning temperature controller of IMO precision control (UK)

The E5K controller is a low-cost PID temperature controller with a self-tuning feature. The user is required to select the self-tuning mode when the process temperature is some way from the desired temperature. During the transient that occurs while the process approaches its final temperature, the PID controller is automatically tuned to its optimum settings.

12.14 Robustness

12.14.1 Initial discussion

Robustness describes the ability of a system to maintain a 'reasonable' performance despite significant changes in parameters, gains, phases, and despite large disturbances, inadequate models and even, for multivariable systems, the partial or complete loss of sensors, signal paths and actuators.

Clearly, robustness and sensitivity are closely related since sensitivity is measured in terms of a partial derivative relating (typically) an output variation with a disturbance variation, whereas robustness is concerned with describing a bound on the disturbance within which reasonable system behavior can be guaranteed.

To give a practical illustration of a robustness requirement we quote from Sandell (1981), part of the United States robustness specification for the closed-loop flight control systems (FCS) of military aircraft.

Stability margins are required for FCS to allow for variations in system dynamics. Three basic types of variations exist:

Mathematical modelling and data errors in defining the nominal system and plant.

Variations in dynamic characteristics caused by changes in environmental conditions, manufacturing tolerances, aging, wear, noncritical material failures, and off-nominal power supplies.

Maintenance induced errors in calibration, installation and adjustment.

In multiple-loop systems, variations should be made with all gain and phase values in the feedback paths held at nominal values except for the path under investigation.

12.14.2 Robustness of single-loop systems

In order to ensure robustness in the face of a whole armoury of possible variations, uncertainties and disturbances, it would seem almost essential to downgrade control performance to bring it far away from the boundary of instability. Such a safety margin as is indicated by a wide gain and phase margin (Figure 12.15) would seem to, and will be found to ensure practical robustness.

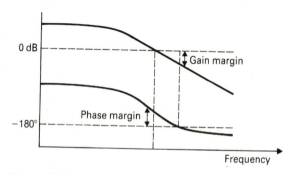

Figure 12.15 Bode plot.

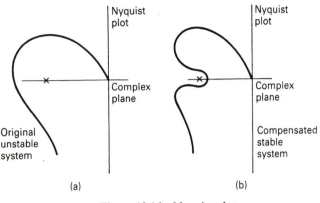

(a)   (b)

Figure 12.16   Nyquist plots.

A designer's instinct will lead to a selective degradation of performance at critical frequencies, leaving the performance undegraded elsewhere (Figures 12.16(a) and (b)).

Such a strategy clearly depends on the invariance of several factors to ensure that the compensation by notch filter shown in Figure 12.16(b) is maintained in practice, where a requirement for robustness may militate against the use of such highly selective compensation.

12.14.3 Robustness of multi-loop systems

Robustness for single-loop systems is a relatively simple concept but for multivariable systems it is the subject of a very considerable literature.

Let us first note that the United States military aircraft requirement for multiple-loop control systems quoted in the section above is not really adequate, since in real life a system must remain operational when some or all loops experience simultaneous perturbation.

In considering the robustness of a multi-loop system it is important to consider the possible effect of simultaneous, reinforcing perturbations on the overall stability.

A common approach is to use the model of Figure 12.17 in which a diagonal perturbation matrix P contains n perturbations $P_1 \ldots P_n$ on its leading diagonal with interconnection to the process represented by an $n \times n$ square transfer function matrix $G(s)$.

Robustness of this system depends on the properties of the matrix $G_{zx}(s)$. This is the matrix that relates the vectors z and x.

The robustness of the system is governed by *the greatest singular value* of the matrix $G_{zx}(j\omega)$. The greatest singular value of any matrix G is denoted $\bar{\sigma}(G)$

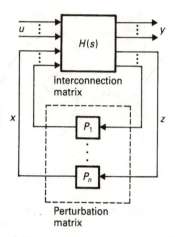

Figure 12.17   Linear system with perturbation.

and it is defined by the relation

$$\bar{\sigma}(G) = +(\lambda_{max})^{1/2} \tag{12.33}$$

where λ_{max} is the largest eigenvalue of the set of eigenvalues of the matrix $G_{zx}(j\omega)G_{zx}(j\omega)^{\dagger}$ in which the superscript $\dagger$ indicates the complex conjugate of the transpose of $G_{zx}(j\omega)$.

The system will remain stable provided that each perturbation satisfies the inequality

$$\log|p_i(j\omega)| < -\log(\bar{\sigma}G_{zx}(j\omega)). \tag{12.34}$$

Notice that the inequality needs to be satisfied for all positive frequency ω and the requirement can be well illustrated using a magnitude versus frequency plot as illustrated in Figure 12.18, see Safonov (1980) and Postlethwaite (1982).

Another approach due to Lehtomaki (1988) obtains the robustness results in terms of the minimum and maximum values of the system's return difference transfer-function matrix. Again this robustness test requires an inequality to be

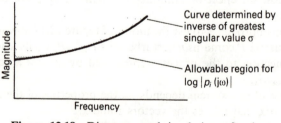

Figure 12.18   Diagram used in design of robust multi-loop systems.

satisfied for all real frequencies and again the graphical interpretation of the test represents a generalization of the gain and phase margin concepts of a single-loop system.

Sandell (1979) gives very good illustrative examples of the LQG approach being applied to the design of a robust flight control system for a helicopter.

Davison and Ferguson (1981) pioneered a parameter optimization approach to the robust design problem for multi-loop feedback control systems. In their approach, they set out to choose design parameters that would ensure rapid response and minimize interaction between loops whilst satisfying constraints on the position of closed-loop eigenvalues and individual loop gain (to prevent saturation) while maximizing tolerance to plant parameter perturbations and the effect of individual or group actuator or sensor failures.

The design problem is formulated in terms of a performance index that must be maximized within constraints by choice of design parameters.

Having given an overview of the available approaches to robust control design, we next go into more detail on the mathematical background and the technique of one particular approach—the H^∞ robust design method.

12.14.4 The H^∞ approach to robust control design
An aside on norms

The magnitude of a vector x, of a function f, or of an operator L is measured by its *norm* denoted $\|x\|$, $\|f\|$, $\|L\|$ respectively.

Any quantity whatever can serve as a norm, provided that it satisfies the three axioms.

$\|x\| = 0$ if and only if $x = 0$

$\|ax\| = |a|\,\|x\|$ for any x and for any scalar a

$\|x + y\| \leqslant \|x\| + \|y\|$ for any x, y in the space where the norm is defined

(It follows from these axioms that $\|-x\| = |-1|\cdot \|x\| = \|x\|$ for all x).

The 'distance between two functions' may be measured by the norm

$$\|f_1 - f_2\| = \left(\int (f_1(t) - f_2(t))^p \, dp \right)^{1/p}, \tag{12.35}$$

with equal mathematical validity provided that p satisfies the requirement $1 \leqslant p \leqslant \infty$. (If $p < 1$, it will be found that the quantity is no longer a norm since the triangle inequality (third axiom) above fails).

For different choice of p, the resulting norm is denoted $\|\ \|_p$.

It is easily shown that the greater the value of p, the 'stronger' the resulting norm.

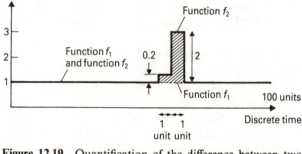

Figure 12.19   Quantification of the difference between two functions (not to scale).

For instance, if the norm is to be used to investigate the convergence of a sequence of functions $\{f_i\}$ towards a function f. It will be found that convergence in norm $\| \ \|_p$ implies convergence in norm $\| \ \|_q$ provided that $p > q$, but not conversely.

As another illustration of the effect of choice of p, consider the quantification of the difference between the two functions f_1 and f_2 shown in Figure 12.19. In the figure, the two functions are identical except for their divergence at the shaded section. Using the norm $\| \ \|_p$, defined by

$$\| \ \|_p = \left(\frac{1}{100} \sum_{i=1}^{100} |f_1 - f_2|^p \right)^{1/p} \tag{12.36}$$

we find

$\| \ \|_1 = 0.022$

$\| \ \|_2 = 0.2$

$\| \ \|_8 = 1.12$

$\| \ \|_{20} = 1.59$

$\| \ \|_{100} = 1.91$

$\| \ \|_{300} = 1.97$

and $\| \ \|_p \to 2$ as $p \to \infty$. That is,

$$\| \ \|_\infty = \sup_i |f_1 - f_2|,$$

see Figure 12.20. It can be seen that if two functions are near in $\| \ \|_\infty$, they are certainly near throughout in the general sense of the word and they are near in $\| \ \|_p$ norm, $p < \infty$. The value of a $\| \ \|_\infty$ norm rather than the usual $\| \ \|_2$ norm can be appreciated when guarantees of stability are being sought.

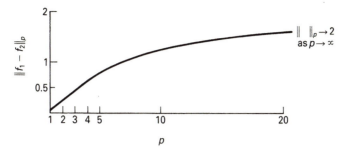

Figure 12.20   The variation of $\| \ \|_p$ with p .

THE NORM OF AN OPERATOR

Let $L: U \rightarrow Y$ be an operator on the space U with a range in the space Y. The norm of the operator L, denoted $\|L\|$ is defined by

$$\|L\| = \sup_{\substack{u \in U \\ u \neq 0}} \frac{\|y\|_p}{\|u\|_p}.$$

In control theory, L is the system, u, y are input and output variables and p will often be set to $p = 2$.

The operator norm, defined as above is what is called by control engineers the H^∞ norm of the system transfer-function matrix. The operator norm of a transfer function G is given by

$$\|G\| = \frac{\|y\|_2}{\|u\|_2} = \frac{I \int y(t)^T y(t) \, dt}{I \int u(t)^T u(t) \, dt} = \sup_\omega \bar{\sigma}[G(j\omega)] \qquad (12.37)$$

where p has been set to 2 and the integrals are taken over some suitable interval I. $\bar{\sigma}$ denotes, as before, the largest singular value of $G(j\omega)$.

Note that no p value appears against the norm of the operator G, since operator norms cannot have p values other than $p = \infty$ which is not therefore stated. However $\|G\|$ is the so-called H^∞ norm for G.

Comment   The H^∞ norm operates in the H^∞ normed space, consisting of all complex-valued functions F of a complex argument s, which are analytic and bounded in the open right complex half-plane.

The attraction of the H^∞ space is that it allows multivariable linear control problems to be defined and manipulated within the powerful framework of operator theory with the H^∞ norm being easily related to a cost function that is to be minimized.

THE H^∞ DESIGN PROCEDURE

In the following simple demonstration of H^∞ design, the process G receives two sets of inputs u_1, u_2, produces two sets of outputs y_1 and y_2 and the object

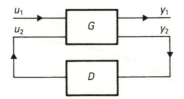

Figure 12.21 The configuration
used to demonstrate H^∞ design.

is to design a robust feedback controller D, see Figure 12.21. If we write

$$\begin{bmatrix} y_1 \\ y_2 \end{bmatrix} = G \begin{bmatrix} u_1 \\ u_2 \end{bmatrix} \tag{12.38}$$

and let H denote the transfer-function matrix of the complete closed loop, then

$$H = G_{11} + G_{12}D(1 - G_{22}D)^{-1}G_{21} \tag{12.39}$$

where the G_{ij} are the elements of the partitioned matrix G.

The design procedure is continued by partitioning H into two parts, one relating to performance and one to stability and attaching weighting functions that are functions of frequency to yield an augmented H matrix, H', i.e.

$$H' = \begin{bmatrix} W_{11}(s) & H_1 & W_{12}(s) \\ W_{21}(s) & H_2 & W_{22}(s) \end{bmatrix}. \tag{12.40}$$

Note that the weighting functions W_{ij} are functions of frequency and the judicious choice of these by the designer will allow interactive shaping of the character of the system to reach a compromise between performance and stability margin.

Conventional wisdom requires that the weighting matrices $W_1(s)$, $W_2(s)$ should have singular values with complementary frequency responses (Figure 12.22). Such an arrangement allows good low-frequency system performance combined with good high-frequency tolerance of modelling errors.

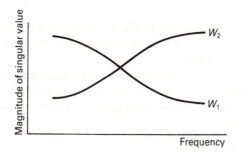

Figure 12.22 Typical weighting functions
used in H^∞ design.

A very extensive literature exists and a selection is given in the bibliography at the end of the book. Hammond (1991), Dorato (1990) and Doyle (1986) are particularly relevant. Note also that a robust toolbox exists for control systems design within the Matlab set of software.

12.15 Some practical points on process identification and modelling

12.15.1 Introduction

Here we consider:

Undermodelling (i.e. the case where process mechanisms are ignored or under-represented)—how the problem will not be revealed unless sufficiently exciting input signals are used.

How, in the case of undermodelling, parameters in the fitted model will be input-signal-dependent.

Overmodelling—the inclusion of too many process parameters/effects
    modelling using orthogonal functions
    principal components analysis
    model reduction

Experimental identification of process nonlinearities

Identification in closed loop

Time-varying processes

Linearization about a standard trajectory.

12.15.2 Undermodelling

Suppose that a process is linear and of nth order and that an mth-order linear model is chosen $(m < n)$ (i.e. the process is undermodelled), then it is quite possible for an apparently perfect fit to be obtained between model and process despite the 'undermodelling'—*unless* input signals sufficiently rich to excite all the modes of the process are used.

Some simple (idealized) illustrations: let $n = 2$, $m = 1$, i.e. a first-order model is used to represent a second-order process. Suppose that the input to the process is a 10 Hz sinusoid, then a perfect fit between model and process can be obtained by suitable choice of model parameter.

Suppose next that, in another experiment, a 100 Hz sinusoid is input to the same process as above, then again a perfect fit between model and process can be obtained—despite the undermodelling!—but for a different model parameter than that giving a perfect fit at 10 Hz. In other words, in the case that a process is undermodelled, perfect fits can be obtained for different single-frequency inputs but the model parameters are quite different in the two cases.

What if the two sinusoids at 10 Hz and 100 Hz are input together? In that case, it will no longer be possible to obtain a perfect fit between model and process. A compromise solution will be automatically found by experimental identification, with the weighting of the fitting being proportional to the relative energies of the input signal components.

Thus in the case of undermodelling, the model parameters are input-signal-dependent.

Corollary If inputs are not sufficiently rich in excitation signals then an inadequate model will appear adequate. If the model, identified under relatively quiescent conditions, is to be used for more adventurous control purposes then model inadequacies may emerge at a later important stage. In other words, models obtained based on data from normal plant operating conditions may not be extrapolatable to represent the new conditions that will apply once a control system is built up around the model.

Wherever possible, additional test signals should be injected to enrich the spectrum of the input data. The test signal frequencies need to be matched to the natural frequencies of the modes that are suspected to be within the process.

INCORRECT CONTROLLER DESIGN RESULTING FROM UNDERMODELLING
In the event that a process is undermodelled but well fitted to recorded data (due to insufficiently wide-ranging input signals) then theoretically-derived control algorithms, using a process model as part of their raw material, will necessarily produce significantly different control laws compared with design using a correct model.

12.15.3 Overmodelling (inclusion of too many effects) in a linear model

Suppose that a curve 'near in shape' to the curve $y = \sin x$ is to be approximated near the origin by a power series:

$$y = ax + bx^3 + cx^5 + \cdots. \tag{12.40}$$

Experimental curve fitting to a particular case will produce best values for the parameters $\{a, b, c, \ldots\}$.

Suppose next that the curve under study is modified only slightly to take on a different shape and again best numerical values are found for $\{a, b, c, \ldots\}$. In general, it will be found that slight shape changes in the curve (for curve, read process data) will cause large, non-obvious changes in the parameters ($\{a, b, c, \ldots\}$) (for $\{a, b, c, \ldots\}$ read model parameters).

If a less ambitious model $y = ax$ is fitted to the original and then to the modified curve, the two values of a found for the two cases will be similar—the low-order model is found to be robust.

This artificial example illustrates an effect of overmodelling that, in the event of overmodelling, there is a wide variation of parameters between models fitted to nearly identical data sets.

12.15.4 Modelling using orthogonal functions

If an arbitrary curve is fitted by a Fourier series, say

$$y = p \sin x + q \sin 3x + r \sin 5x + \cdots \tag{12.41}$$

by choice of best $p, q, r, \ldots$ parameters and the curve is slightly changed, refitting of the $p, q, r, \ldots$ parameters will not exhibit the ill-conditioning described in Section 3.1. The improved behavior results from the orthogonality of the functions $\sin x, \sin 3x, \ldots$ used in the fitting. This effect arises because each function in an orthogonal set is separately a best possible fit to the experimental curve—a healthy situation. For instance, in orthogonal fitting, increasing the order of the model by an extra term does not disturb the best values already found in the previous, lower-order, model.

12.15.5 Principal components analysis

Principal components analysis attempts to move from the axes arbitrarily provided by a set of measured data to a set of new orthogonal axes, spanning the same parameter space, to provide all the benefits of orthogonality described in Section 3.2.

The method is to perform an eigenvalue, eigenvector analysis on the experimental data with the eigenvectors playing the role of the axes in parameter space and the magnitudes of associated eigenvalues indicating the significance of each particular axis.

A disadvantage of the method is the loss of physical meaning that occurs—it is no longer possible perhaps to speak of dissolved oxygen or pH value—only of parameter a_n!

12.15.6 Model reduction

A model of high dimension can be reduced to lower dimension by suitable extension of principal components analysis. Here, the magnitude of eigenvectors is used to indicate the relative significance of effects extracted from the data. It is quite straightforward using model reduction techniques of the sort outlined to feed in a 15-input–15-output set of data to a model reduction technique and, with some hand-holding of the algorithm, to emerge with a 'most significant' 4-input–4-output dynamic model that best approximates the mass of data.

Model reduction methods named for Davidson and Marshall respectively can be found in the literature. Principal components analysis is well described by Hyvarinen. All three approaches and full references are given in Reference L4.

12.15.7 Identification of nonlinearities in the process

Nonlinearities are amplitude-dependent effects and there is no systematic way to identify them. In an ideal scenario for identification of a nonlinear process, the same input sequence would be fed to the process repeatedly, being identical on each run except for magnitude. In many processes, only a small range of magnitude is physically possible because of essential process constraints and simplistic attempts to identify nonlinearities will fail.

12.15.8 Process identification using data obtained from a system that is already operating under closed loop conditions

1. Feedback alone (i.e. without allowing for a controller) alters the weighting of signals seen by the process, compared with the weighting of input signals to the combination. The effect on an undermodelled process needs to be worked through carefully.
2. Closed-loop identification may be degraded by dependencies between input and output signals that invalidate assumptions of statistical independence in the identification methods.

12.15.9 Time-varying processes

A moving 'window' in the process data of sufficient length for good identification but short enough to be sensitive to sudden or gradual shifts in process mechanism is needed. If a systematic change can be seen in model parameters, the way is open to quantify the time variation, possibly in closed form.

12.15.10 'Subtracting out' a standard trajectory/linearizing about a standard trajectory

Many batch processes in industry operate to a highly repetitive schedule with the result that graphical displays of behavior all have a family likeness that can be represented as a mean 'standard trajectory' surrounded by an envelope. In many cases, the control aim will be to manipulate the control behavior not too drastically from the standard trajectory. In such a case, a model that represents the behavior about the standard trajectory as a datum may be a useful tool. In some cases, the standard trajectory can be viewed largely as a function of initial conditions and repetitive programmed actions that are the same on every batch, whereas the deviations from the standard trajectory can be viewed as dependent on in-batch steering actions that will differ from batch to batch.

12.16 Some final realism

12.16.1 Principles of adaptive control

The concept of adaptive control is very simple. The practice of adaptive control is in two parts:

Quantitative identification of the process.

Design of the controller

Both of these operations have to be undertaken on-line, automatically and reliably in real time.

Here we list some of the difficulties that make achievement of adaptive control more difficult than might have been expected.

12.16.2 Application of adaptive control

Adaptive control is, logically, best applied to those processes where no fixed controller can operate successfully. Such processes tend, in the main to be: not well understood, nonlinear, continuous time and time-varying. The accepted approaches to adaptive control attempt to encapsulate the characteristics of these continuous time, complex, variable, nonlinear processes in a low-order, fixed-parameter discrete-time difference equation.

Unsurprisingly, this aspect (attempting to model a complex process by a simple difference equation) can lead to difficulties.

12.16.3 Closed-loop processes

Processes that are operating under closed loop are difficult to identify for two reasons.

1. Under good control, there is little variation in process output. If the process output is insufficiently rich in information, identification is difficult.

2. Under closed-loop control, input signals are functions of previous output signals. This dependence complicates identification.

12.16.4 Identification

Identification will be aided if perturbation signals are deliberately injected during process operation. A compromise must be reached whereby perturbations are chosen to aid identification without undue degradation of the normal operation.

12.16.5 Real processes

Real processes may be non-minimum phase and/or have time-varying delays. Both features complicate adaptive control.

Many processes are subject to sudden changes in operating regime as well as experiencing slow long-term drifts. Classical adaptive control is better able to compensate for the second type of change than for the first.

12.16.6 Necessary compromises

In all approaches to real-time identification, it is necessary to compromise between obtaining rapid, up-to-date results based on small data sets and slower outdated results based on large data sets. Such compromises need to be soundly based on statistical considerations. Current identification techniques used in adaptive control schemes are often theoretically weak from the point of view of their scientific statistical basis.

12.16.7 Prespecified trajectories

Many important processes need to be steered along prespecified desired trajectories. Adaptive control techniques to achieve such a result are not yet well developed.

12.17 Conclusions

The field of adaptive and robust control, applied particularly to multi-loop systems, remains a very active research area.

It is pointed out that systems based on fuzzy logic, pattern recognition and qualitatively-based situation recognition are under intensive development.

These are not felt to be yet within the remit of the book. However, many of these developments will increasingly compete with the methods of this chapter. In any case, there are few 'pure' adaptive control problems waiting to be solved and almost always a judicious combination of techniques will need to be customized to satisfy the inevitable complexities of any particular group of applications.

Because of the intensive ongoing research in the topic of this chapter, an extensive separate bibliography has been provided.

Miniprojects

12.1 There is an assumption in the Model Reference Adaptive Control (MRAC) approach that a system can be forced to behave like a given fixed model by the imposition of suitable input signals. Investigate the truth of this assumption. In particular:

(a) Investigate whether a process that has become too oscillatory in its step response can be corrected in this respect by suitable input signals generated from the differences between process and reference model outputs.

(b) Allowing for all possible ways of generating MRAC forcing signals from the differences between process and reference model outputs, attempt to demonstrate complete equivalence between an MRAC configuration and the 'on-line identification followed by controller redesign' types of adaptive control. (*Hint*: under what circumstances does the difference (do the differences) between process and reference models contain sufficient information to completely model the process?)

12.2 The whole topic of adaptive control, it might be argued, is a branch of statistics. For instance, we may have some initial confidence in a particular process model. On-line data indicate, again within some confidence measure, that the process model has changed.

Design a simple single-loop adaptive controller in which initial confidence in the model and confidence in the on-line-derived model are based from first principles on simple statistical tests, such as may be found in reference W1.

Go on to consider ways of resetting the model, length of data windows and the operational details—all from simple statistical first principles.

12.3 Some years before the ready availability of on-line computer power, the author built an adaptive servomechanism that works as follows. Let $e(t)$ be the error in the closed loop then, by Hölder's inequality

$$\left| \int_0^T e(t)\, dt \right| \leq \int_0^T |e(t)|\, dt$$

with equality during a transient only for an overdamped system. The difference between the two sides of the inequality, suitably normalized, was found to be a realiable quantifier Q of the degree of oscillatoriness. By comparing Q with a desired value for Q and using the difference to vary loop gain, the servomechanism response could be kept reasonably constant despite changing loads. Investigate this approach and comment on the potential of this approach and developments of it for real application.

The Structure and Operation of a Peripheral Interface Adaptor (PIA)

(An 8-bit PIA chip is described here. Note that the structure and programming approach for the 16-bit 68000 series PIA chip is basically similar.)

A PIA is a programmable interface chip. Here we describe the outline of the Motorola MC6820, 40-pin PIA chip. The chip has the following features:

An 8-bit data bus directly connectable to the microprocessor.

Two 8-bit data buses for connection to peripherals. Any of the 16 lines can be allocated as input or output by software control.

Program-controlled interrupt facility.

Logic sufficient to control most types of data transfer.

Before proceeding further we indicate, in Figure A1, how the PIA might fit into a control loop.

As shown in Figure A1, the PIA is fully connected to the system data bus and it has access to some of the address and control lines. Data is transferred between the PIA and the data bus in blocks on one byte.

The PIA has two halves, normally designated A and B. They differ only in respect of their output buffering—the output lines from the B half can be tri-stated (i.e. they can be put into a high-impedance state when not being used or driven).

Each half of the PIA has:

an 8-bit output register, OR;

an 8-bit data direction register, DDR;

an 8-bit control register, CR

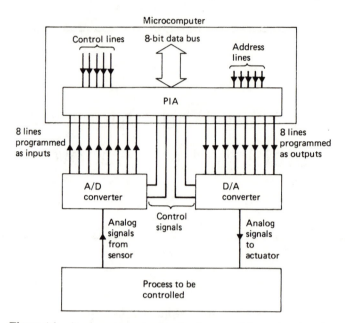

Figure A1   A microcomputer in a basic control loop showing the
function of the registers.

The output register is a buffer store for data that is being transferred to a peripheral device.

Each bit in the DDR register controls the direction of data transfer in one of the 8 lines on the peripheral side of the PIA.

The control register contents determine the PIA characteristics and allow it to respond to timing and interrupt requests.

Each of the three registers can be addressed so that a normal instruction in the microprocessor program can accomplish transfers of the contents of the registers.

A1 Addressing the PIA chip

There are five lines from the address bus to the PIA (Figure A2). Three of the lines are chip-select lines. In small configurations, a three-bit number is often sufficient to identify uniquely which device is being addressed (recall that the PIA has only a partial connection to the address bus). The fourth line from the address bus selects either the A or B side of the PIA. The fifth line selects either the control register CR or the data-direction register DDR/output register OR. (The routing of the selection between registers DDR and OR depends on the contents of the register CR.)

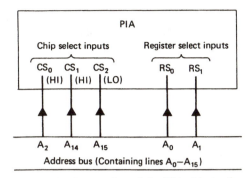

Figure A2   Connection of the PIA chip to the
address bus.

Specifically, the PIA will be selected whenever inputs CS_0, CS_1, CS_2 are high, high, low respectively.

If the connections to the address bus are as in Figure A2, then the PIA is selected from any address of the form:

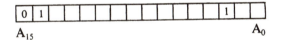

$$A_{15} \qquad\qquad\qquad\qquad\qquad\qquad\qquad\qquad A_0$$

(the blank locations are arbitrary).

Bits A_0 and A_1 then determine which particular register in the PIA is addressed, as shown in Table A1.

Table A1

Address line	Value	Action	Value	Action
$A_1(RS_1)$	0	Side A of the PIA is selected	1	Side B of the PIA is selected
$A_0(RS_0)$	0	Select *either* data direction register DDR *or* output register OR	1	Select control register CR
Bit 2 in register CR	0	Select register DDR	1	Select register OR

Thus, as an example, the hexadecimal address 5007_{16} selects the control register of side B of the PIA. (Notice that a large number of other addresses would select exactly the same register but this causes no problems.)

A2 PIA control lines

PIA ENABLE (E)

This is the only timing signal from the microprocessor to the PIA. The signal is normally obtained from the system clock ϕ_2.

READ/WRITE SIGNAL (R/W)

A 0 on this line enables the transfer of data from MPU to PIA when the E signal occurs provided that the PIA has been selected.

Conversely a 1 on this line enables a transfer of data from the PIA to the data bus on the occurrence of the signal E.

RESET

A 0 on this line resets all bits in the PIA registers to 0.

INTERRUPT REQUEST OUTPUT (IRQA, IRQB)

Through these two outputs (one associated with each half of the PIA) the PIA is able to request the MPU to break normal operation to service an interrupt.

INTERRUPT REQUEST INPUTS (CA1, CA2, CB1, CB2)

The inputs allow peripherals to request interrupt service. In each side of the PIA, four bits control the logic of the interrupt servicing to allow priorities to be established and the sources of the interrupts to be identified.

The control lines CA2, CB2 can also serve as logical outputs to control peripherals. The function of CA2 and CB2 is determined by the content of the respective control registers.

The allocation of functions in the control registers is as shown in Table A2.

Figures A3, A4 are general diagrams showing the overall functions of the PIA chip and its interconnection into the microcomputer configuration.

Table A2   The functions of the control register CRA (CRB is similar)

7	6	5	4	3	2	1	0
IRQA1 Interrupt outputs to MPU	IRQA2	Select CA2 as input (0) or output (1)	Control of signal CA2		Address control OR/DDR	Control of input CA1	

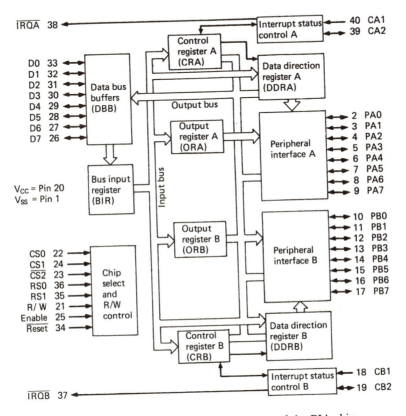

Figure A3   The overall functional diagram of the PIA chip.

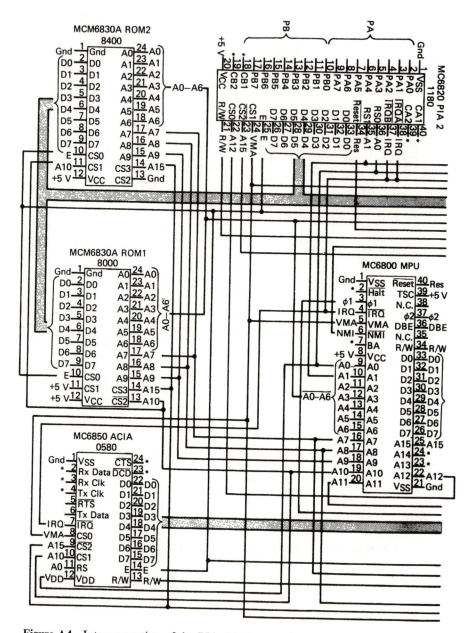

Figure A4   Interconnection of the PIA chip into a microcomputer configuration.

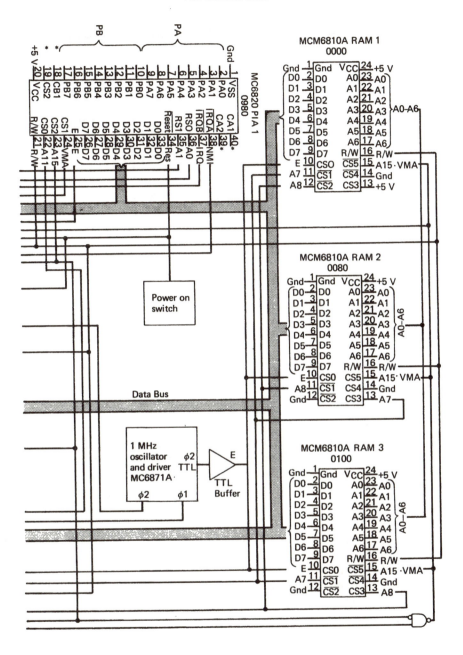

Personal Computers on the VMEbus for Control Implementation

IBM-compatible computers, based on the Intel Processors 80286, 80386 and 80486 dominate the personal computer market and their PC/AT architecture forms a platform for a huge range of software, much of it highly suitable for real-time control applications. However, the physical construction of personal computers is best suited to office or perhaps laboratory environments and there are some doubts about their suitability for hostile industrial installation and their ability to be configured and closely interconnected to meet stringent reliability requirements.

One solution is to connect PC/AT systems to a VMEbus system. The VMEbus system complements the PC/AT architecture nicely, providing a robust, reliable, bus system with high memory address space and high-performance multi-processor capabilities.

The combination provides a PC/AT software environment in a VMEbus hardware confirmation as shown in Figure B1.

Refer back also to case histories C and D in Chapter 11. In both cases, hardware solutions similar to that described above were implemented.

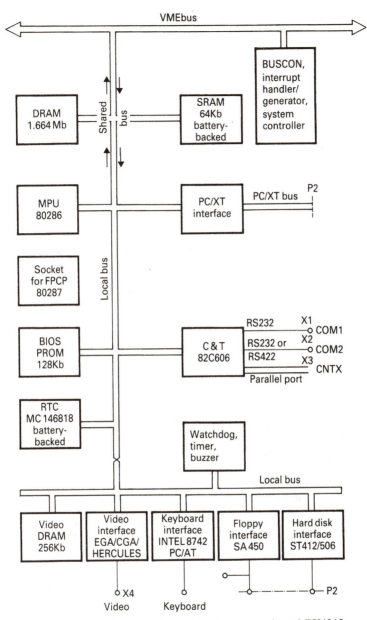

Figure B1   Philips VMEbus PC/AT processor board PX4010.

Tables of Transform Pairs

Time function $f(kT)$	$\mathscr{Z}$ transform $F(z) = \sum\limits_{k=0}^{\infty} f(kT)z^{-k}$
$\alpha f_1(kT) + \beta f_2(kT)$ where α, β are scalar multipliers	$\alpha F_1(z) + \beta F_2(z)$
$e^{-akT}f(kT)$	$F(e^{aT}z)$
1 for $k \geqslant 0$ 0 otherwise	$\dfrac{z}{z-1}$
1 for $k = 0$ 0 otherwise	1
1 for $k = p$ 0 otherwise	z^{-p}
kT	$\dfrac{Tz}{(z-1)^2}$
$(kT)^2$	$\dfrac{T^2 z(z+1)}{(z-1)^3}$
$(kT)^3$	$\dfrac{T^3 z(z^2 + 4z + 1)}{(z-1)^4}$
e^{-akT}	$\dfrac{z}{z - e^{-aT}}$

Time function $f(kT)$	$\mathscr{Z}$ transform $F(z)$
a^k	$\dfrac{z}{z-a}$
$1-e^{-akT}$	$\dfrac{z(1-e^{-aT})}{(z-1)(z-e^{-aT})}$
$e^{-akT}-e^{-bkT}$	$\dfrac{z(e^{-aT}-e^{-bT})}{(z-e^{-aT})(z-e^{-bT})}$
$\sin(akT)$	$\dfrac{z\sin aT}{z^2-(2\cos aT)z+1}$
$\cos(akT)$	$\dfrac{z(z-\cos aT)}{z^2-2(\cos aT)z+1}$
$kTe^{-akT}=kTc^k$ where $c=e^{-aT}$	$\dfrac{Tz\,e^{-aT}}{(z-e^{-aT})^2}=\dfrac{Tzc}{(z-c)^2}$
$f(kT+T)$	$zF(z)-zf(0)$
$f(kT+pT)$	$z^pF(z)-z^pf(0)-\cdots-zf[(p-1)T]$

$$f(0) = \lim_{z\to\infty}F(z)$$

$$\lim_{k\to\infty}f(kT) = \lim_{z\to1}(z-1)F(z)$$

(provided that the poles of $(z-1)F(z)$ are inside the unit circle)

Laplace transform	Corresponding $\mathscr{Z}$ transform
$\dfrac{1}{s}$	$\dfrac{z}{z-1}$
$\dfrac{1}{s^2}$	$\dfrac{Tz}{(z-1)^2}$
$\dfrac{1}{s^3}$	$\dfrac{T^2z(z+1)}{2(z-1)^3}$
$\dfrac{1}{s^4}$	$\dfrac{T^3z(z^2+4z+1)}{6(z-1)^4}$

Laplace transform	Corresponding $\mathscr{Z}$ transform
$\dfrac{1}{s+a}$	$\dfrac{z}{z-\mathrm{e}^{-aT}}$
$\dfrac{1}{(s+a)^2}$	$\dfrac{Tz\,\mathrm{e}^{-aT}}{(z-\mathrm{e}^{-aT})^2}$
$\dfrac{1}{(s+a)^3}$	$\dfrac{T^2}{2}\mathrm{e}^{-aT}\dfrac{z(z+\mathrm{e}^{-aT})}{(z-\mathrm{e}^{-aT})^3}$
$\dfrac{a}{s(s+a)}$	$\dfrac{z(1-\mathrm{e}^{-aT})}{(z-1)(z-\mathrm{e}^{-aT})}$
$\dfrac{a}{s^2(s+a)}$	$\dfrac{Tz}{(z-1)^2}-\dfrac{(1-\mathrm{e}^{-aT})z}{a(z-1)(z-\mathrm{e}^{-aT})}$
$\dfrac{a^2}{s(s+a)^2}$	$\dfrac{z}{z-1}-\dfrac{z}{z-\mathrm{e}^{-aT}}-\dfrac{\mathrm{e}^{-aT}aTz}{(z-\mathrm{e}^{-aT})^2}$
$\dfrac{b-a}{(s+a)(s+b)}$	$\dfrac{z(\mathrm{e}^{-aT}-\mathrm{e}^{-bT})}{(z-\mathrm{e}^{-aT})(z-\mathrm{e}^{-bT})}$
$\dfrac{(b-a)s}{(s+a)(s+b)}$	$\dfrac{(b-a)z^2-(\mathrm{e}^{-aT}b-\mathrm{e}^{-bT}a)z}{(z-\mathrm{e}^{-aT})(z-\mathrm{e}^{-bT})}$
$\dfrac{ab}{s(s+a)(s+b)}$	$\dfrac{z}{z-1}+\dfrac{b}{a-b}\dfrac{z}{z-\mathrm{e}^{-aT}}$
	$-\dfrac{a}{a-b}\dfrac{z}{z-\mathrm{e}^{-bT}}$
$\dfrac{a}{s^2+a^2}$	$\dfrac{z\sin aT}{z^2-2z\cos aT+1}$
$\dfrac{s}{s^2+a^2}$	$\dfrac{z^2-z\cos aT}{z^2-2z\cos aT+1}$
$\dfrac{s}{(s+a)^2}$	$\dfrac{z(z-\mathrm{e}^{-aT}(1+aT))}{(z-\mathrm{e}^{-aT})^2}$
$\dfrac{1}{s-(\ln a)/T}$	$\dfrac{z}{z-a}$
e^{-ksT}	z^{-k}

Bibliography—General

A1   Ahmed, N. U., *Elements of Finite-dimensional Systems and Control Theory*, Longman, 1988.

A2   Ahmed, N. U., *Semigroup Theory with Applications to Systems and Control*, Longman, 1991.

A3   Amouroux, M. (ed.), *Control of Distributed Parameter Systems*, Pergamon Press, 1990.

A4   Anderson, T. (ed.), *Dependability of Resilient Computers*, Blackwell Scientific Publications, 1989.

A5   Anderson, T. (ed.), *Safe and Secure Computing Systems*, Blackwell Scientific Publications, 1989.

A6   Anderson, R. L., *Robot Ping-pong Player: Experiment in Real-time Intelligent Control*, MIT Press, 1988.

A7   Ausländer, D. M. and Tham, C. H., *Real Time Software for Control*, Prentice-Hall, 1989.

B1   Bastin, G. and Dochain, D., *On-line Estimation and Adaptive Control of Bioreactors*, Elsevier, 1990.

B2   Bennett, S., *Real Time Computer Control: An Introduction*, Prentice-Hall, 1988.

B3   Bennett, S. and Linkens, D. A. (eds.), *Computer Control of Industrial Processes*, Peter Peregrinus, Stevenage, UK, 1982.

B4   Bennett, S. and Virk, G. S. (eds.), *Computer Control of Real Time Processes*, Peter Peregrinus, Stevenage, UK, 1990.

B5   Bertil, T., 'New PID parameter tuning methods for industrial applications', *First IFAC Symposium on Design Methods for Control Systems*, Zürich, 1991.

B6   Blakelock, J. H., *Automatic Control of Aircraft and Missiles*, Wiley, 1991.

B7   Bollinger, J. D. and Duffie, U. A., *Computer Control of Machines and Processes*, Addison-Wesley, 1988.

B8   Bolton, W., *Industrial Control and Instrumentation*, Longman, 1991.

B9   Boullart, L. (ed.), *Industrial Process Control Systems: Reliability, Availability and Maintainability*, Pergamon Press, 1989.

B10   Brogan, W. L., *Modern Control Theory*, Prentice-Hall, 1991.

B11   Brooks, R. W., Millette, P. and Mitchell, T., *Use of Realtime and Relational Databases for CIM Implementation in Computer Applications in Chemical Engineering*, Elsevier Science Publishers, Amsterdam, 1990.

B12   Bryson, A. E. and Ho, Y-C., *Applied Optimal Control: Optimization Estimation and Control*, Hemisphere Publishing Corporation, USA, 1988.

B13   Butkovskiy, A. G., *Phase Portraits of Control Dynamical Systems*, Kluwer Academic Publishers, 1991.

C1   Carr, J. J., *Data Acquisition and Control: Microcomputer Applications for Scientists and Engineers*, TAB Books, USA, 1988.

C2   Cattermole, K. W., *Principles of Pulse Code Modulation*, Iliffe, UK, 1969.

C3   Chen, G. (ed.), *Distributed Parameter Control Systems: Trends and Applications*, Dekker, 1990.

C4   Clements, R. R., *Statistical Process Control and Beyond*, RE Krieger Publishing Co., USA, 1988.

C5   Christensen, G. S., *Optimal Control of Distributed Nuclear Reactors*, Plenum Publishing Company, 1990.

C6   Clayton, G. B., *Data Converters*, Macmillan, London, 1982.

C7   Coulbeck, B. and Orr, C-H., *Computer Applications in Water Supply*, Research Studies Press, 1988.

D1   Dahlin, E. B., 'Designing and tuning digital controllers', *Instrumentation Control Systems*, **41**(6), 77–83, 87–91, 1968.

D2   Dale-Harris, L., *Introduction to Feedback Systems*, Wiley, New York, 1961.

D3   Daniels, B. K. (ed.), *International Federation of Automatic Control, Safety of Computer Control Systems: Workshop Proceedings*, Pergamon Press, 1990.

D4   Datta, B. N. (ed.), *Linear Algebra in Signals, Systems and Control*, Society for Industrial & Applied Maths, USA, 1988.

D5   Debs, A. S., *Modern Power Systems Control and Operation*, Kluwer-Nijhoff, USA, 1988.

D6   Dehnad, K. (ed.), *Quality Control, Robust Design and the Taguchi Method*, Brooks-Cole, 1988.

D7   Desrochers, A. A., *Modelling and Control of Automated Manufacturing Systems*, IEEE Computer Society Press, USA, 1989.

D8   Dote, Y., *Servomotor and Motion Control Using Digital Signal Processors*, Prentice-Hall, 1990.

D9   Drouin, M., *Control of Complex Systems: Methods and Technology*, Plenum Publishing Company, 1990.

E1   Efstathiou, J., *Expert Systems in Process Control*, Longman, 1989.

E2   Ehrenberger, W. D. (ed.), *International Federation of Automatic Control, Safety of Computer Control Systems: Workshop Proceedings*, Pergamon Press, 1988.

F1   Fargeon, C., *Digital Control Systems*, Chapman & Hall, 1989.

F2   Findeisen, W., Bailey, F. N., Brdyś, M., Malinowski, K., Tatjewski, P. and Woźniak, A., *Control and Coordination in Hierarchical Systems*, International Series on Applied Systems Analysis, Wiley, Chichester, 1980.

F3   Fleming, P. J., (ed.), *Parallel Processing in Control: The Transputer and Other Architectures*, Peter Peregrinus, Stevenage, UK, 1988.

F4   Franklin, G. F., Powell, J. D. and Workman, M. L., *Digital Control of Dynamic Systems*, Addison-Wesley, 1990.

F5   Furuta, K., *State Variable Methods in Automatic Control*, Wiley, 1988.

G1   Genser, R. (ed.), *International Federation of Automatic Control, Safety of Computer Control Systems: Workshop Proceedings*, Pergamon Press, 1990.

G2   Giorgi, B. (ed.), *Personal Computers in Industrial Control*, Butterworth, 1988.

G3   Goff, K. W., 'Dynamics in direct digital control', *Journal of the Instrumentation Society of America*, I 45–49, II 44–54, 1966.

G4   Gopal, M., *Digital Control Engineering*, Halsted Press, 1988.

G5   Grimble, M. J. and Johnson, M. A., *Optimal Control and Stochastic Estimation: Theory and Applications*, Wiley, 1988.

G6   Gupta, M. M. and Yamakawa, T. (ed.), *Fuzzy Logic in Knowledge-based Systems, Decision and Control*, North-Holland Publishing Company, 1988.

H1   Harris, C. J. (ed.), *Application of Artificial Intelligence to Command and Control Systems*, Peter Peregrinus, Stevenage, UK, 1988.

H2   Herman, S. L. and Alerich, W. N., *Industrial Motor Control*, Delmar Publishing Company, USA, 1990.

I1   Isidori, A. (ed.), *Nonlinear Control Systems Design: Symposium Proceedings*, Pergamon Press, 1990.

J1   Joseph, B., *Real-time Personal Computing For Data Acquisition and Control*, Prentice-Hall, 1988.

K1   Kalani, G., *Microprocessor Based Distributed Control Systems*, Prentice-Hall, 1989.

K2   Keats, B. and Hubele, N. F. (eds.), *Statistical Process Control in Automated Manufacturing*, Dekker, 1988.

K3   Keviczky, L., *Mathematics and Control Engineering of Grinding Technology*, Kluwer Academic Publishers, 1989.

K4   Khoo, M. C. K. (ed.), *Modelling and Parameter Estimation in Respiratory Control*, Plenum Publishing Company, 1990.

K5   Knowles, J. B., *Simulation and Control of Electrical Power Stations*, Research Studies Press, 1990.

K6   Knuth, E. and Rodd, M. G. (eds.), *International Federation of Automatic Control, Distributed Data Bases in Real-time Control Proceedings: Workshop Proceedings*, Pergamon Press, 1990.

K7   Koenig, D. M., *Control and Analysis of Noisy Processes*, Prentice-Hall, 1991.

K8   Kreyszig, E., *Advanced Engineering Mathematics*, 6th edition, John Wiley, 1988.

K9   Kučera, V., *Analysis and Design of Discrete Linear Control Systems*, Prentice-Hall, 1989.

K10 Kuo, B. C., *Automatic Control Systems*, Prentice-Hall, 1990.

L1   Lain, P. K., *Digital System Design Using PLC Devices*, Prentice-Hall, 1990.

L2   Landau, I. D., *Systems Identification and Control Using PIM + Software*, Prentice-Hall, 1989.

L3   Leigh, J. R., *Functional Analysis and Control Theory*, Volume 156 in the International Series, Mathematics in Science and Engineering, Academic Press, 1980.

L4   Leigh, J. R., *Modelling and Simulation*, IEE Topics in Control Series, 1, Peter Peregrinus, Stevenage, UK, 1983.

L5   Leigh, J. R., *Essentials of Nonlinear Control Theory*, IEE Topics in Control Series, 2, Peter Peregrinus, Stevenage, UK, 1983.

L6   Leigh, J. R. and Thoma, M. (eds.), *Special Issue of IEE Proceedings, Part D, Control in Bioprocessing*, **133**, D, No. 5, 1986.

L7   Leigh, J. R. (ed.), *Modelling and Control of Fermentation Processes*, Peter Peregrinus, Stevenage, UK, 1987.

L8   Leigh, J. R., *Temperature Measurement and Control*, Peter Peregrinus, Stevenage, UK, 1988.

L9   Leigh, J. R., *Control Theory: A Guided Tour*, Peter Peregrinus, Stevenage, UK, 1992.

L10 Leondes, C. T. (ed.), *Control and Dynamic Systems: Advances in Theory and Applications*, Academic Press, some 40 volumes published at least annually since 1964.

L11 Luyben, W. L., *Process Modelling, Simulation and Control for Chemical Engineers*,

McGraw Hill Book Company, 1990.

M1   Macleod, I. M. and Heher, A. D. (eds.), *Software for Computer Control: Symposium Proceedings*, Pergamon Press, 1989.

M2   McAvoy, T. J. (ed.), *Model-Based Process Control*, Pergamon Press, 1989.

M3   McClean, D., *Automatic Flight Control Systems*, Prentice-Hall, 1990.

M4   McGhee, J. (ed.), *Knowledge-based Systems for Industrial Control*, Peter Peregrinus, Stevenage, UK, 1990.

M5   McGreavy, C. (ed.), *Dynamics and Control of Chemical Reactors and Distillation*, Pergamon Press, 1988.

M6   Meirovitch, L., *Dynamics and Control of Structures*, Wiley, 1990.

M7   Middleton, R. and Goodwin, G. C., *Digital Control and Estimation*, Prentice-Hall, 1990.

M8   Min, J. L. and Schrage, J. J., *Designing Analogue and Digital Control Systems*, E. Horwood, 1988.

M9   Minton, S., *Learning Search Control Knowledge: An Explanation Based Approach*, Kluwer-Nijhoff, USA, 1988.

M10 Motus, L. and Narita, S. (eds.), *International Federation of Automatic Control, Distributed Computer Control Systems: Workshop Proceedings*, Pergamon Press, 1990.

N1   Naidu, D. S., *Singular Perturbation Methodology in Control Systems*, Peter Peregrinus, Stevenage, UK, 1988.

N2   Nelson, R. C., *Flight Stability and Automatic Control*, McGraw Hill Book Company, 1990.

N3   Newell, R. B. and Lee, P. L., *Applied Process Control: A Case Study*, Prentice-Hall, 1989.

N4   Nishimura, T. (ed.), *Automatic Control in Aerospace: International Symposium*, Pergamon Press, 1990.

O1   Ogata, K., *Discrete Time Control Systems*, Prentice-Hall, 1987.

O2   Omstead, D. R., *Computer Control of Fermentation Processes*, CRC Press, 1989.

O3   O'Reilly, J. O. (ed.), *Multivariable Control for Industrial Applications*, Peter Peregrinus, Stevenage, UK, 1987.

P1   Parr, E. A., *Industrial Control Handbook*, BSP Professional: Butterworth–Heinemann, May 1989.

P2   Pedrycz, W., *Fuzzy Control and Fuzzy Systems*, Research Studies Press, 1989.

P3   Phillips, C. L. and Nagle, H. T., *Digital Control System Analysis and Design*, Prentice-Hall, 1989.

P4   Phillips, C. L. and Orauc, B. T., *Control Systems Programmes: IBM Personal Computer Version*, Prentice-Hall, 1990.

P5   Plannstiel, D. and Isermann, R., 'Modelling, simulation and digital control of the combustion of a furnace', *Proceedings of the IEE Conference, Control 91*, Edinburgh, 1991.

P6   Pontryagin, L. S., Boltyanskii, V. G., Gamkrelidze, R. V. and Mishchenko, E. F., *The Mathematical Theory of Optimal Processes*, Pergamon Press, 1964.

P7   Prett, D. M. (ed.), *Process Control: Workshop Proceedings*, Butterworth, 1990.

R1   Ray, W. H., *Advanced Process Control*, Butterworth, 1990.

R2   Raynal, M. and Helary, J-M, *Synchronization and Control of Distributed Systems and Programmes*, Wiley, 1990.

R3   Rijnnsdorp, J. E. (ed.), *Dynamics and Control of Chemical Reactors, Distillation Columns and Batch Processes*, Pergamon Press, 1990.

R4   Rijnnsdorp, J. E., *Integrated Process Control and Automation*, Elsevier, Amsterdam, 1991.

R5   Rodd, M. G. and D'Epinay, T. L. (eds.), *Distributed Computer Control Systems*, Pergamon Press, 1989.

S1   Seborg, D. E., *Process Dynamics and Control*, Wiley, 1989.

S2   Shannon, C. E. and Weaver, W., *The Mathematical Theory of Communication*, University of Illinois Press, Urbana, USA, 1972.

S3   Shinskey, F. G., *Process Control Systems*, McGraw Hill Book Company, 1988.

S4   Sigmund, K. and Kurzhanski, A. S. (eds.), *Evolution and Control in Biological Systems*, Kluwer Academic Publishers, USA, 1989.

S5   Siljak, D. D., *Decentralized Control of Complex Systems*, Academic Press, 1990.

S6   Slotine, J-J. E. and Weiping, L., *Applied Nonlinear Control*, Prentice-Hall, 1990.

S7   Sommerville, I., *Software Engineering*, 3rd Edition, Addison-Wesley, 1989.

S8   Spong, M. W. and Vidyasagar, M., *Robot Dynamics and Control*, Wiley, 1989.

S9   Stock, M., *Artificial Intelligence in Process Control*, McGraw Hill Book Company, 1988.

S10 Stoecker, W. F. and Stoecker, P., *Microcomputer Control of Thermal and Mechanical Systems*, Van Nostrand Reinhold, 1988.

S11 Sutton, R., *Modelling Human Operators in Control System Design*, Research Studies Press, John Wiley, 1990.

T1   Takahashi, Y., Rabins, M. J. and Ausländer, D. M., *Control and Dynamic Systems*, Addison-Wesley, Reading, Massachusetts, 1970.

T2   Thaler, G. J., *Automatic Control Systems*, West Publishing Company, USA, 1988.

T3   Tooley, M. H., *Bus Based Industrial Control*, Heinemann, 1988.

T4   Triebel, W. A. and Singh, A., *68000 and 68020, Microprocessors, Hardware, Software and Interfacing Techniques*, Prentice-Hall, 1991.

T5   Truxal, J. G., *Automatic Feedback Control System Synthesis*, McGraw Hill Book Company, New York, 1955.

T6   Tzafestas, S. G. and Pal, J. K. (eds.), *Real Time Microcomputer Control of Industrial Processes*, Kluwer Academic Publishers, 1990.

V1   Vardulakis, A. I. G., *Linear Multivariable Control: Algebraic Analysis and Synthesis*, Wiley, 1991.

W1   Wardlaw, A. C. *Practical Statistics*, John Wiley, 1985.

W2   Warwick, K. and Pugh, A. (eds.), *Robot Control: Theory and Applications*, Peter Peregrinus, Stevenage, UK, 1988.

W3   Warwick, K. and Rees, D. (eds.), *Industrial Digital Control Systems*, Peter Peregrinus, Stevenage, UK, 1988.

W4   Warwick, K. and Tam, M. T., *Fail-safe Control Systems: Applications and Emergency Management*, Chapman & Hall, 1990.

W5   Wetherill, G. B. and Brown, D. W., *Statistical Process Control: Theory and Practice*, Chapman & Hall, 1990.

W6   Wheeler, C., *Microprocessors and Industrial Control*, Hutchinson Educational, 1988.

Z1   Zalewski, J. and Ehrenberger, W. D. (eds.), *Hardware and Software for Real Time Process Control*, North-Holland Publishing Company, 1988.

Z2   Ziegler, J. G. and Nichols, N. B., 'Optimum settings for automatic controllers', *Transactions of the ASME*, **64**(11), 759, 1942.

Z3   Zikic, A. M., *Practical Digital Control*, E. Horwood, 1989.

Z4   Zinober, A. S. I. (ed.), *Deterministic Control of Uncertain Systems*, Peter Peregrinus, Stevenage, UK, 1990.

References—Adaptive, Self-Tuning and Robust Control

In chapter 12, we have introduced H^∞, DMC, MRAC, and many other techniques that are still under very active development. Further information on these approaches will be found in the following bibliography.

Other techniques that we did not find space to include, for example, the Internal Model Control (IMC) approach, the Horowitz Quantitative Feedback Theory (QFT), the μ method, and the Extended Self-Tuning Regulator (ESTR) approaches will be found described in references cited in the following list.

Ackerman, J. and Wilde, A., 'Robustness of sampled data control systems', *European Control Conference*, Grenoble, France, 1991.

Agarwal, M. and Seborg, D., 'Self-tuning controllers for nonlinear systems', *Automatica*, **23**, 209–214, 1987.

Åström, K. J., 'Theory and applications of adaptive control—A survey', *Automatica*, **19**, 471–486, 1983.

Åström, K. J., 'Design and implementation of digital and adaptive controllers', *First IFAC Symposium on Design Methods of Control Systems*, Zürich, 1991.

Åström, K. J. and Hagglund, T., *Automatic Tuning of PID Controllers*, chapter 5 'Auto-tuning', 105–132, ISA, 1989.

Åström, K. J. and Wittenmark, B., *Adaptive Control*, Addison-Wesley, 1989.

Bada, A. T., 'Robust brake control for a heavy-duty truck', *IEEE Proceedings D Control Theory and Applications*, **134**(1), 1–8, 1987.

Bai, E. W. and Dasgupta, S., 'Robust control design of sampled systems', *International Journal of Systems Science*, **21**(5), 985–992, 1990.

Bailey, F. N., Panzer, D. and Gu, G., '2 Algorithms for frequency domain design of robust control systems', *International Journal of Control*, **48**(5), 1787–1806, 1988.

Bamani, A. H. and Iskanderani, A. I., 'Robust approximate non-interacting control design for a class of non-linear stochastic systems', *Optimal Control Applications & Methods*, **10**(3), 275–283, 1989.

Banda, S. S., Yeh, H. H. and Heise, S. A., 'Robust control of uncertain systems with combined H-infinity and LQG optimizations', *International Journal of Systems Science*, **22**(1), 85–96, 1991.

Barmish, B. R. and Khargonezar, P. P., 'Robust stability of feedback control systems with uncertain parameters and unmodeled dynamics', *Mathematics of Control Signals and Systems*, **3**(3), 197–210, 1990.

Bastin, G. and Dochain, D., *On-line Estimation and Adaptive Control of Bioreactors*, Elsevier, Amsterdam, 1990.

Beale, S. and Shafai, B., 'Robust control system design with a proportional integral observer', *International Journal of Control*, **50**(1), 97–111, 1989.

Bendotti, P. and Msaad, M., 'Generalised predictive adaptive control for autopilot design', *European Control Conference*, Grenobe, France, 1991.

Bengtsson, G. and Egerdt, B., 'Experiences with self-tuning control in the process industry', *IEEE Decision and Control Conference*, 132–140, 1982.

Berger, C. S., 'Robust pole placement algorithm for adaptive control', *IEEE Proceedings D Control Theory and Applications*, **135**(6), 493–498, 1988.

Berger, C. S., 'Robust control of discrete systems', *IEEE Proceedings D Control Theory and Applications*, **136**(4), 165–170, 1989.

Bernstein, D. S. and Hollot, C. V., 'Robust stability for sampled data control systems', *Systems & Control Letters*, **13**(3), 217–226, 1989.

Bhat, J., Chidambaram, M. and Madhavan, K. P., 'Robust control of batch reactors', *Chemical Engineering Communications*, **87**, 195–204, 1990.

Bitmead, R. R., *Adaptive Optimal Control: The Thinking Man's GPC*, Prentice-Hall, Englewood Cliffs, New Jersey, 1990.

Bontsema, J., Curtain, R. F. and Schumacher, J. M., 'Robust control of flexible structures—A case study', *Automatica*, **24**(2), 177–186, 1988.

Bristol, E. H., 'Pattern recognition: An alternative to parameter identification in adaptive control', *Automatica*, **13**, 197–202, 1977.

Bristol, E. H., 'The design of industrially useful adaptive controllers', *ISA Transactions*, **22**(3), 17–25, 1983.

Bucholt, F., Kummel, M., 'A multivariable self-tuning regulator to control a double effect evaporator', *Automatica*, **17**(5), 737–743, 1981.

Bueno, S. S. and Favier, G., 'Self tuning PID controllers: A review', *First IFAC Symposium on Design Methods of Control Systems*, Zürich, 1991.

Campo, P. J., Morari, M., 'Robust control of processes subject to saturation nonlinearities', *Computers & Chemical Engineering*, **14**(4–5), 343–358, 1990.

Carmon, A., 'Applying self-tuning control to plant', *Control and Instrumentation*, 81–83, 1986.

Chalam, V. V., *Adaptive Control System: Techniques and Applications*, Marcel Dekker, New York, 1987.

Chen, B. S., Lin, C. L. and Hsiao, F. B., 'Robust observer based control of a vibrating beam', *Proceedings of the Institution of Mechanical Engineers Part C—Journal of Mechanical Engineering Science*, **205**(2), 77–89, 1991.

Chen, B. S., Wang, S. S. and Lu, H. C., 'Robust stability of perfect model matching control system', *International Journal of Systems Science*, **20**(5), 889–906, 1989.

Chen, G. and Defigueiredo, R. J. P., 'On robust stabilization of nonlinear control systems', *Systems & Control Letters*, **12**(4), 373–379, 1989.

Chen, Y. H., 'Decentralized robust control system design for large scale uncertain systems', *International Journal of Control*, **47**(5), 1195–1205, 1988.

Chen, Y. H., 'Modified adaptive robust control system design', *International Journal of Control*, **49**(6), 1869–1882, 1989.

Chen, Y. H., 'Robust control system design—Non-adaptive versus adaptive', *International Journal of Control*, **51**(6), 1457–1477, 1990.

Chen, Y. H., 'Adaptive robust control system design—Using information related to the bound of uncertainty', *Control Theory and Advanced Technology*, **7**(1), 31–53, 1991.

Chen, Y. H. and Pandey, S., 'Robust control strategy for take off performance in a windshear', *Optimal Control Applications & Methods*, **10**(1), 65–79, 1989.

Cheung, Louis, S., 'Performance of two industrial self-tuning controllers', *ISA Control Exposition*, 686–689, 1987.

Cheung, Louis, S., 'A new automated optimal tuning strategy for a PID controller', *Proceedings of ISA 87 International Conference and Exhibition*, 1487–1495, 1987.

Chidambaram, M. and Rao, Y. S. N. M., 'Robust control of a semibatch chemical reactor', *Journal of Loss Prevention in the Process Industries*, **3**(3), 330–332, 1990.

Chizeck, H. J., Crago, P. E. and Kofman, L. S., 'Robust closed loop control of isometric muscle force using pulsewidth modulation', *IEEE Transactions on Biomedical Engineering*, **35**(7), 510–517, 1988.

Chou, J. H., 'Control systems design for robust role assignment in a specified circular design', *Control Theory and Advanced Technology*, **7**(2), 237–245, 1991.

Choura, S. and Jayasuriya, S., 'Robust finite time settling control of distributed parameter systems', *International Journal of Control*, **52**(6), 1425–1453, 1990.

Clarke, D. W., *Introduction to Self-Tuning Controllers, Self-Tuning and Adaptive Control: Theory and Applications*, Peter Peregrinus, London, 1981.

Clarke, D. W., 'Generalised predictive control', *Colloquium on 'Advances in Adaptive Control'*, IEE Computing and Control Division, London, 1986.

Clarke, D. W., Mohtadi, C. and Tuffs, P. S., 'Generalised predictive control—Part 1., The basic algorithm', *Automatica*, **231**(2), 137–148, 1987.

Cluett, W. R., Shah, S. L. and Fisher, D. G., 'Robust design of adaptive control systems using conic sector theory', *Automatica*, **23**(2), 221–224, 1987.

Colgate, J. E. and Hogan, N., 'Robust control of dynamically interacting systems', *International Journal of Control*, **48**(1), 65–88, 1988.

Correa, G. O., 'A system identification problem motivated by robust control', *International Journal of Control*, **50**(2), 575–602, 1989.

Cutler, C. R. and Ramaker, B. L., 'Dynamic matrix control—A computer control algorithm', *ACC*, WP5-B, 1980.

Davison, E. J., 'The robust control of a servomechanism problem for linear time-invariant multivariable systems', *IEEE Transactions on Automatic Control*, **21**, 25–34, 1976.

Davison, E. J. and Ferguson, I. J., 'The design of controllers for the multivariable robust servomechanism problem using parameter optimisation methods', *IEEE Transactions on Automatic Control*, **26**, 93–110, 1981.

Dawson, D. M., Qu, Z., Lewis, F. L. and Dorsey, J. F., 'Robust control for the tracking of robot motion', *International Journal of Control*, **52**(3), 581–595, 1990.

Delasen, M., 'A robust indirect discrete adaptive control approach based on passivity results for nonlinear systems', *Computers & Mathematics with Applications*, 15(5), 389–403, 1988.

Doraiswami, R. and Bordry, F., 'Robust 2-time level control strategy for sampled-data servomechanism problem', *International Journal of Systems Science*, 18(12), 2261–2277, 1987.

Dorato, P. (ed.), *Robust Control*, IEEE Press, 1987.

Dorato, P. and Yedavalli, R. K. (eds.), *Recent Advances in Robust Control: Selected Conference Papers*, IEEE Press, 1990.

Dorling, C. M. and Zinober, A. S. I., 'Robust hyperplane design in multivariable variable structure', *International Journal of Control*, 48(5), 2043–2054, 1988.

Doyle, J. C., 'Quantitative feedback theory (QFT) and robust control', *Proceedings of the ACC*, FA-10, 1691–1698, 1986.

Dumont, G. A., 'Self-tuning control of a chip refiner motor load', *Automatica*, 18(3), 307–314, 1982.

Economou, C., Morari, M. and Palsson, B., 'Internal model control.5. Extension to nonlinear systems', *Industrial Engineering and Chemical Process Design and Development*, 25, 403–411, 1986.

Egardt, B., *Stability of Adaptive Controllers*, Springer-Verlag, Berlin, 1979.

Elshal, S. M., 'Digital modelling and robust control of the glucose homeostasis system', *International Journal of Systems Science*, 20(4), 575–586, 1989.

Elshal, S. M., 'Microcomputer based robust digital control for the glucoregulatory system', *International Journal of Systems Science*, 22(7), 1279–1293, 1991.

Elshal, S. M. and Mahmoud, M. S., 'Microcomputer based robust control for slow time delay processes—Pole placement approach', *International Journal of Systems Science*, 20(12), 2395–2401, 1989.

Epton, J., 'Adaptive and self-tuning control', *Control and Instrumentation*, 61–64, 1989.

Feuer, A. and Goodwin, G. C., 'Integral action in robust adaptive control', *IEEE Transactions on Automatic Control*, 34(10), 1082–1085, 1989.

Foo, Y. K., 'Quantitative design of robust multivariable control systems', *IEE Proceedings D Control Theory and Applications*, 135(6), 404, 1988.

Fossen, T. I. and Balchen, J. G., 'Modelling and non-linear self-tuning robust trajectory control of an autonomous underwater vehicle', *Modelling Identification and Control*, 9(4), 165–177, 1988.

Fradkov, A. L., 'A survey of the works on the comparison of adaptive algorithms', *Proceedings of the IFAC Workshop*, 25–28, Tbilisi, USSR, 1989.

Francis, B. A. and Wonham, W. M., 'The internal model principle of control theory', *Automatica*, 12, 457–465, 1976.

Fu, L. C., 'New approach to robust model reference adaptive control for a class of plants', *International Journal of Control*, 53(6), 1359–1375, 1991.

Fung, K. Y. and Tam, H., 'Robust confidence-intervals for comparing several treatment groups to a control group', *Statistician*, 37(4–5), 387–399, 1988.

Furuta, K., 'Alternative robust servo-control system and its digital-control', *International Journal of Control*, 45(1), 183–194, 1987.

Garcia, C. and Morari, M., 'International model control—1. A unifying review and some new results', *Industrial Engineering and Chemical Process Design and Development*, 21, 308–323, 1982.

Gawthrop, P. J., 'An introduction to continuous time self-tuning control', *Colloquium on 'Advances in Adaptive Control'*, IEE Computing and Control Division, London, 1986.

Gawthrop, P. J., *Continuous-time Self-tuning Control*, Research Studies Press, 1990.

Giri, F., Dion, J. M., Dugard, L. and Msaad, M., 'Robust pole placement direct adaptive control', *IEEE Transactions on Automatic Control*, **34**(3), 356–359, 1989.

Goodwin, G. C. (ed.), *Robust Adaptive Control: Workshop Proceedings*, Pergamon Press, 1989.

Goodwin, G. C., Hill, D. J. and Palaniswami, M., 'A perspective on convergence of adaptive control algorithms', *Automatica*, **20**, 519, 1984.

Grimble, M. J., 'An LQG approach to self-tuning control', *Fourth Workshop on 'Self-Tuning and Adaptive Control'*, IEE Computing and Control Division, Oxford, 1987.

Grimble, M. J., 'H-infinity robust controller for self-tuning control applications, 1, Controller design', *International Journal of Control*, **46**(4), 1429–1444, 1987.

Grimble, M. J., 'H-infinity robust controller for self-tuning, control applications, 2, Self-tuning and robustness', *International Journal of Control*, **46**(5), 1819–1840, 1987.

Grimm, W. M., Lee, P. L. and Callaghan, P. J., 'Practical robust predictive control of a heat exchange network', *Chemical Engineering Communications*, **81**, 25–53, 1989.

Gupta, M. M. (ed.), *Adaptive Methods for Control System Design*, IEEE Press, New York, 1986.

Gutman, P. O., Levin, H., Neumann, L., Sprecher, T. and Venezia, E., 'Robust and adaptive control of a beam deflector', *IEEE Transactions on Automatic Control*, **33**(7), 610–619, 1988.

Halme, A. and Visala, A., 'Combining symbolic and numerical information in modelling the state of biotechnological processes', *Proceedings of the First European Control Conference*, France, 1991.

Hammond, P. H. (ed.), *Robust Control, System Design Using H-infinity and Related Methods*, Institute of Measurement and Control, London, 1991.

Hawk, Jr., W. M., 'A self-tuning self-contained PID controller', *Proceedings of the ACC*, 838–842, 1983.

Hill, D. J., Wen, C. Y. and Goodwin, G. C., 'Stability analysis of decentralised robust adaptive control', *Systems & Control Letters*, **11**(4), 277–284, 1988.

Hsia, T. C. S., 'A new technique for robust control of servo systems', *IEEE Transactions on Industrial Electronics*, **36**(1), 1–7, 1989.

Hunt, K. J., Grimble, M. J., Chen, M. J. and Jones, R. W., 'Industrial LQG self-tuning controller design', *IEEE CDC Proceedings*, Athens, Greece, 1986.

Ioannou, P. and Sun, J., 'Theory and design of robust direct and indirect adaptive control schemes', *International Journal of Control*, **47**(3), 775–813, 1988.

Ioannou, P. A. and Tao, G., 'Dominant richness and improvement of performance of robust adaptive control', *Automatica*, **25**(2), 287–291, 1989.

Isermann, R. and Lachman, K. H., *Adaptive Digital Control Systems*, Prentice Hall, 1990.

Jacobs, O. L. R., 'A survey of adaptive control', *Third Workshop on 'The Theory and Application of Adaptive Control'*, IEE Computing and Control Division, Oxford, 1985.

Jacubasch, A., Kuntze, H. B., Arber, C. and Richalet, J., 'Application of a new concept of fast and robust position control of industrial robots', *Robotersysteme*, **3**(3), 129–138, 1987.

Jaworska, I. and Tzafestas, S., 'Improvement of systems reliability using robust control theory', *International Journal of Systems Science*, **22**(3), 587–593, 1991.

Johnson, M. A. (ed.), *Adaptive Systems in Control and Signal Processing*, Pergamon Press, 1990.

Johnson, W. J. D. and Abdallah, C. T., 'Robust control of accelerators', *Nuclear Instruments & Methods in Physics Research Section A—Accelerators Spectrometers Detectors and Associated Equipment*, **304**(1–3), 364–367, 1991.

Kanno, S. and Chubachi, T., 'Automatic robust landing control system design for Ente plane using equivalent nonlinear elimination method', *Transactions of the Japan Society for Aeronautical and Space Sciences*, **31**(93), 134–145, 1988.

Kanno, S. and Chubachi, T., 'Large motion robust flight control of aircraft by equivalent nonlinear elimination', *Transactions of the Japan Society for Aeronautical and Space Sciences*, **31**(91), 48–60, 1988.

Karim, M. N. and Lee, G. K. F., 'On the design of robust control systems for distillation columns', *Chemical Engineering Communications*, **68**, 81–98, 1988.

Kautsky, J., Nichols, N. K., Chu, E. K. W., 'Robust pole assignment in singular control systems', *Linear Algebra and its Applications*, **121**(Aug), 9–37, 1989.

Kaya, A. and Titus, S., 'A critical performance evaluation of four single loop self-tuning control products', *1988 American Control Conference*, 1659–1664, 1988.

Keel, L. H., Bhattacharyya, S. P. and Howze, J. W., 'Robust control with structured perturbations', *IEEE Transactions on Automatic Control*, **33**(1), 68–78, 1988.

Keviczky, L. and Banyasz, Cs. 'A completely adaptive PID regulator', *IFAC Symposium on Identification and System Parameter Estimation*, **1**, 91–97, 1988.

Keyes, M. A. and Kaya, A., 'Evolution of adaptive control algorithms and products: A critical review and evaluation', *Proceedings of the IFAC Workshop*, 1–8, Tbilisi, USSR, 1989.

Keyes, M. A. and Kennedy, J. P., 'Adaptive control techniques for control of suspension PVC processes', *ISA Annual Conference*, 1971.

Khargonezar, P. P., Petersen, I. R. and Zhou, K. M., 'Robust stabilisation of uncertain linear systems—Quadratic stabilizability and H infinity control theory', *IEEE Transactions on Automatic Control*, **35**(3), 356–361, 1990.

Kim, K. H., and Clarke, D. W., 'Robust adaptive pole-placement control using the pseudo-plant method', *International Journal of Systems Science*, **20**(11), 2043–2061, 1989.

Kraus, F. J. and Truol, W., 'Robust stability of control systems with polytopical uncertainty—A Nyquist approach', *International Journal of Control*, **53**(4), 967–983, 1991.

Kreisselmeier, G., 'A robust indirect adaptive control approach', *International Journal of Control*, **43**(1), 1986.

Kreisselmeier, G. and Anderson, B. D. O., 'Robust model reference adaptive control', *IEEE Transactions on Automatic Control*, **31**(2), 1986.

Kumar, V. R., Kulkarni, B. D. and Deshpande, P. B., 'On the robust control of nonlinear systems', *Proceedings of the Royal Society of London Series A Mathematics and Physical Sciences*, **433**(1889), 711–722, 1991.

Kummel, M. (ed.), *International Federation of Automatic Control, Adaptive Control of Chemical Processes: Workshop Proceedings*, Pergamon Press, 1989.

Kummel, M. and Andersen, H. W., 'Controller adjustment for improved nominal performance and robustness.2. Robust geometric control of a distillation column', *Chemical Engineering Science*, **42**(8), 2011–2023, 1987.

Kuo, C. Y., and Wang, S. P. T., 'Nonlinear robust industrial robot control', *Journal of Dynamic Systems Measurement and Control Transactions*, **111**(1), 24–30, 1989.

Kuo, C. Y. and Wang, S. P. T., 'Nonlinear robust hybrid control of robotic manipulators', *Journal of Dynamic Systems Measurement and Control Transactions*, **112**(1), 48–54, 1990.

Lane, J. D., 'Description of a modular self-tuning control system', *Proceedings of the ACC*, 2052–2057, 1989.

Lane, J. D., 'A closed loop deadtime/parameter estimator for self-tuning process control', *American Controls Conference*, 1989.

Lee, T. H. and Narendra, K. S., 'Robust adaptive control of discrete time systems using persistent excitation', *Automatica*, **24**(6), 781–788, 1988.

Lehtomaki, N. A., Sandell, N. R. Jr. and Athans, M. 'Robustness results in linear quadratic Gaussian based multivariable control designs'. *IEEE Transactions on Automatic Control*, **26**, 75–93, 1981.

Lewin, D. R. and Morari, M., 'Robex—An expert system for robust control synthesis', *Computers & Chemical Engineering*, **12**(12), 1187–1198, 1988.

Lin, C., Chang, J. R. and Jenc, S. C., 'Robust control of a boiling water reactor', *Nuclear Science and Engineering*, **102**(3), 283–294, 1989.

Ljung, L. and Söderström, T., *Theory and Practice of Recursive Identification*, MIT Press, Cambridge, Massachusetts, 1983.

Locatelli, A., Scattolini, R. and Shiavoni, N., 'On the design of reliable robust decentralised regulators for linear systems'. *Large Scale Systems*, **10**, 95–113, 1986.

Logemann, H. and Owens, D. H., 'Robust high-gain feedback control of infinite dimensional minimum phase systems', *IMA Journal of Mathematical Control and Information*, **4**(3), 195–220, 1987.

Lototsky, V. A., 'Evaluation of adaptive control strategies in industrial applications', *IFAC Workshop*, Tbilisi, USSR, International Federation of Automatic Control, Pergamon Press, Oxford, 1989.

Lou, S. X. C. and Kager, P. W., 'A robust production control policy for VLSI wafer fabrication', *IEEE Transactions on Semi-Conductor Manufacturing*, **2**(4), 159–164, 1989.

Lozanoleal, R., Collado, J. and Mondie, S., 'Model-reference robust adaptive control without *a priori* knowledge of the high frequency gain', *IEEE Transactions on Automatic Control*, **35**(1), 71–78, 1990.

Lunze, J., *Robust Multivariate Feedback Control*, Prentice-Hall, 1988.

Madiwale, A. N., Haddad, W. M. and Bernstein, D. S., 'Robust H-infinity control design for systems with structured parameter uncertainty', *Systems & Control Letters*, **12**(5), 393–407, 1989.

Mandler, J. A., Morari, M. and Seinfeld, J. H., 'Robust multivariable control system design for a fixed bed reactor', *Industrial & Engineering Chemistry Fundamentals*, **25**(4), 1986.

Manousiouthakis, V. and Arkun, Y., 'Hybrid approach for the design of robust control systems', *International Journal of Control*, **45**(6), 2203–2220, 1987.

Mareels, I. M. Y. and Bitmead, R. R., 'Bifurcation effects in robust adaptive control', *IEEE Transactions on Circuits and Systems*, **35**(7), 835–841, 1988.

McClaren, M. D. and Slater, G. L., 'Robust multivariable control of large space structures using positivity', *Journal of Guidance Control and Dynamics*, **10**(4), 393–400, 1987.

McClean, D. and Asiam-mir, S., 'Reconfigurable flight control systems', *Proceedings of the IEE Conference, Control 91*, Edinburgh, 1991.

McClusky, E. G. and Thompson, S., 'Rule-based adaptive PID control using a pattern recognition approach', *Proceedings of the IEE Colloquium on 'Expert Systems in Process Control'*, London, 1988.

Mendozabustos, S. A., Penlidis, A. and Cluett, W. R., 'Robust adaptive process control of a polymerization reactor', *Computers & Chemical Engineering*, **14**(3), 251–258, 1990.

Middleton, R. H., Goodwin, G. C., Hill, D. J. and Mayne, D. Q., 'Design issues in adaptive control', *IEEE Transactions on Automatic Control*, AC-33, 50–58, 1988.

Milanese, M. and Tempo, R. (ed.), *Robustness in Identification and Control: International Workshop Proceedings*, Plenum Publishing Company, 1989.

Minter, J. B. and Fisher, D. C., 'A comparison of adaptive controllers: Academic vs. industrial', *1988 American Control Conference*, 1653–1658, 1986.

Moden, P. E. and Nybrant, T., 'Adaptive control of rotary drum dryers', *Proceedings of the 6th IFAC/IFIP Conference*, Pergamon Press, 1986.

Montgomery, P. A., Williams, D. and Swanick, B. H., 'Control of a fermentation process by an on-line adaptive technique', *IFAC Conference, Modelling and Control of Biotechnological Processes*, 111–119, Holland, 1985.

Morari, M., 'Robust process control', *Chemical Engineering Research & Design*, **65**(6), 462–479, 1987.

Morari, M., 'Robust process control', *Abstracts of Papers of the American Chemical Society*, **197**(Apr), 57, 1989.

Morari, M. and Zafiriou, E., *Robust Process Control*, Prentice Hall, Englewood Cliffs, 1989.

Morris, A. J., 'Multivariate self-tuning process control', *Optimal Control Applications & Methods*, 3, 363–387, 1982.

Mudge, S. K. and Patton, R. J., 'Analysis of the technique of robust eigenstructure assignment with application to aircraft control', *IEE Proceedings D Control Theory and Applications*, **135**(4), 275–281, 1988.

Mufti, I. H., 'An improved adaptive control for robust adaptation', *IEEE Transactions on Automatic Control*, **34**(3), 339–343, 1989.

Mukhopadhyay, V., 'Digital robust control law synthesis using constrained optimization', *Journal of Guidance Control and Dynamics*, **12**(2), 175–181, 1989.

Myron, T. J., 'Self-tuning PID control—An expert system approach', *Proceedings of the IFAC Automatic Control in Petroleum, Petrochemical and Desalination Industries*, 77–80, Kuwait, 1986.

Nachtigal, C. L., 'Adaptive controller simulated process results', *Proceedings of the ACC Conference*, 1434–1439, 1986.

Narendra, K. S. and Annaswamy, A. M., 'Robust adaptive control in the presence of bounded disturbances', *IEEE Transactions on Automatic Control*, **31**(4), 1986.

Narendra, K. S. and Annaswamy, A. M., *Stable Adaptive Systems*, Prentice Hall, Englewood Cliffs, 1988.

Newman, W. S., 'Robust near time optimal control', *IEEE Transactions on Automatic Control*, **35**(7), 841–844, 1990.

Nwokah, O. D. I., 'Quantitative design of robust multivariable control systems', *IEE Proceedings D Control Theory and Applications*, **135**(1), 57–66, 1988.

Ohshima, M., Hashimoto, I., Takamatsu, T. and Ohno, H., 'Robust stability of model predictive control', *Kagaku Kogaku Ronbunshu*, **14**(4), 517–524, 1988.

Okada, T., Kihara, M. and Ikeda, M., 'Robust control system design synthesis with observers', *Journal of Guidance Control and Dynamics*, **13**(2), 337–342, 1990.

Owens, D. H. and Chotai, A., 'Approximate models in multivariable process control—An inverse Nyquist array and robust tuning regulator interpretation', *IEE Proceedings D Control Theory and Applications*, **133**(1), 1986.

Park, C. and David, A. J., 'An adaptive controller for heating and cooling systems: modelling implementation and testing', *ASME Winter Annual Meeting*, 1984.

Piovoso, M. J. and Williams, J. M., 'Self-tuning control of pH', *ISA 84—International Conference & Exhibition*, 705–723, 1984.

Postlethwaite, I., *Robustness in Multivariable Control System Design, Design of Modern Control Systems*, Peter Peregrinus, New York, 1982.

Reed, J. S. and Ioannou, P. A., 'Instability analysis and robust adaptive control of robotic manipulators', *IEEE Transactions on Robotics and Automation*, **5**(3), 381–386, 1989.

Rivera, D. E., Skogestad, S. and Morari, M., 'Internal model control; PID controller design', *I and E. C. Research*, **25**, 252–265, 1986.

Rocke, D. M., 'Robust control charts', *Technometrics*, **31**(2), 173–184, 1989.

Rotstein, H., Bandoni, J., Desages, A. and Romagnoli, J., 'Mathematical programming with linear uncertain constraints—Application to robust control', *Computers & Chemical Engineering*, **14**(4–5), 373–379, 1990.

Safonov, M. G. *Stability and Robustness of Multivariable Feedback Systems*, MIT Press, Cambridge, Massachusetts, 1980.

Safonov, M. G., Chiang, R. Y. and Flashner, H., '*H*-Infinity robust control synthesis for a large space structure', *Journal of Guidance Control and Dynamics*, **14**(3), 513–520, 1991.

Sandell, N. R., Jr., 'Robust stability of systems with application to singular perturbations', *Automatica*, **15**, 467–70, 1979.

Sandell, N. R., Jr., Gully, S. W., Lee, W. H. and Lehtomaki, N. A., Multivariable stability margins for vehicle flight control systems'. *Proceedings of the IEEE Conference on Decision and Control*, IEEE, New York, 1981.

Sastry, S. and Bodson, M., *Adaptive Control: Stability, Convergence and Robustness*, Prentice Hall, Englewood Cliffs, 1989.

Schmitendorf, W. E., 'Methods for obtaining robust tracking control laws', *Automatica*, **23**(5), 675–677, 1987.

Schoukens, J. and Pintelon, R., *Identification of Linear Systems*, Pergamon Press, 1991.

Seborg, D. E., Shah, S. L. and Edgar, T. F., 'Adaptive control strategies for process control: A survey', *AIChE Journal*, **32**(6), 881–913, 1986.

Sevaston, G. E. and Longman, R. W., 'Systematic design of robust control systems using small gain stability concepts', *International Journal of Systems Science*, **19**(3), 439–451, 1988.

Shimizu, K. and Matsubara, M., 'Singular value analysis for the robust control system—Design of distillation systems', *Journal of Chemical Engineering of Japan*, **18**(6), 1985.

Shoureshi, R., Momot, M. E. and Roesler, M. D., 'Robust control for manipulators with uncertain dynamics', *Automatica*, **26**(2), 353–359, 1990.

Sikora, R. F., 'Self-tuning control of moisture content', *Pulp & Paper Canada*, **85**(5), 51–54, 1965.

Siljak, D. D., 'Parameter space methods for robust control design—A guided tour', *IEEE Transactions on Automatic Control*, **34**(7), 674–688, 1989.

Singh, S. N., 'Robust nonlinear attitude control of flexible spacecraft', *IEEE Transactions of Aerospace and Electronic Systems*, **23**(3), 380–387, 1987

Skogestad, S. and Morari, M., 'Robust performance of decentralized control systems by independent designs', *Automatica*, **25**(1), 119–125, 1989.

Skogestad, S., Morari, M. and Doyle, J. C., 'Robust control of ill-conditioned plants—High purity distillation', *IEEE Transactions on Automatic Control*, **33**(12), 1092–1105, 1988.

Sobel, K. M., Banda, S. S. and Yeh, H. H., 'Robust control for linear systems with structured state space uncertainty', *International Journal of Control*, **50**(5), 1991–2004, 1989.

Söderstrom, T. and Stoica, P., *System Identification*, Prentice Hall, 1989.

Spurgeon, S. K. and Patton, R. J., 'Robust variable structure control of model reference systems', *IEE Proceedings D Control Theory and Applications*, **137**(6), 341–348, 1990.

Stepan, J., 'The uncertainty problem in control theory 2 The internally robust procedures', *Kybernetika*, **26**(2), 122–133, 1990.

Stoten, D. P., *Model Reference Adaptive Control of Manipulators*, Research Studies Press, 1990.

Sun, J. G., 'On numerical methods for robust pole assignment in control system design', *Journal of Computational Mathematics*, **5**(2), 119–134, 1987.

Sun, J. G., 'On numerical methods for robust pole assignment in control system design, 2', *Journal of Computational Mathematics*, **5**(4), 352–363, 1987.

Tao, G. and Ioannou, P. A., 'Robust adaptive control of bilinear plants', *International Journal of Control*, **50**(4), 1153–1167, 1989.

Tao, G. and Ioannou, P. A., 'Robust stability and performance improvement of discrete time multivariable adaptive control systems', *International Journal of Control*, **50**(5), 1835–1855, 1989.

Tao, G. and Ioannou, P. A., 'Robust adaptive control of plants with unknown order and high-frequency gain', *International Journal of Control*, **53**(3), 559–578, 1991.

Tao, G. and Ioannou, P. A., 'Robust adaptive control—A modified scheme', *International Journal of Control*, **54**(1), 241–256, 1991.

Taylor, R. G., 'Adaptive regulation of nonlinear systems with unmodelled dynamics', *IEEE Transactions on Automatic Control*, 405–412, 1989.

Tesi, A. and Vicino, A., 'Robust absolute stability of lure control systems in parameter space', *Automatica*, **27**(1), 147–151, 1991.

Triantafyllou, M. S. and Grosenbaugh, M. A., 'Robust control for underwater vehicle systems with time delays', *IEEE Journal of Oceanic Engineering*, **16**(1), 146–151, 1991.

Tsay, S. C., 'Robust control for linear uncertain systems via linear quadratic state feedback', *Systems & Control Letters*, **15**(3), 199–205, 1990.

Tso, S. K. and Shum, H. Y., 'Robust model reference adaptive control algorithm for a class of plants with bounded input disturbances', *Electronics Letters*, **26**(15), 1220–1221, 1990.

Tzou, Y. Y. and Wu, H. J., 'Multimicroprocessor-based robust control of an AC induction servo motor', *IEEE Transactions on Industry Applications*, **26**(3), 441–449, 1990.

Utkin, V. I., *Sliding Modes in Control Optimisation*, Springer-Verlag, Berlin, 1991.

Verghese, G. C., Fernandez, R. B. and Hedrick, J. K., 'Stable, robust tracking by sliding mode control', *Systems & Control Letters*, **10**(1), 27–34, 1988.

Virk, G. S. and Tahir, J. M., 'A fault-tolerant optimal flight control system', *Proceedings of the IEE Conference, Control 91*, Edinburgh, 1991.

Wang, D. C. and Leondes, C. T., 'Robust tracking control of non-linear systems with uncertain dynamics 1', *International Journal of Systems Science*, **20**(12), 2619–2641, 1989.

Wang, D. C. and Leondes, C. T., 'Robust tracking control of non-linear systems with uncertain dynamics 2', *International Journal of Systems Science*, **20**(12), 2643–2661, 1989.

Wang, K. and Ljung, L., 'A discussion of adaptive stabilization and robust adaptive control', *Systems & Control Letters*, **12**(1), 53–56, 1989.

Wang, S. D., Kuo, T. S., Lin, Y. H., Hsu, C. F. and Juang, Y. T., 'Robust control design for linear systems with uncertain parameters', *International Journal of Control*, **46**(5), 1557–1567, 1987.

Warwick, K., 'Self-tuning control in the state space', *Third Workshop on 'The Theory and Applications of Adaptive Control'*, IEE Computing and Control Division, Oxford, 1985.

Warwick, K. (ed.), *Implementation of Self-Tuning Controllers*, Peter Peregrinus, UK 1988.

Weeden, D. A., Jr. and Myron, T. J., 'Artificial intelligence applied to control tuning', *ISA/POWID Conference*, 1985.

Wellstead, P. E., 'Pole assignment self-tuning control', *Third Workshop on 'The Theory and Application of Adaptive Control'*, IEE Computing and Control Division, Oxford, 1985.

Wen, J. T. Y. and Balas, M. J., 'Robust adaptive control in Hilbert space', *Journal of Mathematical Analysis and Applications*, **143**(1), 1–26, 1989.

Widrow, B., 'Adaptive inverse control', *Proceedings of the 2nd IFAC Workshop on Adaptive Systems in Control and Signal Processing*, 1–5, Lund, Sweden, 1986.

Williams, D., Yousefpour, P. and Wellington, E. M. H., 'On-line adaptive control of a fed-batch fermentation of saccharomyces cerevisiae', *Biotechnology and Bioengineering*, **28**, 631–645, 1986.

Wit, C. A. C. De, *Adaptive Control for Partially Known Systems: Theory and Applications*, Elsevier, 1988.

Wu, H. S., 'Decentralized robust control for a class of large scale interconnected systems with uncertainties', *International Journal of Systems Science*, **20**(12), 2597–2608, 1989.

Wu, W. T., Jang, Y. J. and Chu, Y. T., 'Online stability index for robust control', *International Journal of Systems Science*, **22**(3), 495–505, 1991.

Yang, W. C. and Tomizuka, M., 'Discrete time robust control via state feedback for single input systems', *IEEE Transactions on Automatic Control*, **35**(5), 590–598, 1990.

Yanushevsky, R. T., 'Approach to robust control systems design', *Journal of Guidance Control and Dynamics*, **14**(1), 218–220, 1991.

Ydstie, B. E., 'Extended horizon adaptive control', *Proceedings of the 9th IFAC World Congress*, Budapest, 1984.

Yedavalli, R. K., 'Robust control design for aerospace applications', *IEEE Transactions on Aerospace and Electronic Systems*, **25**(3), 314–324, 1989.

Yeh, H. H., Banda, S. S., Heise, S. A. and Bartlett, A. C., 'Robust control design with real parameter uncertainty and unmodelled dynamics', *Journal of Guidance Control and Dynamics*, **13**(6), 1117–1125, 1990.

Yoerger, D. R. and Slotine, J. J. E., 'Robust trajectory control of underwater vehicles', *IEEE Journal of Oceanic Engineering*, **10**(4), 1985.

Young, G. E. and Rao, S., 'Robust sliding mode control of a nonlinear process with uncertainty and delay', *Journal of Dynamic Systems Measurement and Control—Transactions of the ASME*, **109**(3), 203–208, 1987.

Young, P. C., 'An instrumental variable method for real-time identification of a noise process', *Automatica*, **6**, 271–287, 1970.

Zafiriou, E., 'Robust model predictive control of processes with hard constraints', *Computers & Chemical Engineering*, **14**(4–5), 359–371, 1990.

Zafiriou, E. and Morari, M., 'Internal model control—Robust digital controller synthesis for multivariable open-loop stable or unstable processes', *International Journal of Control*, **54**(3), 665–704, 1991.

Zeng, F. Y. and Dahhou, B., 'Model reference adaptive estimation and control applied to a continuous fermentation process', *IFAC Symposium on Advanced Control of Chemical Processes*, Toulouse, France, 1991.

Zhang, Q. and Tomizuka, M., 'Multivariable direct adaptive control of thermal mixing process', *Dynamics and Control of Thermofluid Process and Systems*, ASME, 19–26, New Orleans, Louisiana, 1984.

Zohdy, M. A., Loh, N. K. and Abdulwahab, A. A., 'A robust optimal model matching control', *IEEE Transactions on Automatic Control*, **32**(5), 410–414, 1987.

Glossary of Symbols

Some symbols can take on meanings different from those given in the list—such meanings are defined locally in the text.

$\mathcal{E}$	Expected value
$\mathcal{F}$	Fourier transform
$\mathcal{L}$	Laplace transformation
$\mathcal{Z}$	$\mathcal{Z}$ transformation
$\mathbb{R}^n$	Real n-dimensional space
A, B, C	System matrices
G, H	Transfer functions
I	The identity matrix
$L_1(t_1, t_2)$	The space of functions whose absolute value is integrable over the internal (t_1, t_2)
T	The sampling interval in a discrete time system
w, w'	See the index
s, z	The complex variables associated with Laplace and $\mathcal{Z}$ transformations respectively
Σ	An abstract system
Φ	The transition matrix
ζ	Damping factor
λ	System eigenvalue
σ	The real part of s or standard deviation
ω	Angular frequency
ω_b	Bandwidth
ω_n	Undamped natural frequency
ω_s	Sampling frequency
A^T	The transpose of matrix A
$f^*(t)$	The time function $f(t)$ after ideal sampling

$f * g$	Convolution of the functions, f, g
$\dot{f}$	The time derivative of f
det ( )	Determinant
dim ( )	Dimension
ln	Natural logarithm
log	Logarithm to base 10
$n!$	Factorial n
$\rightarrow$	Implies
$\langle , \rangle$	Inner product
$\mid \; \mid$	Absolute value
$f(t)\vert_{t = t_1}$	The value of $f(t)$ when $t = t_1$
$\angle$	Denotes the angle in polar coordinates
$\Vert \; \Vert$	Norm
$\triangleq$	Equal to by definition
$\forall$	For all, for every

Index